D0905045

Q
127
.U6
S3146
1986

Science and society
in early America

DISCARD

$30.00

DATE			
		^	

BUSINESS/SCIENCE/TECHNOLOGY
DIVISION

© THE BAKER & TAYLOR CO.

Science and Society in Early America:
Essays in Honor of Whitfield J. Bell, Jr.

Science and Society in Early America:
Essays in Honor of
Whitfield J. Bell, Jr.

EDITED BY

RANDOLPH SHIPLEY KLEIN

AMERICAN PHILOSOPHICAL SOCIETY

INDEPENDENCE SQUARE 1986

Publication of this volume has been aided
in part by the Exxon Corporation

ON THE COVER: *The Great Alachua-Savana, in East Florida,*
pen and ink by William Bartram, ca. 1774,
at the American Philosophical Society.

Library of Congress Catalog Card No.: 85-71740
International Standard Book Number: 0-87169-166-3
US ISSN: 0065-9738

Copyright 1986 by the American Philosophical
Society for its Memoirs series, Volume 166

Contents

BST R

$ 30.00

President Jonathan E. Rhoads's Remarks on Presentation of Essays in Honor of Whitfield J. Bell, Jr. and the Society's Franklin Medal

W HITFIELD J. BELL, JR.'s service to the Society has been long and distinguished. About a year ago he wrote that it was his wish to resign as Executive Officer not later than 1 January 1984. We met this deadline by a very narrow margin when the search committee discovered that Herman H. Goldstine would be willing to take on this important responsibility. We could not let the departure of one who has done so much for all of us pass unnoticed and tonight I have two items to present—one a promise and the other not. First for the promise—

Scholarship has always meant a great deal to Whitfield J. Bell, Jr. Upon his election to the Society in 1964, he was praised for being "a rare combination of outgoing, enthusiastic teacher with a warm interest in people and the quiet, painstaking scholar." While at Dickinson College where he became Boyd Lee Spahr Professor of History, Whit was a very popular teacher while he regularly produced useful articles on the history of medicine and other subjects.

He left Dickinson to teach at The College of William and Mary, and while there was active with the Institute for Early American Culture and was visiting editor of the *William and Mary Quarterly*, the premier journal for early American history. In addition to bringing together a useful issue on the history of science, he also wrote *Early American Science: Needs and Opportunities for Study* which abounded with useful insights and identified many subjects which others developed into significant books and articles. It also contained a useful bibliography for the relatively young field.

Promoting useful knowledge occurs in many ways, and Whit soon became a pioneer in another area. Until the 1950s, historical editing of American materials consisted of collecting a large number of an individual's letters, arranging them chronologically, and hoping that the printer made few errors. In many instances an editor silently deleted material considered distasteful or unflattering. Our member Julian Boyd initiated a new concept of historical editing. When Whit Bell joined Leonard Labaree to edit *The Papers of Benjamin Franklin*, he helped broaden the scope of what an historical editor did and endowed the calling with scholarly respectability which it previously lacked. Together they meticulously collected and accurately reproduced all the letters *to* as well as from Franklin, and also included his other writings. This proved valuable in itself, but Labaree and Bell went beyond this to provide extraordinarily useful footnotes which identified individuals and events mentioned in the text and explained references which would have eluded almost any scholar. Frequently they also provided lengthy introductions to individual items. Critical reviews praised the series not only because of the importance of Franklin and Franklin's lively style; they also exclaimed over the great accomplishment of the editors as well.

Whit then came to the American Philosophical Society, first as Associate Librarian, then as Librarian, and finally as Executive Officer. He remained the warm, friendly man so many had come to know, and also the creative and energetic promoter of scholarly projects. He built up the Society's holdings, compiled useful guides to the manuscript collections, and helped launch the *Joseph Henry Papers* (an important editorial project for the history of science). With others he recognized that the Society would soon outgrow its present facilities, and played an active role in acquiring the Farmers and Mechanics Bank building at 427 Chestnut Street and raising funds to help pay for it. In addition to administrative responsibilities, he also found time to write a book on *John Morgan* and other works. Among the important projects begun soon after his arrival at the Society, was a biographical dictionary of the early members of this society.

As chairman of the Research Committee, Whit also evaluated great numbers of research grant proposals and his insights and broad knowledge helped the committee choose the best for funding. He also took steps to improve the publications program and delighted in putting together the wonderful selection of papers which members enjoy when they attend the Society's meetings.

As Whit prepared to leave his administrative duties and take up the life of a scholar on a full-time basis, his friends at the Society (his fellow workers as well as members), thought that an especially fitting gift for someone who has given so much to others and promoted scholarship in so many ways, would be a scholarly tribute. After several months of preliminaries, we can reveal what we hope was a well-kept secret. The Society is proud to announce that it will publish a book in its *Memoir* series, *Essays in Honor of Whitfield J. Bell, Jr.* Mrs. Jean T. Williams, Dr. Bell's secretary for many years, first suggested the tribute. Although Dr. Carter, Dr. Strayer, Dr. Oppenheimer, Dr. Goldstine, Mr. Miller, Mrs. Williams, Miss Le Faivre and others helped with the idea, Dr. Klein gave life to the project by urging it, shaping it, and recruiting the authors.

The book will contain twelve articles including footnotes and illustrations, a biographical sketch of Whitfield J. Bell, Jr., along with a list of his honors and awards and complete bibliography, and of course a solid index. I hope that this brief appreciation of Dr. Bell's many contributions to the American Philosophical Society held at Philadelphia for Promoting Useful Knowledge may serve as a kind of introduction to the papers that will follow. I think *Essays in*

Honor of Whitfield J. Bell, Jr. is a splendid idea and that Dr. Klein, Whit's former student at the University of Pennsylvania and colleague, is the right person to edit it.

This will contain scholarly contributions of high quality, for a majority of the authors are members of this Society, and the others are well-equipped to make significant contributions as well. Edwin Wolf of the Library Company of Philadelphia and Esmond Wright of the University of London will reflect Whit's fascination with Franklin. Whit's former student, Marvin Wolfgang, will write about the founding of the Pennsylvania Prison Society (an organization founded by Bishop William White, a member of the APS). Bentley Glass and Jane Oppenheimer will contribute essays on the history of science. William Abbot of the *Papers of George Washington* represents Whit's interest in historical editing, as Joseph Ewan does Whit's interest in the history of science and book collecting. Other authors include long-time friends such as Brooke Hindle, Silvio A. Bedini, and Thomas C. Cochran.

Whit, although the idea of a book in your honor met with overwhelming applause, even your friends could not produce it by this evening. We are presenting you an advance mockup of the actual book, which we hope to present in 1985. It was hand crafted by Willman Spawn, the Society's conservator, and contains the title page and table of contents. In due time we will have a suitably modest ceremony at which the authors and some of your other friends can present the real scholarly tribute to a scholar who has contributed so much to scholarship.

To mark the two hundredth anniversary of the birth of Benjamin Franklin in 1906 the United States Congress authorized a commemorative medal. It was designed by Louis and Augustus St. Gaudens. One copy in gold was presented, "under the direction of the President of the United States," to the Republic of France; and of the 150 copies struck in bronze, 100 were for distribution as the

President might direct, and 50 were given to the American Philosophical Society for its use.

The Society has awarded the Franklin Medal very infrequently. This evening we present this rare medal to a rare individual, Whitfield J. Bell, Jr.

Philadelphia, 19 April 1984

Acknowledgments

With pleasure I express my gratitude to those who helped make this book possible. I thank the Society for its support, especially Jonathan E. Rhoads, whose positive response to the proposal and whose suggestions along the way meant a great deal. And of course I add my thanks to all he mentioned in his "Remarks." I especially relied upon Edward C. Carter II and Jane M. Oppenheimer for sound advice, and repeatedly benefited from it. It's also appropriate to thank the authors, for they were a pleasure to work with and helped make this editor's job enjoyable. Herman H. Goldstine gave important help when it was needed. I also benefited from conversations with Jean T. Williams, Carl F. Miller, and Maryann Klein. Susan M. Babbitt's careful reading of copyedited material enhanced accuracy, and Carole N. Le Faivre handled production with skill.

Special thanks goes to the Exxon Research and Engineering Company for its subvention of $6,000 and to John Peckham of Meriden-Stinehour, long-time friend of Whit Bell and printer of some of the Society's most attractive publications in the past. His generous financial contribution helped produce a well-crafted book at a reasonable price.

Most of all I thank my good friend Whit Bell to whom this book is dedicated.

R.S.K.

Whitfield J. Bell, Jr.

FOR years visiting scholars and others were often welcomed personally to the American Philosophical Society by its Executive Officer, a man who played an important role in the nation's oldest and perhaps most prestigious learned society which was founded by Benjamin Franklin in 1743. Whitfield J. Bell, Jr., is a man whose warmth immediately puts people at ease even as they marvel over the depth and breadth of his knowledge about Franklin, the Society, early America, and the history of medicine and science. So polished is his ability with words and in recreating the past, that when you read his works or meet him at lunch you would swear he had been talking to Franklin, Thomas Bond, John Morgan, Benjamin Rush, or John Vaughan that very morning. Although it would be easy to write a great deal about Whitfield J. Bell, Jr.'s acccomplishments and contributions to useful knowledge, a lengthy essay trumpeting achievements seems inappropriate for a gentleman to whom modesty means so much. I offer instead a sketch.

As an original member of the editorial team which worked with *The Papers of Benjamin Franklin*, Whit became acquainted with Franklin as few others have. In addition to the five volumes which came out during his tenure at New Haven, Whit published a selection of Franklin letters (with Leonard W. Labaree) and introductions to a number of Franklin pieces such as "The Morals of Chess," "The Old Mistresses' Apologue," and Franklin's essay on literary style. He has also probed Franklin's relationship with medicine, philanthropy, and provisioning the British army during the French and Indian War, as well as a number of other topics.

Whit has made major contributions to our knowledge of medi-

9

cine in early America. *John Morgan: Continental Doctor* provided an
ideal way to deal with the founding of the nation's first medical
school, the medical problems associated with the War for Independence, and the nuances of the development of the practice of
medicine in America. A review in the *William and Mary Quarterly*
observed that "it provides a brilliant view of the late colonial period
and of medicine of that time. . . . He has contributed a work with
broad appeal for general students of eighteenth-century America
as well as for the specialist in the history of science or medicine." A
steady flow of articles on medical history has enlightened scholars
over the years, and an impressive tribute to their importance occurred in 1975 when a group of his friends gathered fifteen of the
most important ones together in *The Colonial Physician*. With char-
acteristic humility, Whit informed the publisher it would not sell.
No one is perfect—it did, and this scholarly publication has actually paid the author royalties ever since. A very favorable review in
the *William and Mary Quarterly* began "occasionally, imagination
and modern printing technology combine to produce from previously published essays a book transcending in value the elements
of which it is composed. This is such an occasion." The reviewer
concluded that the author was "that rare historian who has writen
a series of related essays that offer an important complex of understanding."

Ironically, Whit's mentor, Richard Harrison Shryock, a great
historian of medicine, did not guide his doctoral students into
medicine at the time Whit studied at the University of Pennsylvania. In Whit's case, the attraction flowed naturally from the
original topic. That topic was "Science and Humanity in Philadelphia, 1775–1790." Many find his excellent doctoral dissertation
a treasure, and rely upon it more than many officially published
works.

Whit Bell's approach to historical topics is often biographical;
hence the world he recreates retains the authenticity of actual
people in thought and action, rather than becoming a place over-

whelmed by impersonal forces, trends, movements, and groups. All the while he maintains focus upon the larger picture with its social and intellectual context, and his insights are clear; his understanding is neither lost amid useless details nor distorted by myopic vision. To emphasize a biographical approach need not limit one to a great-lives approach, which can be a distortion in itself. Whit remarked upon the importance of people of "the second rank." "These men's lives have meaning," he explained, and then elaborated that "it is the large body of unimaginative, undistinguished men of the middle sort who keep alive the ideas other men conceive and hold together the institutions that other men create. . . . They originate little; they transmit much."

An interest in people coincided well with his fascination with the world of the eighteenth century and science. In the 1960s he embarked upon an ambitious undertaking, the creation of a biographical dictionary of all the members of the American Philosophical Society elected before 1800. To date he has written over one hundred and eighty biographical sketches. Those who have seen them look forward to the publication of this useful work which duties as Executive Officer caused him to lay aside. In recent years his artistry has preserved the memory of important twentieth-century members such as Catherine Drinker Bowen, Lyman H. Butterfield, George W. Corner, Leonard W. Labaree, and Richard H. Shryock. Quite naturally he was asked to provide the introduction to a republication of James Thacher's useful *American Medical Biography*. Bell wisely added a bibliography to it. His reflections on Thacher and other medical biographers, namely Stephen W. Williams and Samuel D. Gross, appeared in the *Bulletin of the History of Medicine* and *The Colonial Physician*.

Whit Bell enjoys research, writing, and teaching undergraduates. Although he has taught graduate students, the guidance of doctoral students rarely lured him from other activities. His influence is felt in other ways. Always a man to encourage others and share his insights and knowledge with them, Whit has often published useful works which have opened new pathways. *Early American Science: Needs and Opportunities for Studies*, published by the Institute for Early American History and Culture, is a fine example. It not only attracted many to what had been a neglected field, but remains an important reference work, as its republication in 1971 demonstrates. His Presidential address to the American Association for the History of Medicine provided many keen observations on the strengths and limitations of past approaches to the history of medicine as did his reflections on noted medical historians including pioneers such as Fielding Garrison and Joseph M. Toner. Other suggestions appear in "The Writing of American Medical History before Professionalization." As Librarian of the Society, Whit was instrumental in launching a joint program with Bryn Mawr College and the University of Pennsylvania in graduate instruction in the history of science. This program of major importance attracted attention well beyond the Philadelphia area. The first seminar met at the Society and included distinguished scholars and geneticists L. C. Dunn, Bentley Glass, Theodosius Dobzhansky, Joseph L. Fruton, I. Michael Lerner, and George W. Beadle. In addition to classroom activities the scholars made suggestions regarding acquisitions for the library and the library staff, in turn, provided students with much guidance to the Society's rich holdings.

Another means of encouraging others lies in preparing useful guides and reference works. Whit often observed that such works, or careful scholarly editions of historical papers, make a more useful and lasting contribution to knowledge than many of the articles churned out in the publish-or-perish atmosphere of some universities and colleges. Whit put it succinctly when he wrote

> a library, like a museum and some other educational institutions, has an obligation not only to those who come to it but to those who remain beyond its walls. This Library, without postponing its immediate tasks, exhibits selected holdings to

the public, has prepared bibliographies, guides, and a newsletter, encourages the scholarly publication of its manuscript collections, and has joined actively in a conventional university education program.

With Murphy D. Smith he prepared a *Guide to the Archives and Manuscript Collections of the American Philosophical Society*. It helped increase awareness and use of the Society's valuable holdings. An article of suggestions for research in the local history of medicine in the United States was followed by several bibliographies of the history of medicine published in the *Bulletin of the History of Medicine*. His annual reports as Librarian of the Society from 1966 to 1981 deftly introduced readers to new collections, books, paintings, and some sense of their potential. As Librarian he also enlisted others to prepare guides to the Society's pamphlets, the Thomas Paine collection, and other holdings. In addition to serving on the administrative board of *The Papers of Benjamin Franklin* since 1969, Dr. Bell deserves recognition for initiating and helping launch *The Joseph Henry Papers* (the first secretary of the Smithsonian Institution and an important scientist), and he has served since its inception in 1965 on the committee of sponsors for that project. He also vigorously supported the Darwin Papers Project, which is still jointly sponsored by the Society, Cambridge University, and the American Council of Learned Societies, and provided space in the Library for its American operations. He has also been an active and valuable member of the editorial board of *The Papers of Benjamin Henry Latrobe* which is now housed in the Library.

Beyond his work with the *Franklin Papers* Whit has edited *Mr. Franklin: A Selection of his Personal Letters* (with Leonard Labaree) and a number of shorter works ranging from "The Clinical Notes of John Archer, M.B., 1768," to "Nicholas Collin's Appeal to American Scientists" (*William and Mary Quarterly*), to "Diary of George Bell: A Record of Captivity in a Federal Military Prison." Insights into historical editing based on experience appear in articles such as "Franklin's Papers and *The Papers of Benjamin Franklin*," "Editors and Great Men," "Papers for the History of Science," and "Editing a Scientist's Papers" (*Isis*). He also edited a special volume of the *William and Mary Quarterly* devoted to the history of science in early America.

Editorial work of another sort stems from his wonderful command of the English language. Despite his busy schedule, Whit always makes time for those who ask his opinion of their writings. He is a master craftsman when it comes to revisions and clarification. If necessary, he can help an author rewrite a sentence or whole section so that it exactly replaces existing copy instead of causing expensive resetting of type. In such instances, or those requiring less rigor, he always manages to improve a piece without imposing his style upon it. And as with all advice he so freely offers, he maintains a pleasant manner that enables an author to think and write more clearly, yet never feel a sting of criticism.

A guide of another sort was *A Rising People*. This exhibit and its catalog had their origins in the bicentennial of the Declaration of Independence. While some scholars merely complained that the nation's birthday was becoming a mindless party, Whit joined with the directors of the Library Company of Philadelphia and the Historical Society of Pennsylvania to put on an exhibit which featured about one thousand manuscripts, books, pamphlets, broadsides, maps, paintings, and furniture. For nine months it was open at no charge, seven days a week for twelve hours a day, and drew great praise. A happy historian recorded that it is "no exaggeration to assert, as many who saw it told us, that in content 'A Rising People' was the most important historical exhibition presented anywhere in this country or in England in 1976—an incomparable celebration of the founding of the nation by three institutions . . . which preserve so much the authentic history of the nation." It was a matter of quiet pride that this celebration, which cost several thousand dollars, was organized and carried through without so much as an initialled memorandum by the three institutional sponsors.

Collecting manuscripts and making them available to scholars came naturally to one so at ease with people. At first he ferreted out material for his own research and then Franklin material in England and Scotland as well as in the United States. Perusal of the footnotes of his writings frequently reveals that he had obtained access to papers in private hands. Later he persuaded people such as Sol Feinstone, the Cope family, and the Macholds to present their collections including 120 Washington letters, papers of Arctic traveler Elisha Kent Kane, and important Shippen manuscripts, to appropriate institutions. The reports of the Committee on Library teem with reference to gifts and acquisitions. Such happy occurrences do not just happen, although the Librarian was too modest to indicate his role in this valuable process of growth which serves current scholars and will benefit scholarship for generations. Sometimes when gifts were not forthcoming, Whit had the good sense to realize that more than words of encouragement were needed. For example, the Benjamin Smith Barton papers had eluded the Society despite efforts of past Presidents and Librarians. When Whit explained that they were "unquestionably of the greatest importance to the history of science in early America," the Society made a significant purchase.

As his former student, I gladly record that few teachers are Whit Bell's equal. That view is common among those who have taken his courses at Dickinson College, William and Mary, Yale, and the University of Pennsylvania. His love of his subject radiates throughout the class, and his encouragement to explore on your own the massive reading lists doubtless prompted and inspired many to probe far beyond requirements. In addition to promoting independent work and thought, he has a wonderful ability to help people sort out options and opportunities which may confuse them. This comes in part, no doubt, from a genuine interest in people and from respectfully dealing with students as fellow seekers of truth, rather than in a manner that in any way smacks of superiority. Unlike many scholars with great knowledge, Whit can always convey it on an appropriate level—whether speaking to students, historical societies, clubs, or professional associations. And unlike some professors, he was regarded with warmth and affection, even though he was not deemed an easy grader. Part of this came through the clear indications that he never confused academic achievement with personal worth. I knew him when I was a student at Penn; however, I know that at Dickinson the brilliant, those seeking "a gentleman's 'C,'" and those in between or below were welcome at his home in Carlisle. The social intercourse he promoted, often enhanced by his famous cider and gingerbread, fostered an open atmosphere in class which made it easier for ideas to flow.

Fostering an atmosphere where minds can meet and easily exchange ideas is not limited to a college, university, or formal institutions. When Whit came to Philadelphia from the Franklin Papers, he was instrumental in encouraging those with interests in early America to get together. For many years, professional and amateur historians alike gathered once a month for lunch and talked about their research. In addition to the content of individual conversations which had both social and intellectual value, the gathering of people from the APS, Park Service, Historical Society, Historical Commission, local universities and colleges, and those with no affiliation fostered an important sense of being part of a larger endeavor, in touch wih the main currents of historical scholarship.

Because of Dr. Rhoads's remarks on Whitfield J. Bell, Jr.'s contribution to the American Philosophical Society during a tenure of twenty-two years as Associate Librarian, Librarian, and Executive Officer, I shall not elaborate further. During parts and in some cases all of that period, he also served as presidential appointee to the National Historical Publications Research Commission, trustee of Dickinson College, the Rosenbach Foundation (president,

1974), Winterthur Museum, Hagley Museum–Eleutherian Mills Historical Library, University of Pennsylvania Press, and the Rockefeller Archives Center. For more than a decade he also served on the Council of the Institute of Early American History and Culture and more recently helped with the birth of the Philadelphia Center for Early American Studies. Dr. Bell is recipient of several honorary degrees, including one from his alma mater, Dickinson College, which later honored him at his fiftieth class reunion this past June as Alumnus of the Year.

One of the appealing features of a tribute to a person rather than to his memory is not only that the one so honored is present, but also we can cheer on—rather than lament—work yet unfinished. Although Whit gave much of his life and energy to the Society, the attractions of devoting full time to scholarly research and writing could not be reined in indefinitely. When he retired, it was not a retreat from challenges or in response to a wish to rusticate at his home in Carlisle. Although the pleasures of that town call him west from time to time, the library which he added to the house enables him to continue the scholarly activities which delight him in Philadelphia. He is completing a bicentennial history of the College of Physicians of Philadelphia and then can turn to other projects including the biographical dictionary of early members of the Society. It is an honor to join the other authors and the Society in recognizing Whitfield J. Bell, Jr., and wish him well as he pursues interests which give him such pleasure and delight.

RANDOLPH SHIPLEY KLEIN

Selected Bibliography of Whitfield J. Bell, Jr.

BOOKS

Edited. *Bulwark of Liberty: Early Years at Dickinson. The Boyd Lee Spahr Lectures in Americana.* I. New York: Fleming H. Revell, 1950. 174 p.

Two Hundred Years in Cumberland County. With D. W. Thompson and others. Carlisle, Pa.: Hamilton Library and Historical Association, 1951, 388 p.

Early American Science: Needs and Opportunities for Study. Williamsburg, Va.: Institute of Early American History and Culture, 1955. 85 p. (Reprinted by Russell and Russell, New York, 1971.)

Mr. Franklin: A Selection from His Personal Letters. With Leonard W. Labaree. New Haven: Yale University Press, 1956. 61 p.

The Papers of Benjamin Franklin. With Leonard W. Labaree and others. New Haven: Yale University Press, 1959–. [Associate editor for volumes 1–5, 1959–1962.]

John Morgan: Continental Doctor. Philadelphia: University of Pennsylvania Press, 1965. 301 p.

The Art of Philadelphia Medicine. Philadelphia: Philadelphia Museum of Art, 1965, 131 p.

Guide to the Archives and Manuscript Collections of the American Philosophical Society, Memoirs, vol. 66, 1966. 182 p.

The Colonial Physician & Other Essays. New York: Science History Publications, 1975. 229 p.

A Rising People: The Founding of the United States. With James E. Mooney and Edwin Wolf 2nd. Philadelphia: American Philosophical Society, 1976. xi, 292 p.

18

ARTICLES

"Belles Lettres Society Through 150 Years." *Dickinson Alumnus* 13, No. 1 (1936): 13–19.

"James Henry Morgan." *Dickinson Alumnus* 4, No. 3 (1937): 18–19, 22–33.

"Some Aspects of the Social History of Pennsylvania, 1760–1790." *Pennsylvania Magazine of History and Biography* 62 (1938): 281–308.

Edited. "Diary of George Bell: A Record of Captivity in a Federal Military Prison." *Georgia Historical Quarterly* 22 (1938): 169–184.

"The Relation of Herndon and Gibbon's Exploration of the Amazon to North American Slavery." *Hispanic American Historical Review* 19 (1939): 494–503.

"Dr. James Smith and the Public Encouragement for Vaccination for Small Pox." *Annals of Medical History*, 3d ser., 2 (1940): 500–517.

"Washington County, Pennsylvania, in the Eighteenth Century Anti-Slavery Movement." *Western Pennsylvania Historical Magazine* 25 (1942): 135–142.

"Philadelphia Medical Students in Europe, 1750–1800." *Pennsylvania Magazine of History and Biography* 67 (1943): 1–29.

"Thomas Anburey's 'Travels through America': A Note on Eighteenth Century Plagiarism." *Bibliographical Society of America, Papers* 37 (1943): 3–16.

"Suggestions for Research in the Local History of Medicine in the United States." *Bulletin of the History of Medicine* 17 (1945): 460–476.

"Holmes and History." *Baker Street Journal* 2 (1946): 447–456.

"The Scientific Environment of Philadelphia, 1775–1790." *American Philosophical Society, Proceedings* 92 (1948): 6–14.

"John Morgan." *Bulletin of the History of Medicine* 22 (1948): 543–561.

"Problems and Promises of a Course in State History." *American Heritage* (February 1949).

"A Box of Old Bones: A Note on the Identification of the Mastodon, 1766–1806." *American Philosophical Society, Proceedings* 93 (1949): 169–177.

"Thomas Parke, M.B., Physician and Friend." *William and Mary Quarterly*, 3d ser., 6 (1949): 569–595.

Edited. "Bibliography of the History of Medicine of the United States and Canada—1948." *Bulletin of the History of Medicine* 23 (1949): 494–517.

The Founding of a Lodge. Carlisle, Pa. Cumberland Star Lodge, No. 197, F. and A. M. (1950): 36 p.

Charles Francis Himes, Local Historian. Carlisle, Pa.: Hamilton Library and Historical Association, 1950, 12 p.

"Some American Students of 'That Shining Oracle of Physic,' Dr. William Cullen of Edinburgh, 1755–1766." American Philosophical Society, *Proceedings* 94 (1950): 275–281.

"Boyd Lee Spahr: An Appreciation." In *Bulwark of Liberty: Early Years at Dickinson. The Boyd Lee Spahr Lectures in Americana* 1 (1950): 17–28.

Edited. "Bibliography of the History of Medicine of the United States and Canada—1949." *Bulletin of the History of Medicine* 24 (1950): 541–558.

"Thomas Parke's Student Life in England and Scotland, 1771–1773." *Pennsylvania Magazine of History and Biography* 75 (1951): 237–259.

"The Reverend Mr. Joseph Morgan, an American Correspondent of the Royal Society, 1732–1739." American Philosophical Society, *Proceedings* 95 (1951): 254–261.

"John Morgan, Founder of the Medical School." *General Magazine and Historical Chronicle* 53 (1951): 213–223.

Edited. "Bibliography of the History of Medicine of the United States and Canada—1950." *Bulletin of the History of Medicine* 25 (1951): 464–489.

"Carlisle to Pittsburgh: A Gateway to the West, 1750–1815." *Western Pennsylvania Historical Magazine* 35 (1952): 157–166.

Edited. "Bibliography of the History of Medicine of the United States and Canada—1951." *Bulletin of the History of Medicine* 26 (1952): 452–476.

"Thomas Cooper as Professor of Chemistry at Dickinson College, 1811–1815." *Journal of the History of Medicine and Allied Sciences* 8 (1953): 70–87.

"Medical Students and their Examiners in Eighteenth Century America." College of Physicians of Philadelphia, *Transactions and Studies,* 4th ser., 21 (1953): 14–24.

Edited. "Bibliography of the History of Medicine of the United States and Canada—1952." *Bulletin of the History of Medicine* 27 (1953): 451–481.

"Physicians and politics in the Revolution: The Case of Adam Kuhn, with a Note on Philip Turpin." College of Physicians of Philadelphia, *Transactions and Studies,* 4th ser., 22 (1954): 25–31.

"Scottish Emigration to America: A Letter of Dr. Charles Nisbet to Dr. John Witherspoon, 1784." *William and Mary Quarterly,* 3d ser., 11 (1954): 276–289.

"A Note on Franklin and His Lightning Rod." In *A Poetical Epistle from Charles Woodmason to Benjamin Franklin.* Richmond, Va.: William Byrd Press, 1954, 1–6.

"Dr. Adam Cunningham: Physick and Traffick in Old Virginia." Cleveland Medical Library, *Bulletin,* n.s., 1 (1954): 67–69.

Edited. "Bibliography of the History of Medicine of the United States and Canada—1953." *Bulletin of the History of Medicine* 28 (1954): 442–470.

"Adam Cunningham's Atlantic Crossing, 1728." *Maryland Historical Magazine* 50 (1955): 195–202.

The Amateur Historian [address at the 177th anniversary and the 78th annual commemoration service of the battle and massacre of Wyoming]. Wilkes-Barre, Pa.: Wyoming Commemorative Association, 1955, 16 p.

"Franklin's Papers and *The Papers of Benjamin Franklin.*" *Pennsylvania History* 22 (1955): 3–19.

"Benjamin Franklin and the German Charity Schools." American Philosophical Society, *Proceedings* 99 (1955): 381–387.

"An Eighteenth Century Medical Manuscript: The Clinical Notebook of John Archer, M.B., 1768." [University of Pennsylvania] *Library Chronicle* 22 (1956): 1–8.

"The Other Man on Bingham's Porch [John Montgomery]." In '*John and Mary's College:' The Boyd Lee Spahr Lectures in Americana* 2. New York: Fleming H. Revell, 1956, 33–59.

"'All Clear Sunshine:' New Letters of Franklin and Mary Stevenson Hewson." American Philosophical Society, *Proceedings* 100 (1956): 521–536.

"Benjamin Franklin: Doctor and Patient." *Philadelphia Medicine* 52 (1956): 517–520.

"The Father of All Yankees." In *The American Story: The Age of Exploration to the Age of the Atom*, edited by Earl Schenck Miers. Great Neck, N.Y.: Channel Press, 1956, 67–74.

Edited. "Nicholas Collin's Appeal to American Scientists." *William and Mary Quarterly*, 3d ser., 13 (1956): 519–550.

Editorial Note to [Benjamin Franklin's] *The Old Mistresses' Apologue*. Philadelphia: Rosenbach Foundation, 1956.

"John Redman, Medical Preceptor, 1722–1808." *Pennsylvania Magazine of History and Biography* 81 (1957): 157–169. Reprinted, with Redman's "Letter on Inoculation, 1760," and an introduction, in College of Physicians of Philadelphia, *Transactions and Studies*, 4th ser., 25 (1957): 103–115.

"Benjamin Franklin as an American Hero." Association of American Colleges, *Bulletin* 43 (1957): 121–132.

'*The Noble Chieftain and the Illustrious Sage:' Washington and Franklin*. (Morristown, N.J.: Washington Association of New Jersey, 1957), 13–28.

Introduction to Benjamin Franklin's Essay on Literary Style. Philadelphia: privately printed, 1957.

"Medical Practice in Colonial America." *Bulletin of the History of Medicine* 31 (1957): 442–453.

"Franklin and the 'Wagon Affair,' 1755." With Leonard W. Labaree. American Philosophical Society, *Proceedings* 101 (1957): 551–558.

"The Goal of Independence." *McNeese Review* 9 (1957): 81–90.

"The Stuart W. Jackson Lafayette Collection." *Yale University Library Gazette* 33 (1958): 49–56.

"A Tribute to John Bartram, with a Note on Jacob Engelbrecht." With Ralph L. Ketcham. *Pennsylvania Magazine of History and Biography* 83 (1959): 446–451.

"Autographs and Archives in the American Philosophical Society." *Manuscripts* 11 (1959): 39–47; reprinted with revisions and notes in American Philosophical Society, *Proceedings* 103 (1959): 761–767.

"Editors and Great Men." Kent State University, *Aspects of Librarianship*, no. 23 (1960).

"Michel-Guillaume de Crèvecoeur." In *Unforgettable Americans*, edited by John A. Garraty. New York, 1960, 71–74.

"The Medical Institution of Yale College." *Yale Journal of Biology and Medicine* 33 (1961): 169–183.

Edited. Francis Hopkinson's Account of the Federal Procession, *Old South Leaflet*, nos. 230–31 (1962).

"Benjamin Franklin and the Practice of Medicine." Cleveland Medical Library Association, *Bulletin*, n.s., 9 (1962): 51–62.

"Franklin among the Philanthropists." National Council on Community Foundations (1962).

"The Federal Processions of 1788." *New York Historical Society Quarterly* 46 (1962): 5–39.

"Editing a Scientist's Papers." *Isis* 53 (1962): 14–19.

"The American Philosophical Society as a National Academy of Sciences, 1780–1846." Tenth International Congress of the History of Science, *Proceedings* (Paris, 1962), 165–175.

"John Morgan." *Encyclopedia Britannica* 15 (1963): 801.

"Henry Stevens, his Uncle Samuel, and the Franklin Papers." Massachusetts Historical Society, *Proceedings* 72 (1963): 143–211.

"Old West, Dickinson College." *Dickinson Alumnus* 41, no. 1 (1963): 8–10; also in *Congressional Record*, 21 May 1963.

"Astronomical Observatories of the American Philosophical Society, 1769–1843." American Philosophical Society, *Proceedings* 108 (1964): 7–14.

"The Middle States Tradition in American Historiography: Introduction." American Philosophical Society, *Proceedings* 108 (1964): 145–146.

"The Court-Martial of Dr. William Shippen, Jr., 1780." *Journal of the History of Medicine* 19 (1964): 218–238.

"Dr. James Rush on his Philadelphia, Edinburgh, and London Teachers." *Journal of the History of Medicine* 19 (1964): 419–421.

"The Worlds of Benjamin Franklin." National Microfilm Association, *Proceedings*, Thirteenth Annual Meeting, Philadelphia (1964), 31–40.

"Historical Introduction" to "John Morgan's Medical Thesis, 'Pus Production,' Edinburgh, 1763." College of Physicians of Philadelphia, *Transactions and Studies*, 4th ser., 32 (1965): 123–125.

Introduction to *An Account of Dr. John Morgan by Dr. Benjamin Rush*. Philadelphia: privately printed, 1965.

"Patriot-Improvers: Some Early Delaware Members of the American Philosophical Society." *Delaware History* 11 (1965): 195–207.

"Huxley, Tyndall, and American Gold: A Letter of Oliver Wendell Holmes, 1872." *Journal of the History of Medicine* 20 (1965): 164–165.

"The Mother of Franklin's Son Again." Historical Society of Pennsylvania, *140th Annual Meeting, March 31, 1965* [3–4].

"Preserving the Archives." *Pennsylvania Hospital Bulletin* 21, no. 4 (Summer 1965): 1–6.

"North American and West Indian Medical Graduates of Glasgow and Aberdeen to 1800." *Journal of the History of Medicine* 20 (1965): 411–414.

"The American Philosophical Society and Medicine." *Bulletin of the History of Medicine* 40 (1966): 112–123.

Report of the Committee on Library, 1966–1981. Philadelphia: American Philosophical Society, 1967–1982. (Also in American Philosophical Society *Year Book*, 1966–1981.)

"Medicine: Foster Mother of the Sciences." *Journal of the American Medical Association* 196 (4 April 1966): 50–54.

"James Hutchinson (1752–1793): Letters from an American Student in London." College of Physicians of Philadelphia, *Transactions and Studies*, 4th ser., 34 (1966): 20–25.

"Autobiographical Verses by John Rogers, M.D." *Journal of the History of Medicine* 21 (1966): 413–414.

"The Forgotten Woman in Franklin's Life." *Sunday Bulletin Magazine*.

Introduction and bibliographical appendix, James Thacher, *American Medical Biography*. Reprint, New York, 1967.

"Papers for the History of Science." *Quarterly Journal of the Library of Congress* 24 (1967): 163–170.

"The Cabinet of the American Philosophical Society." In *A Cabinet of Curiosities: Five Episodes in the Evolution of American Museums*, ed. Walter M. Whitehill. Charlottesville: University Press of Virginia, 1967, 1–34.

"Lives in Medicine: The Biographical Dictionaries of Thacher, Williams, and Gross." *Bulletin of the History of Medicine* 42 (1968): 101–120.

"An American Medical Plutarch." *Journal of the American Medical Association* 204 (1 April 1968): 11–14.

"James Hutchinson (1752–1793): A Physician in Politics." In *Medicine, Science and Culture: Historical Essays in Honor of Owsei Temkin*, ed. L. G. Stevenson and Robert P. Multhauf. Baltimore, 1968, 265–283.

"The American Philosophical Society." *ACLS Newsletter* 19 (1968): 1–4.

"Mosquitoes as Therapists." *Journal of the History of Medicine* 23 (1968): 108.

"Body-Snatching in Philadelphia." *Journal of the History of Medicine* 23 (1968): 108–110.

"Rattlesnakes and Hummingbirds: Philadelphia's Resources for the History of Science." *Papers* of the Bibliographical Society of America 64 (1970): 13–27.

"Local Scientific Societies in Nineteenth Century America." *The Pennsylvanian* 21 (May 1970): 1–9.

"Everett T. Tomlinson, New Jersey Novelist of the American Revolution." In *New Jersey in the American Revolution, Political and Social Conflict: Papers presented at the First Annual New Jersey History Symposium . . . December 6, 1969* (Trenton, 1970), 62–72; revised edition, Trenton: New Jersey Historical Commission, 1974, 77–88.

"John Bartram." *Dictionary of Scientific Biography* 1 (1970): 486–488.

"William Bartram." *Dictionary of Scientific Biography* 1 (1970): 488–490.

"A Portrait of the Colonial Physician." *Bulletin of the History of Medicine* 44 (1970): 497–517.

Edited. *Dear Friend at Home: Letters written by Nathan Guilford on a Journey to Kentucky in 1814.* North Hills, Pa.: Bird & Bull Press, 1970, 49–88.

Introduction to *The Complete Poor Richard's Almanacs.* Barre, Mass.: Imprint Society, 1970: v–xxii.

"William Shippen, Jr.'s Introductory Lecture, 1762." *Journal of the History of Medicine* 25 (1970): 478–479.

"For Mutual Improvement in the Healing Art: Philadelphia Medical Societies of the 18th Century." *Journal of the American Medical Association* 216 (5 April 1971): 125–129.

"Jeremiah Dixon." *Dictionary of Scientific Biography* 4 (1971): 131–132.

"The King of France's Picture." *University Hospital Antiques Show 1971* [Catalogue], 190–191.

"Benjamin Smith Barton, M.D. (Kiel)." *Journal of the History of Medicine and Allied Sciences* 26 (1971): 197–203.

"Sarah (Franklin) Bache." In *Notable American Women, 1607–1950,* edited by Edwin T. James (1971), 1: 75–76.

"Promoting Useful Knowledge: The How and Why of the American Philosophical Society Library." Pennsylvania Library Association *Bulletin* 27 (1972): 141–146.

"As Others Saw Us: Notes on the Reputation of the American Philosophical Society." American Philosophical Society, *Proceedings* 116 (1972): 269–278.

"[Richard H. Shryock]." *American Historical Review* 77 (1972): 1203–1205.

"Richard Harrison Shryock." *Bulletin of the History of Medicine* 46 (1972): 499–503.

"The Amateur Historian." *New York History* 53 (1972): 265–282.

Introduction to *Benjamin Franklin: A Biography in His Own Words*, ed. Thomas Fleming. New York: Harper and Row, 1972, 6–11.

"Richard Harlan." *Dictionary of Scientific Biography* 6 (1972): 119–121.

"The Old Library of the Pennsylvania Hospital." *Bulletin of the Medical Library Association* 60 (1972): 543–560.

"Memoir of Richard Harrison Shryock (1893–1972)." College of Physicians of Philadelphia, *Transactions and Studies*, 4th ser., 40 (1973): 202–204.

"Joseph M. Toner (1825–1896) as a Medical Historian." *Bulletin of the History of Medicine* 47 (1973): 1–24; abridged in *Journal of the American Medical Association* 224 (2 April 1973): 107–111.

"Isaac Lea." *Dictionary of Scientific Biography* 8 (1973): 103–104.

Introduction to John Bartram et al., *A Journey from Pennsylvania to Onondaga in 1743*. Barre, Mass.: Imprint Society, 1973.

[Unsigned.] "Dr. Franklin: 'Travelling is one Way of Lengthening Life,'" in Oliver W. Holmes, *Shall Stage Coaches Carry the Mail? With a Supplement of Documentary Accounts of Travel . . .* Washington, 1972, 32–39.

"Nathan Smith Davis: An Autobiographical Letter, Previously Unpublished." *Journal of the American Medical Association* 224 (14 May 1973): 1014–1016.

"Painted Portraits and Busts in the American Philosophical Society." *Antiques* 104 (1973): 878–894.

"Richard H. Shryock: Life and Work of a Historian." *Journal of the History of Medicine* 29 (1974): 15–31.

"Addenda to Watson's *Annals of Philadelphia*: Notes by Jacob Mordecai, 1836." *Pennsylvania Magazine of History and Biography* 98 (1974): 131–70.

"The Writing of American Medical History before Professionalization." College of Physicians of Philadelphia, *Transactions and Studies* 4th ser., 42 (1974): 49–60.

"Samuel George Morton." *Dictionary of Scientific Biography* 9 (1974): 540–41.

"Catherine Drinker Bowen (1897–1973)." American Philosophical Society *Year Book 1974* (1975): 115–20.

"I. Minis Hays, Secretary, Librarian, and Benefactor of the American Philosophical Society." American Philosophical Society, *Proceedings* 119 (1975): 401–411.

The Bust of Thomas Paine. Philadelphia: Friends of the Library of the American Philosophical Society, 1974.

"W. B. McDaniel, 2d (1897–1975)." *Bulletin of the History of Medicine* 49 (1975): 429–31.

"Charles Willson Peale." *Dictionary of Scientific Biography* 10 (1974): 438–39.

"Practitioners of History: Philadelphia Medical Historians before 1925." *Bulletin of the History of Medicine* 50 (1976): 73–92.

(With George W. Corner.) "The American Philosophical Society—From Franklin to Burmese Earthworms." In *Philadelphia's Publishers and Printers: An Informal History*, ed. R. Kenneth Bussy. Philadelphia Book Clinic, 1976, 6–10.

"Betsy Copping Corner, 1888–1976." *Journal of the History of Medicine* 31 (1976): 224–225.

"Oliver Wendell Holmes Declines an Invitation." *Journal of the History of Medicine* 31 (1976): 369–370.

"American Philosophical Society." *Dictionary of American History* 1 (1976): 110–111.

"Caspar Wistar." *Dictionary of Scientific Biography* 14 (1976): 456–457.

"American Medicine and National Independence 1765–1789." *Pathologist* 31, no. 5 (1977): 269–273.

The Declaration of Independence: Four 1776 Versions. Philadelphia: American Philosophical Society, 1976. 23 p.

'. . . the best Company in the World . . .': Lord Le Despencer and Benjamin Franklin at West Wycombe Park.* Philadelphia: The Friends of the Library of the American Philosophical Society, 1977. 32 p.

"History of Anatomical Instruction in New England in a Letter of Benjamin Waterhouse, 1808." *Journal of the History of Medicine* 33 (1978): 215–17.

"Martha Brand (1755?–1814): An Early American Physician." *Journal of the History of Medicine* 33 (1978): 218–19.

'Towards a National Spirit': Collecting and Publishing in the Early Republic to 1830. Sixth Annual Maury A. Bromsen Lecture in Humanistic Bibliography. Boston: Boston Public Library, 1978. 37 p.

With L. H. Butterfield. Introduction to *My Dearest Julia: The Love Letters of Dr. Benjamin Rush to Julia Stockton*. Philadelphia: Rosenbach Foundation, xi–xvii.

"Lessing J. Rosenwald." American Philosophical Society *Year Book 1978* (1979): 103–06.

"Medicine in Boston and Philadelphia: Comparisons and Contrasts, 1750–1820." In *Medicine in Colonial Massachusetts, 1620–1820.* Publications of the Colonial Society of Massachusetts 57 (1980): 159–83.

"Leonard Woods Labaree (1897–1980)." American Philosophical Society *Year Book 1980* (1981): 603–07.

"Charles Coleman Sellers (1903–1980)." American Philosophical Society *Year Book 1980* (1981): 622–25.

"George Washington Corner (1889–1981)." *Journal of the History of Medicine* 37 (1982): 77–78.

"George W. Corner (1889–1981): Historian." *Bulletin of the History of Medicine* 56 (1982): 93–98.

"Private Physicians and Public Collections: Medical Libraries in the United States before 1900." In *A Celebration of Medical History*, ed. Lloyd G. Stevenson. Baltimore: Johns Hopkins University Press, 1982, 85–102.

"Leonard Woods Labaree." Massachusetts Historical Society, *Proceedings* 92 (1980): 156–60.

"Views from Another East Berlin." With George F. Kennan. American Philosophical Society, *Proceedings* 126 (1982): 472–80.

"Lyman Henry Butterfield." Massachusetts Historical Society, *Proceedings* 94 (1982): 107–110.

"Oliver Wendell Holmes." Massachusetts Historical Society, *Proceedings* 94 (1982): 81–85.

"Lyman Henry Butterfield, An Appreciation." *William and Mary Quarterly*, 3d ser., 40 (1983): 166–67.

"George W. Corner (1889–1981)." With Jane M. Oppenheimer. American Philosophical Society *Year Book 1982* (1983): 460–68.

"Julian P. Boyd (1903–1980)." American Philosophical Society *Year Book 1982* (1983): 441–48.

"Lyman H. Butterfield (1909–1982)." American Philosophical Society *Year Book 1982* (1983): 449–53.

"Josiah Wedgwood, Benjamin Franklin, and Their Mutual Friends." *Proceedings of the Fifteenth Annual Wedgwood International Seminar*, Philadelphia, 1970 (published 1982), 54–74.

Introduction to *A Catalogue of the Manuscripts and Archives of the Library of the College of Physicians of Philadelphia*, ed. Rudolf Hirsch. Philadelphia, 1983, xv–xviii.

"Lyman Henry Butterfield." American Antiquarian Society, *Proceedings* 93 (1983): 47–51.

Introduction

THESE twelve essays reflect Dr. Whitfield J. Bell, Jr.'s interests not only as a distinguished scholar of Benjamin Franklin and of the cultural and scientific life of early America, but also as Librarian and Executive Officer of the American Philosophical Society. A majority of the authors are distinguished as members of the Society, and the others are well-equipped to make significant contributions to this solid piece of scholarship of high quality. In addition to broad themes which link many of the essays, connections to the Society's history provide a cohesive bond. The essays appear in chronological order.

Several essays are biographical. Esmond Wright of the University of London, deals with Benjamin Franklin the Old England Man, who loved England, yet helped destroy the fragile china vase, the British Empire. The author discusses much of what made Franklin feel so at home in England and effectively uses Franklin's own words to express those sentiments and also his important feelings as an American. The Revolution sparked many humanitarian reforms, such as the founding of the Prison Society in Pennsylvania which Marvin Wolfgang portrays as a positive consequence of independence. In particular he examines concern with the plight of prisoners and the burning issues of the period from 1787 to 1829, namely, whether prisoners should be held in solitary confinement, and whether such prisoners should be allowed to engage in labor. The nation's oldest prison society was the driving force behind "the model for cultural diffusion throughout Europe and Asia." Independence involved conflict and exerted major influences on the first medical school graduates of the College of

Philadelphia (University of Pennsylvania), who had to deal almost immediately with the medical side of warfare. As Randolph S. Klein shows, their lives reveal much about science and professional careers in the early republic. Their roles in combatting smallpox, hospital fever and mismanagement during the war, serving the public on the national, state, or local level, and establishing a solid medical legacy will introduce the reader to some of Dr. Bell's "old friends." Promoting science in another way was Benjamin Henry Latrobe, the talented engineer and architect, who receives expert treatment from the editor of the Latrobe papers, Edward C. Carter II. This author explains the relationship among Latrobe, the American Philosophical Society, and the advancement of science and art. Clearly Latrobe gained in reputation from the association which facilitated entry to more influential circles. The Society in turn benefited from his considerable abilities and energy. Bentley Glass and Jane M. Oppenheimer deal with eighteenth- and nineteenth-century biological scientists. Dr. Oppenheimer evaluates the work of Louis Agassiz, the first embryologist in America, who helped establish the National Academy of Sciences. Agassiz brought embryology to America from Europe and relied upon his European training when producing an extensive series of works which Dr. Oppenheimer discusses. Dr. Glass successfully probes the details of the lives of many biologists and discovers patterns which explain why some were elected to the American Philosophical Society, the nation's first scientific organization, while others were overlooked or scorned. Useful tables of distinguished foreign biologists of the eighteenth and nineteenth centuries, including thumbnail sketches and summaries of their accomplishments, accompany his analysis.

Another approach to history lies in Silvio Bedini's study of the observatory behind Independence Hall where eighteenth-century members of the Society observed the transit of Venus and helped earn the young organization an international reputation for excellence. His story involves a fascinating account of efforts to fix the exact location of the first astronomical observatory erected in the British colonies. "That Awfull Stage" later served as the rostrum from which the Declaration of Independence was first publicly proclaimed. The search involved numerous individuals and groups over the ensuing years. Thomas Cochran's comparative analysis of the Industrial Revolution in England and America reveals a major misconception in the literature, and provides a perceptive antidote to the cotton-textile myth of the Industrial Revolution. Professor Cochran's geo-cultural approach makes it clear that the rise of industrialism came about for many causes which operated differently in different areas. His perceptive analysis of the development in America clearly explains why using the British pattern as a standard inhibits understanding. Brooke Hindle's profusely illustrated article demonstrates that three-dimensional survivals or material culture of that Revolution provide the key to a different access to understanding, which lies beyond the written record's ability to relate. After identifying some of the promise and limitations of such sources, he skillfully develops his insights by examining artifacts including the *John Bull* engine and its adaption to American uses, Oliver Evans's automated grist mill, and John Augustus Roebling's bridges.

Whit Bell never neglected the opportunity to promote useful research tools. Joseph Ewan's "Catalog of Books Belonging to Benjamin Smith Barton" and Edwin Wolf's "Benjamin Franklin's Medical Books" are only the most recent examples. Barton's was the largest library of natural history in Jefferson's time. Today most of his books can be found at the Library of the Pennsylvania Hospital, American Philosophical Society, and Library Company of Philadelphia. An amusing squabble among heirs accompanies an analysis and list of the Franklin collection. Although few of the actual volumes which Franklin owned were located, most of the titles can be found among the rich holdings of the Historical Society of Pennsylvania, Library Company of Philadelphia, and American Philosophical Society. Finally, William W. Abbot displays the historical editor's craft in his introduction and annotations to letters which William Byrd III of Westover wrote while recruiting Cherokees for Lord Loudoun's campaign against Ft. Duquesne during the French and Indian War. Indians were deemed essential to the campaign; however, the young Virginian found gathering recruits and keeping them on hand until battle a frustrating experience.

This book begins with an appreciation of Whitfield J. Bell, Jr.'s great contribution to the American Philosophical Society by Jonathan E. Rhoads, who was President of the Society at the time its Executive Officer retired. It also contains a biographical sketch of Dr. Bell and a select bibliography of his writings. A detailed index and numerous appropriate illustrations complement the essays. All involved in this tribute hope that those interested in science and early American society will find it lively and enlightening.

Philadelphia, 29 June 1985   RANDOLPH SHIPLEY KLEIN

Contributors

WILLIAM W. ABBOT, Editor of *The Papers of George Washington*, was on the faculty of the College of William and Mary, Northwestern University, and Rice University before accepting appointment as professor at the University of Virginia. For over a decade he served as book review editor and then editor of the *William and Mary Quarterly*.

SILVIO A. BEDINI, APS, Keeper of the Rare Books of the Smithsonian Institution, was for many years Assistant then Deputy Director of the National Museum of History and Technology (now the National Museum of American History). The author of numerous articles on horological history and early American science, his books include *Early American Scientific Instruments and Their Makers* (1964); (with Werner von Braun and Fred L. Whipple) *Moon, Man's Greatest Adventure* (1970); *The Life of Benjamin Banneker* (1972); *Thinkers and Tinkers, Early American Men of Science* (1975); and *Thomas Jefferson and His Copying Machines* (1984).

EDWARD C. CARTER II, APS, the Librarian of the American Philosophical Society, is also Editor of *The Papers of Benjamin Henry Latrobe*. In 1985 Yale University Press published the critically acclaimed *Latrobe's View of America, 1795–1820*; the ninth and final volume dealing with the father of American professional architecture and engineering will appear in 1988. Before coming to the Society Dr. Carter was professor of history at the Catholic University of America. He is now adjunct professor of history and of the history and sociology of science at the University of Pennsylvania.

35

36   *Science and Society in Early America*

THOMAS C. COCHRAN, APS, Benjamin Franklin Professor Emeritus of History, University of Pennsylvania, is the author of *Age of Enterprise, The American Business System, Social Change in America, The Great Depression and World War II*, and other books. He has served as president of the American Historical Association, Organization of American Historians, and Economic History Association, as well as Pitt Professor at Cambridge and a fellow at St. Anthony's College, Oxford.

JOSEPH EWAN, Ida Richardson Professor Emeritus of Botany, Tulane University, has many research interests including taxonomy, bibliography, and the history of natural history. This former Guggenheim fellow has had several research grants from the Society. Among his books are the *Botanical and Zoological Drawings of William Bartram*, and, with Nesta Ewan, *Biographical Dictionary of Rocky Mountain Naturalists*, and *John Banister and His Natural History of Virginia, 1756–1792*.

BENTLEY GLASS, APS, is Director of the American Philosophical Society's History of Genetics Program, and Distinguished Professor Emeritus of Biology, State University of New York, Stony Brook. This eminent geneticist and member of the National Academy of Sciences is a leader in efforts to improve science education, and to inform the public at large in regard to scientific problems of general concern, such as the extent and nature of radiation hazards, the probable effects of nuclear warfare, problems of disarmament, and kindred subjects. His scientific articles and contributions to the history of science are extensive.

BROOKE HINDLE, APS, Historian Emeritus at the National Museum of American History, Smithsonian Institution, served earlier as its director. Most of his career was spent at New York University where he was professor of history and, for a time, dean of University College of Arts and Sciences. His major works are:

The Pursuit of Science in Revolutionary America, David Rittenhouse, Technology in Early America, and *Emulation and Invention.*

RANDOLPH SHIPLEY KLEIN, Assistant Secretary, American Philosophical Society, was a faculty member of the University of Wisconsin – Stevens Point, Connecticut College, and the University of Pennsylvania, before coming to the Society as Assistant to the Executive Officer. He has written articles on the history of medicine, patriotism, and Revolutionary America. The University of Pennsylvania published his book *Portrait of An American Family, The Shippens of Pennsylvania Across Five Generations.*

JANE M. OPPENHEIMER, APS, Professor Emeritus of Biology and the History of Science, Bryn Mawr College, twice won Guggenheim fellowships. She is a former president of the American Society of Zoologists and has served on the council of the History of Science Society, American Association for the History of Medicine, American Institute of Biological Sciences, and APS. She is the author of *New Aspects of John & William Hunter, Essays in the History of Embryology and Biology,* as well as many articles, and editor with B. H. Willier of *Foundations of Experimental Biology.*

EDWIN WOLF 2ND, APS, Librarian Emeritus, The Library Company of Philadelphia, for thirty years headed the country's oldest private library. This former Guggenheim fellow is past president of the Greater Philadelphia Cultural Alliance, Bibliographic Society of America, and Friends of the University of Pennsylvania Library. Among his many books and articles are *Franklin's Way to Wealth as a Printer, History of the Jews in Philadelphia, The Library of James Logan of Philadelphia,* and *Quarter of a Millennium, Library Company of Philadelphia, 1731–1981.*

MARVIN WOLFGANG, APS, Professor of Criminology and of Law, and Director of the Center for Studies in Criminology and Criminal Law, the Wharton School of the University of Pennsylvania, is president of the American Academy of Political and Social Science and past president of the Pennsylvania Prison Society. This former Fulbright and two-time Guggenheim fellow is author of numerous books and articles including *Patterns in Criminal Homicide, Race and Crime*, with Thorsten Sellin *The Measurement of Delinquency, Crime and Culture*, and with Sir Leon Radzinowicz the three-volume *Crime and Justice*.

ESMOND WRIGHT, Emeritus Professor of American History, University of London, and former Director of the Institute of U.S. Studies, is now Leverhulme Research Fellow at the Institute. He is the author of fifteen books including *George Washington and the American Revolution, Fabric of Freedom, Benjamin Franklin and Independence, American Profiles*, and *The Fire of Liberty*. He is former British editor of the Loyalist Papers Project. In 1986 Harvard University Press will publish his *Franklin of Philadelphia*, the first full-scale biography of Benjamin Franklin in about fifty years.

Benjamin Franklin:
"The Old England Man"

ESMOND WRIGHT

I N the preface to his "Benjamin Franklin at West Wycombe Park" Whit Bell described the documents and books relating to Franklin that Sir Francis Dashwood of Wycombe Park placed on deposit in the library of the American Philosophical Society as evidence of Franklin's "talent for friendship."[1] It is a pleasure to attempt to recount Franklin's "talent for friendship" in a Festschrift that testifies to Whit Bell's own "talent for friendship" for a host of people, and all their varied interests, over a long active and immensely industrious life; and his talent for friendship, like Franklin's, crosses oceans.

The Old England Man

That Benjamin Franklin was an Old England Man needs no stressing. He came first to London in 1724 as an eighteen-year-old apprentice, at the suggestion of Governor Keith—and never put any trust in governors afterwards. He stayed nineteen months, to widen his experience as a printer and to see enough of London life to be driven thereafter to keep a manual for self-denial and self-discipline, and to wish to write *The Art of Virtue.*

Even before his second visit in 1757, he had earned an international reputation, as discoverer of the identification of lightning

---

1. Whitfield J. Bell, Jr., "Benjamin Franklin at West Wycombe Park," The Friends of the Library, American Philosophical Society, 1977.

and electricity, as printer, publisher, and civic planner. He had by that time a host of British friends: the Quaker merchant and botanist Peter Collinson, whose guests he and his son William were on their arrival; another Quaker, the physician and botanist, Dr. John Fothergill; many correspondents of the Royal Society, which had in 1753 awarded Franklin its Copley Medal and which in 1756 would elect him to a Fellowship; William Strachan, the Scottish-born printer and publisher; and Richard Jackson the lawyer, who would succeed him as Pennsylvania agent in 1762, when for two years he returned to Philadelphia.

In the fifteen years of residence (1757–62 and 1764–75) he reigned as uncrowned king at Mrs. Stevenson's home in Craven Street, where he could genially describe himself as "Big Man," "Great Person" and "Dr. Fatsides."[2] The coffee houses of the Strand and the City were close and thus he had a busy social round: erratically at the Dog Tavern or the Pennsylvania Coffee House, Mondays at the George and Vulture, Thursdays at St. Paul's (later the London) Coffee House. He met the great: a visit or two to Lord Despencer's in Wycombe, another to Lord Shelburne's to meet the Abbé Morellet; but he was never of the Establishment. He felt, and it was significant, most at home in Scotland, with Hume and Kames, "Jupiter" Carlyle and "Jolly Jack Phosphorus" (John Anderson). The experiments continued, but with longer and longer interruptions: experiments with his "musical glasses," with heating rooms, with countering colds, with the design of clocks, with phonetics, with the cause of lead poisoning, with a host of topics. And always letters: to a host of friends.

He lived well; he had a coach of his own, hired for twelve guineas a month (the price would go up to £15 on his second trip to Scotland in 1771), and he had two black servants, King and Peter,

---

2. "Big Man, Great Person," in *The Craven Street Gazette*, 22 September 1770, *Papers of Benjamin Franklin* (New Haven: Yale University Press, 1959, ed. Willcox et al.), 17: 220–226 (hereafter *Papers*).

whom he and his son had brought with them from Philadelphia—though King ran away after a year. He sent presents home, two case-loads of them: china and tableware, blankets and tablecloths, a harpsichord costing 42 guineas for Sally and "I also forgot, among the china, to mention a large fine jug for beer, to stand in the cooler. I fell in love with it at first sight; for I thought it looked like a fat jolly dame, clean and tidy, with a neat blue and white calico gown on, good-natured and lovely, and put me in mind of—somebody."[3]

Honors continued: LL.D. of St. Andrews in 1759, a D.C.L. of Oxford in 1762. Certainly Franklin was reluctant in 1762 to leave the home country with its "sensible, virtuous and elegant minds." Writing to "Straney" from Portsmouth, he said: "I cannot, I assure you, quit even this disagreeable place without regret, as it carries me still further from those I love, and from the opportunities of hearing of their welfare. The attraction of reason is at present for the other side of the water, but that of inclination will be for this side. You know which usually prevails. I shall probably make this one vibration and settle here for ever."[4]

"I am going," he wrote to Lord Kames, "from the old world to the new; and I fancy I feel like those who are leaving this world for the next; grief at the parting; fear of the passage; hope of the future. These different passions all affect their minds at once; and these have *tendered* me down exceedingly."[5]

> We had a long passage near ten weeks from Portsmouth to this place [he reported to Strachan from Philadelphia], but it was a pleasant one; for we had ten sail in company and a man-of-war to protect us; we had pleasant weather and fair winds, and frequently visited and dined from ship to ship; we called too at the delightful island of Madeira, by way of half-way house,

3. Benjamin Franklin to Deborah, 19 February 1758, *Papers* 7: 381.
4. Benjamin Franklin to William Strachan, 23 August 1762, *Papers* 10: 149. For Strachan, see J. A. Cochrane, *Dr. Johnson's Printer* (Cambridge, Mass.: Harvard University Press, 1964).
5. Benjamin Franklin to Lord Kames, 17 August 1762, *Papers* 10: 147.

where we replenished our stores and took in many refreshments. It was the time of their vintage, and we hung the ceiling of the cabin with bunches of fine grapes which served as dessert for dinner for some weeks afterwards. The reason of our being so long at sea was that, sailing with a convoy, we could none of us go faster than the slowest, being obliged every day to shorten sail or lay by till they came up. This was the only inconvenience of our having company, which was abundantly made up to us by the sense of greater safety, the mutual good offices daily exchanged, and the other pleasures of society.[6]

And also to Strachan:

I got home well the first of November, and had the happiness to find my little family perfectly well, and that Dr. Smith's report of the diminution of my friends were all false. My house has been full of succession of them from morning to night ever since my arrival, congratulating me on my return with the utmost cordiality and affection. My fellow-citizens, while I was on the sea, had at the annual election chosen me unanimously, as they had done every year while I was in England, to be their representative in Assembly, and would, they say, if I had not disappointed them by coming privately to town, have met me with five hundred horse.[7]

In two years he would, he told Strachan, remove to England, "provided we can persuade the good woman to cross the seas."

"Of all the enviable things England has," Franklin wrote to Polly Stevenson on 25 March 1763, "I envy it most its people. Why should that petty island, which compared to America is but like a stepping-stone in a brook, scarce enough of it above water to keep one's shoes dry; why, I say, should that little island enjoy in almost every neighbourhood more sensible, virtuous and elegant minds than we can collect in ranging a hundred leagues of our vast forests? But 'tis said the arts delight to travel westwards. You have effec-

6. Benjamin Franklin to William Strachan, 7 December 1762, *Papers* 10: 166–7.
7. Benjamin Franklin to William Strachan, 2 December 1762, *Papers* 10: 161.

tually defended us in this glorious war, and in time you will improve us. After the first cares for the necessaries of life are over." Franklin had used almost the same words in his proposal for the American Philosophical Society twenty years before—"We shall come to think of the embellishments. Already some of our young geniuses begin to lisp attempts at painting, poetry and music."[8]

During his third and longest visit even his bitter exchanges with Lord Hillsborough (who from 1768 to 1772 was in effect minister for North American affairs) did not prevent a visit in 1771 to the earl's estate at Hillsborough in County Down—now a center of acute political tension—where the son and heir was charged to act as guide. In the same year he paid two visits to Chilbolton, near Twyford in Hampshire, the handsome Tudor house that his friend Jonathan Shipley, the Bishop of St. Asaph, had fortunately inherited from his wife's family. During the second of these visits the first part of *The Autobiography* was written; and the thank-you letter he wrote on his return to London remains one of the many warm evidences of his uncanny and easy skills with people of all ages, since in it he describes his conversation with Kitty Shipley, the eleven-year-old and the youngest of the five Shipley girls, whom he was accompanying back to her school in Marlborough Street, London, and from whom he coaxes—in the end—a voluble assessment of the men her sisters should marry, and who would best suit her.[9]

Once Dartmouth had replaced Hillsborough, Franklin could write lyrically to his son

> As to my situation here, nothing can be more agreeable, especially as I hope for less embarrassment from the new minister; a general respect paid me by the learned, a number of friends

8. Benjamin Franklin to Polly Stevenson, 25 March 1763, *Papers* 10: 232–3.
9. Benjamin Franklin to Mrs. Anna Mordaunt Shipley, 13 August 1771, *Papers* 18: 201–2. Cf. J. M. Stifler, *My Dear Girl* (New York: George Doran Co., 1927) for the correspondence with Polly Stevenson Hewson and the Shipley girls.

Twyford, at the Bishop of St. Asaph's, where the first part of *The Autobiography* was written, 1771. Courtesy of John Warwick.

and acquaintance among them, with whom I have a pleasing intercourse; a character of so much weight, that it has protected me when some in power would have done me injury, and continued me in an office they would have deprived me of; my company so much desired, that I seldom dine at home in winter, and could spend the whole summer in the country-houses of inviting friends, if I chose it. Learned and ingenious foreigners, that come to England, almost all make a point of visiting me; for my reputation is still higher abroad than here. Several of the foreign ambassadors have assiduously cultivated my acquaintance, treating me as one of their corps, partly I believe from the desire they have, from time to time, of hearing something of American affairs, an object become of importance in foreign courts, who begin to hope Britain's alarming power will be diminished by the defection of her colonies; and partly that they may have an opportunity of introducing me to the gentlemen of their country who desire it.

The King, too, has lately been heard to speak of me with great regard.[10]

The British connection was, however, not a matter merely of comfort, conviviality in coffee-houses, and the enjoyment of occasional contact with the "great men" as well as the "virtuosi" of three kingdoms. He had come in 1757 to persuade the British government to abrogate the proprietary charter of Pennsylvania. He wanted royal government, not the domination of the Penns. The Penn family had ceased to live in Philadelphia, and had become Anglicans. He saw opportunities for America in royal government rather than in government by a proprietor or by anything more democratic. He hoped that the loyalty that he felt for Britain could be translated into a general loyalty, in the concept of a royal colony.

The almost universal "respect for the mother country, and the admiration of everything that is British" on the part of the colonists, Franklin wrote in *The London Chronicle*, was "a natural effect" not only "of their constant intercourse with England, by ships arriving almost every week from the capital," but also, and more importantly, of an ingrained loyalty to a country that, for all free whites, permitted far more liberty than was enjoyed by the colonists of any European power. "Delegates of [British] power" in the colonies might indeed lose the respect of and "give Jealousy to" the colonists either by their corrupt behavior or "by continually abusing and calumniating the People." But such actions did not diminish colonial faith in imperial institutions. "Confidence in the Crown" remained "as great as ever," and Parliament was held in the highest esteem as the ultimate protector of British liberty—in the colonies as well as in Britain.[11]

10. Benjamin Franklin to William, 19 August 1772, *Papers* 19: 258–9.
11. Benjamin Franklin to *The London Chronicle* (editor William Strachan), 16–19 September 1758, 28–30 December 1758, 9 May 1759, 24 November 1767, in *Papers*, passim.

Franklin was rooted in London not only by his task and by seventeen years of happy residence "and everything about us pretty genteel," but by jobs, near-sinecures and by royal approbation.[12] The agency of Pennsylvania came of course from the decision of the Pennsylvania Assembly; and the later agencies for Georgia, New Jersey and Massachusetts could be seen as stamps of colonial approval for the man who in role if not in title was becoming America's ambassador at the Court of St. James. This role, however, did not interfere with his postmastership, which he had obtained by dexterous letter-writing to the right people, and from which appointment as base he could reward with similar jobs his own relatives in familiar patronage fashion. Even if in 1768 he failed to obtain the Colonial Undersecretaryship that he hoped for, his standing had, as early as 1762, helped secure a royal governorship for his son, illegitimate though he was. In family and political as well as personal terms Franklin had a lot to lose when the break came. It was not of his doing, nor was it even in 1775 his quest. He might, equally easily, have become a royal governor himself: his correspondent Cadwallader Colden in New York had become a Lieut.-Governor; in the war that came men of no greater standing (and with less skill and acclaim as scientists and "doers") received British government appointments, and one of them became a count of the Holy Roman Empire. It could as easily have gone that way with Franklin.

Why then did this lover of Britain and of all mankind, and this most unrevolutionary of men, after fifteen years of comfort and esteem, return to his native country and a fevered round of activities, including the drafting and amending of a Declaration of Independence?

The Franklin who arrived in England in 1757 was an experi-

12. Benjamin Franklin to Deborah, January 1758, *Papers* 7: 369.

enced political operator, who had come to the capital of his Empire but as an *Auslander*. His view, and that of Whitehall, were never in harmony. The gap was at one point—the repeal of the Stamp Act—almost bridged; and three years later he almost became part of the Home Establishment. But it was not to be, and some features of the London years steadily widened rather than bridged the rift.

There were features in the imperial relationship that he deplored from the start. On arrival in 1757 he asked why a " 'mother country' . . . would introduce thieves and criminals into the company of her children, to corrupt and disgrace them?" First-hand observation beginning in 1757 provided Franklin with some explanation of why the imperial government could persist in such an "inhuman" policy. The British nation, he wrote his protégé Joseph Galloway in early 1758, seemed to be almost "universally corrupt and rotten from Head to Foot." But it was not primarily a matter of "corruption." Franklin was appalled to find, as his son remarked, "little knowledge of (or indeed Inclination to know) American Affairs" and strong "Prejudices against the Colonies in general." Moreover, many people in power seemed to entertain deep "suspicions and jealousies" of the colonies, suspicions and jealousies nourished by the conviction that the colonies were "not sufficiently obedient," and manifest in fears that they might begin to "feel their own strength" and either enter into competition with the metropolis, or throw off their political dependence and divert the lucrative colonial trade to one of Britain's European rivals. The results were both a widespread feeling within the British political nation that the growth of the colonies should be stinted, lest "the children might in time be as tall as their mother," and strong sentiments for reducing "the People's Privileges" in the colonies. In March 1759, after nearly two years' residence, he wrote to Isaac Norris, "The Prevailing Opinion, . . . among the Ministers and Great Men here is, that the Colonies have too many and too great Privileges; and that it is not only the Interest of the Crown but of the Nation to reduce them . . . and to clip the Wings of (the colonial) Assemblies in their Claims of all the Privileges of a House of Commons."[13]

Within a year after his arrival in England in 1757, he had become convinced that the colonies would be vulnerable to a "general Attack" upon them by the imperial government until their constitutions had been "settled firmly on the foundations of Equity and English Liberty." As he wrote Peter Collinson in April 1764,

> We are in your Hands as Clay in the Hands of the Potter; and so in one more Particular than is generally consider'd: for as the Potter cannot waste or spoil his Clay without injuring himself; so I think there is scarce anything you can do that may be hurtful to us, but what will be as much or more so to you. This must be our chief Security; for Interest with you we have but little: The West Indians vastly outweigh us of the Northern Colonies. What we get above a Subsistence, we lay out with you for your Manufactures. Therefore what you get from us in Taxes you must lose in Trade. The Cat can yield but her Skin. And you must have the whole Hide, if you first cut Thongs out of it, 'tis at your own Expence. The same in regard to our Trade with the foreign West India Islands: If you restrain it in any Degree, you restrain in the same Proportion our Power of making Remittances to you, and of course our Demand for your Goods; for you will not clothe us out of Charity, tho' to receive 100 per Cent for it, in Heaven. In time perhaps Mankind may be wise enough to let Trade take its own Course, find its own Channels, and regulate its own Proportions, &c. At present, most of the Edicts of Princes, Placaerts (Placets), Laws and Ordinances of Kingdoms and States, for that purpose, prove political Blunders. The Advan-

---

13. To Isaac Norris, 19 March 1759, *Papers* 8: 295. The most savage expression of this view was in *Rules by which a great Empire may be reduced to a small one*, in the *Public Advertiser*, 11 September 1773, *Papers* 20: 389–399. But it began earlier: see Benjamin Franklin to Galloway, 17 February 1758, *Papers* 7: 375. Cf. Willard Randall, *A Little Revenge* (Boston: Little Brown, 1984) for a fascinating new portrait of William Franklin.

tages they produce not being *general* for the Commonwealth; but *particular*, to private Persons or Bodies in the State who procur'd them, and *at the Expence of the rest of the People.*[14]

From the start Franklin moved, moreover, almost exclusively in a radical, dissenting, Quaker or Scottish world. His natural setting was in the Club of Honest Whigs. It usually met in the St. Paul's (later the London) Coffee House, on Thursday evenings—not far indeed from Craven Street. This was the group to which in retrospect Franklin looked back with warmest memories: Collinson and Fothergill, Pringle, Price and Priestley, James Burgh and William Rose and John Canton; wherever he went Franklin attracted the questioning, dissenting and scientifically curious. It was to such plain folk with stimulating minds that Franklin responded. In Scotland he had "six weeks of the densest happiness," meeting William Robertson, David Hume, Lord Kames (whose house guest he was in 1771), Sir Alexander Dick, and Adam Smith. Whatever the attitudes of Principal Robertson or of philosopher David Hume, the "tone" of Edinburgh and St. Andrews's student opinion was radical—and it was to them that Franklin directed a generation of Pennsylvania students, especially in medicine. Although the Scottish MP's were a solid phalanx of royal supporters, the two countries were still uneasily conjoined, and political union was sadly threatened by Jacobite invasions in 1715 and 1745. In much of the contemporary discussion of federalism, Scotland—and Ireland—were as often cited as America; the union was still novel, and even fragile, and Bute in 1761 a name for scorn. This Scottish tone did not endear Franklin to those in power.[15]

He had, moreover, few close friendships with any of "the great

14. Benjamin Franklin to Peter Collinson, April 1764, *Papers* 11: 181–2.

15. Benjamin Franklin to Lord Kames, 3 January 1760, *Papers* 9: 5–6. A readable and anecdotal account of father and son in Scotland is in J. B. Nolan, *Benjamin Franklin in Scotland and Ireland* (Philadelphia: University of Pennsylvania Press, 1938).

People" in government, or outside his own rather confined world. He seems to have met Boswell occasionally, Dr. Johnson once only. For the most part he met officials and these only over business matters. As early as January 1766, as he wrote in *The Gazetteer*, he saw evidence of English snobbery.

> Give me leave, Master John Bull, to remind you, that you are related to all mankind; and therefore it less becomes you than anybody, to affront and abuse other nations. But you have mixed with your many virtues a pride, a haughtiness, and an insolent contempt for all but yourself, that, I am afraid, will if not abated, procure you one day or other a handsome drubbing. Besides your rudeness to foreigners, you are far from being civil even to your own family. The Welch you have always despised for submitting to your government: but why despise your own English, who conquered and settled Ireland for you; who conquered and settled America for you? Yet these you now think you may treat as you please, because forsooth, they are a conquered people. Why despise the Scotch, who fight and die for you all over the world? Remember you courted Scotland for one hundred years, and you would fain have had your wicked will of her. She virtuously resisted all your importunities; but at length kindly consented to become your lawful wife. You then solemnly promised to love, cherish, and honour her, as long as you both should live; and yet you have ever since treated her with the utmost contumely, which you now begin to extend to your common children. But, pray, when your enemies are uniting in a Family Compact against you, can it be discreet in you to kick up in your own house a Family Quarrel? And at the very time you are inviting foreigners to settle on your lands, and when you have more to settle than ever you had before, is it prudent to suffer your lawyer, Vindex, to abuse those who have settled there already, because they cannot yet speak "plain English"?—It is my opinion, Master Bull, that the Scotch and Irish, as well as the Colonists, are capable of speaking much plainer English than they ever yet

spoke, but which I hope they will never be provoked to speak.[16]

He had a right to these views, for he had as the son of a small tradesman, one of a family of seventeen, become a scientist of international standing, one who could stand with kings and keep the common touch. But perhaps this was, in the eighteenth century, his special uniqueness?

Nor did his penchant for journalism win him approval in high places. It was, inevitably, his own element, the ladder up which he had himself climbed in America. London seemed made for him: with four daily and six thrice-weekly newspapers, its coffee houses rumor-mills with American affairs a remorselessly recurring theme. The newspaper had been his instrument in Pennsylvania to promote all his activities, and he was adept at telling tales, inventing hoaxes, and so adept that he wrote as much under pseudonym as under his own name. Too often he made enemies; in any case, it was not yet a political ladder up which to climb in London, or for that matter in Paris; even Montesquieu and Voltaire were forced to spend many years in exile.

Again, as agent of—in the end—four colonies, he was thrust into the role of critic, the voice of opposition. He won decisively once, on the repeal of the Stamp Act. He won, on points as it were, against the Penns. He lost totally in 1774. In between, though fertile in expedients, and constructive, as in the suggestion of raising a revenue from a land bank, or on representation in Parliament, or on his western plans, he was listened to—but no more. On each topic he marshaled his arguments impressively, but he was not a lawyer by training; he lacked the lawyer's skill to make his case compellingly, but without personal rancor creeping in. But who knows the emotional storms and inner turmoil that lay behind Franklin's constant

16. Benjamin Franklin in *The Gazetteer*, 14 January 1766. *Papers* 13: 47–48.

smile, his blarney and his efforts at persuading those who he knew felt superior, with whom there was no marriage of minds?

And, again, when his luck ran out it did so decisively in 1774. The repeal of the Stamp Act in 1766 seemed a triumph, like his son's governorship four years before. He spoke by 1770 as agent for four colonies; and in 1770 only one tax remained, the tax on tea, and that as much for the principle as for any revenue it brought. Crisis ceased to be annual—but British troops were now in or near Boston, and the already unpopular customs service was instructed to increase its vigilance. Franklin in London might not have fully sensed the unease and restlessness of the Atlantic coastal ports, but he fed fuel to incipient flames when he obtained and sent to Boston the letters that Governor Hutchinson had years before written to William Whately, Grenville's man of business. He did so, he claimed, to make plain to his Boston paymasters that the mature, vain, but proud governor, who with his family was now monopolizing the tea trade, had never been a friend of colonial ambitions. The letters made clear that London policy, so he believed, originated in fact in a Boston-born Brahmin. And he was unlucky when the petition he brought on the colony's behalf before the Privy Council for the removal of the governor, coincided with the news of the Boston Tea Party. The case, in form that of Massachusetts versus Hutchinson, became in fact that of the British Government versus Franklin. He never forgot nor forgave the bitterness of Solicitor-General Wedderburn's attack upon him, nor its invective: Franklin was a thief, a man without honor, "the true incendiary . . . and abettor," "the first mover and prime conductor," "the actor and secret spring" of the Committee of Correspondence in Boston; the object was to remove Hutchinson—who had discerned their intentions—and to set up a republic; Franklin was indeed already behaving as though he were the minister of an independent and rebellious state. There was here an accumulated dislike that had been building up for twenty years: of the Pennsylvania boss, of the critic of proprietary government, of the colonel in whose honor a suspect militia fired salutes, of the writer of anonymous articles, of the officeholder revealed as conspirator, of the all-too-knowledge-able expert witness of 1766, of the know-all agent-general of colonial assemblies claiming rights of which they had no title. It was on all these resentments that Wedderburn fed. The solicitor-general's effort was a political and personal triumph. The petition for Hutchinson's removal was pronounced "groundless, vexatious, and scandalous"; Franklin was dismissed from his deputy post-master-generalship; and in Parliament the coercive acts were passed to close the port of Boston, to remove the capital to Salem, to curb town meetings and fetter political freedom in Massachu-setts; and if Hutchinson lost his post as governor it was only to be replaced by a soldier, Commander-in-Chief Gage. Boston, obsti-nate and ungovernable as it now seemed, must be brought to heel. Franklin's years of devious politics had reaped a whirlwind. He now stood revealed as Dr. Doubleface, "the Judas of Craven Street."

Franklin sailed from Portsmouth 20 March 1775. "It seems that I am too much of an American," he said. Only a month before he had learned of Deborah's death in Philadelphia (19 December). He was sixty-nine years old; he would only see England once again, when ten years later, he crossed from Le Havre to Southampton to sail home after the war, after the peace was won. But there was now a streak of bitterness behind the bland exterior. And with justice, since it was hard not to see the continuing strain of vindictiveness in the English treatment of him—even the lightning-rods of his devising, erected on government buildings, were patriotically changed from pointed to rounded tips. He never forgave easily: his black list included the Penns, William Smith in Philadelphia, Hillsborough, Wedderburn and, now permanently, his own son, still a royal governor, but in his father's eyes putting filial pieties conspicuously below position and place. In fact, the son handled himself in New Jersey with skill, before being imprisoned as a

Loyalist; and he was denied permission to visit his wife as she lay dying. Between father and son the scars of 1774 never healed. The moderate who never sought independence for his country was thus driven reluctantly towards it. Had he not lived another fifteen years, and become organizer of the French Alliance of 1778, of the Treaty of 1783, and Founding Father in Philadelphia in 1787, it might in 1774–75 have appeared a career in all its range that had culminated in disaster.

As it was, great achievements still lay ahead. And among them not the least important feature was that the Treaty of Peace and the generous terms to which Britain agreed was made possible primarily because of Franklin's English friends: David Hartley, who signed the final treaty, Joseph Priestley, who was Shelburne's librarian at Bowood, and Caleb Whitefoord, his Scottish-born neighbor in Craven Street who was present at the signing of the preliminary treaty of peace in November 1782. It was these friends, some of them visiting him in Passy through the war years—for wars then still had a gentlemanly courtesy about them—who formed the line of approach to Shelburne, the old Chathamite-Imperialist with a liberal and non-mercantilist view of empire, the friend and admirer of Adam Smith. And Shelburne was by chance in power in 1782–83 just long enough to obtain a peace treaty, aided by a Parliamentary recess which gave him a rare few months of freedom. When the hounds returned to Westminster, Shelburne did not last long. The Old England Man had, however, turned to diplomatic advantage his years of friendship in Craven Street. Without them it would have been a more difficult settlement to reach, and if it had been left to Vergennes (or to the proposed European mediators) there is a good chance that a much smaller, seaboard state would have emerged. As it was, the new state was independent—a view on which Franklin never shifted ground after 1774—and extensive and westward-looking. The Old England Man took his revenge on Wedderburn by destroying that frail fragile china vase, the British Empire. Yet his last letter to William

Strachan, 5 March 1785, written a few months before the Scotsman's death, ended with the phrase, "with unchangeable esteem."[17]

The esteem, however, indeed the warmth, did not prevent a mounting pride in America. It was his fortune to be at the crossroads at a dramatic moment; it transformed the Old England Man into the last Anglo-American.

Upon the whole, I have lived so great a part of my life in Britain, and have formed so many friendships in it, that I love it, and sincerely wish it prosperity; and therefore wish to see that Union, on which alone I think it can be secured and established. As to America, the advantages of such a Union to her are not so apparent. She may suffer at present under the arbitrary power of this country; she may suffer for a while in a separation from it; but these are temporary evils that she will outgrow. Scotland and Ireland are differently circumstanced. Confined by the sea, they can scarcely increase in numbers, wealth and strength, so as to over-balance England. But America, an immense territory, favoured by Nature with all advantages of climate, soil, great navigable rivers, and lakes, &c. must become a great country, populous and mighty; and will, in a less time than is generally conceived, be able to shake off any shackles that may be imposed on her, and perhaps place them on the imposers. In the mean time every act of oppression will sour their tempers, lessen greatly, if not annihilate the profits of your commerce with them, and hasten their final revolt; for the seeds of liberty are universally found there, and nothing can eradicate them. And yet, there remains among that people, so much respect, veneration and affection for Britain, that, if cultivated prudently, with kind usage, and tenderness for their privileges, they might be easily governed still for ages, without force, or any considerable expense. But I do not see here a sufficient quantity of the wisdom, that is necessary to produce such a conduct, and I lament the want of it.[18]

17. Benjamin Franklin to Strachan, 5 March 1785, in A. H. Smyth, ed., *Writings of Benjamin Franklin* (New York: Macmillan, 1905–7), 9: 291.
18. Benjamin Franklin to Lord Kames, 25 February 1767, *Papers* 14: 69, 116.

Frustration and Benjamin Franklin's Medical Books

EDWIN WOLF 2ND

F RANKLIN's heirs did not heed gentle Isaac Watts's song:

> Birds in their little nests agree;
> And 'tis a shameful Sight
> When Children of one Family
> Fall out, and chide, and fight.[1]

The falling out and chiding occurred first because nephew Jonathan Williams, Jr., after his return from France, was badly treated in a business affair by Franklin's son-in-law, Richard Bache. That permanently soured their relationship. Then, after Franklin's death, Williams lost the medical books promised him by his aged uncle because of a strict legal construction of the will by grandson William Temple Frankin who, insofar as the library was concerned, was the residuary legatee. In papers of the executors of Franklin's will are some contemporary records of the disagreement about the books. Later, on 22 October 1807, Williams poured out his grievances against Bache and Temple Franklin in a twelve-page letter to William Franklin, then alienated from his son Temple and his step-sister Sarah Bache.[2]

---

1. Isaac Watts, *Divine Songs Attempted in Easy Language for the Use of Children* (Philadelphia: Franklin and Hall, 1750), 18.
2. The Executors' Papers are at the American Philosophical Society (PPAmP), the gift of George Vaux; the Williams letter is part of the William Franklin Papers at the Lilly Library, Indiana University, the excerpts here published by per-

The octogenarian philosopher-diplomat, after his return from France, had made a catalogue—arranged by subject—of those books "I had in France and those I left in Philadelphia being now assembled together." After certain specific bequests to three favored institutions and a few individuals, he bequeathed to his Bache grandsons, Benjamin Franklin and William, and to his nephew, Jonathan Williams, "such and so many of my Books as I shall mark on the said Catalogue with their names." "The residue and remainder of all my Books, Manuscripts and Papers," he continued, "I do give to my Grandson William Temple Franklin."[3]

Williams, who had been closely associated with his relative in France, learned of his uncle's death while he was away from Philadelphia. On 24 September 1790, five months after Franklin died, Williams wrote to "The Gentlemen Executors of D^r. Franklin's will," Henry Hill, John Jay, Thomas Hopkinson and Edward Duffield. In his letter Williams stated:

> Sometime in the summer of 1789 I happen'd to mention to the Doctor my intention of going to Europe in the Fall. He then said to me "Take a copy from my catalogue of all the books that are under the head of *Medicine Surgery &c^a.* for these I have bequeathed to you, and you ought to have a list of them to avoid purchasing duplicates."[4]

His statement seventeen years later was more circumstantial:

> My private interviews with him were never formal, in one of these while lying on his bed, he talked of my intended voyage to London for Mrs Williams, and asked me if I had any inten-

---

mission of William Cagle, Librarian. I used a microfilm of the Indiana MS at PPAmP. The letter was brought to my attention by Dr. Whitfield J. Bell, Jr. in 1958.

3. Benjamin Franklin, MS Will Signed, Philadelphia, 17 July 1788, Register of Wills, Philadelphia, on deposit at PPAmP.

4. Copy of Letter from Jonathan Williams to the Executors, Philadelphia, 24 September 1790, Executors' Papers, PPAmP.

tion of pursuing medical studies (having employed my leisure from the winter of 1785 to 1789 in that way). I said vaguely "perhaps so, as I shall never think of Commerce again." He then told me to go into the other Room and under the Pillow of the Sofa to look for a Paper; I did so & brought his will to him. He then looked at me and read that clause which gives me his Medical, anatomical & chirurgical Books, which were classed in the Catalogue & desired me to mark them with my name: some other expressions descriptive of his doubt of ever seeing me again, & of his approaching dissolution affected me, and I said little, but *did* nothing.[5]

Williams left town soon thereafter. On 28 October 1789 he wrote Mary (Polly) Stevenson Hewson asking her to get the list of medical books for him.[6] She was the daughter of Franklin's London landlady. Widowed, Mary Hewson followed Franklin to Philadelphia and remained close to him until his death. Largely on her evidence Williams laid moral claim to the medical books. According to his later account she took his letter to Franklin.

He then called on Benjamin Franklin Bache his Grandson & ordered him to copy from the Library List the class of Books above mentioned. He did so & brought the Paper to the Doctor who put this Paper in the presence of Benjamin into Mrs Hewson's hands saying "these are the Books I give to Cousin Jonathan."[7]

On 23 October 1790 Temple replied to the claim that Williams had lodged with the executors. Because there were no marks on the catalogue, all the books, lacking that expressed intent of the testator, were his.[8] Five days later he told executor Henry Hill that he

5. Jonathan Williams to William Franklin, Philadelphia, 22 October 1807, William Franklin Papers, Lilly Library.

6. "Extract from Jonathan Williams's letter to M[rs]. Mary Hewson dated Richmond Oct[o]. 28th: 1789," enclosed in Williams's letter of 24 September 1790.

7. Williams, 22 October 1807.

8. Copy of Letter from William Temple Franklin to Jonathan Williams, Philadelphia, 23 October 1790, Executors' Papers, PPAmP.

had made a selection of books for Benny Bache. Further, he had divided the medical books, one half for Billy Bache whose choice of a medical profession his grandfather had been unaware of, the other half for Williams. The Baches had accepted their books gratefully. To buttress his argument that Williams had no legal claim, Temple enclosed the list of medical books that Benny Bache had made out, adding, "Mr. Bache however informs me that his Grandfather did not mention at the time that it was for Mr Williams—& never gave any orders for its being forwarded to him."[9]

Williams acknowledged that Temple had "selected a number of Books, to the amount of 121 volumes" which were offered to him. He recollected in 1807:

> This was indeed an elegant selection, you will judge of it by their being amongst them a very noted book called "*The Mariners compass rectified*," which by the bye was the most *useful* Book in the whole set. In my answer I observed that "the value of the Books to me consisted in their being the *identical ones* intended for me, and none others could replace them."[10]

To settle the matter the executors suggested and Temple agreed that the conflicting claims be left to arbitration. Williams left the choice of the arbitrators to Temple who picked Chief Justice Edward Shippen, Attorney General Jared Ingersoll, and a third person whom Williams could not remember in 1807. They all met at

9. Letter from William Temple Franklin to Henry Hill, [Philadelphia,] 28 October 1790, Executors' Papers, PPAmP. The list of "Medecine & Surgery" he enclosed is the basis for part of the reconstruction of Franklin's medical collection. It is in the hand of Benjamin Franklin Bache; the spelling "Medecine" reflects his European education. The MS list Temple Franklin made out of books for Benjamin Franklin Bache is in PPL.

10. Williams, 22 October 1807. There is in the Library Company of Philadelphia (PPL) a copy of Andrew Wakely's *Mariner's Compass Rectified* (London, 1772), with marks found in books that passed through the hands of the bookseller Dufief who handled part of Franklin's library at the time of its dispersal. There is, however, no Franklin shelf-mark, but it may have been on the missing front endpaper.

Mrs. Hewson's house and Williams was prepared to have her corroborate his claim. Temple, however, then insisted "that the Judge & Lawyer should decide upon the Grounds of a strict professional Opinion, without considering any collateral circumstance, or he would not abide by the decision." According to Williams, the meeting then broke up "& every person present, except Temple, shewed indignation at being made parties to such a farce."[11] The upshot was that Williams got no books except for Pringle's *Observations on the Diseases of the Army*, marked as given to him on the Benny Bache manuscript list.

That list which is with the executors' papers is transcribed here and all but eleven of the items identified. Although Franklin owned even more medical works whose known titles are also here recorded, those included on the sheets copied by Bache and headed by him, "Medecine & Surgery," are the more substantial texts. His list contains 108 entries, including duplicate copies or different editions of five. In most instances, where multiple editions exist (marked below with an asterisk), I have cited an edition in the Library Company as possibly the one Franklin had; in other cases I have used a chronologically logical one found in the catalogues of the British Library, Wellcome Medical Library, or National Library of Medicine. Only seven of Franklin's own copies of the books on the "Medecine & Surgery" list have turned up.[12]

On the other hand 153 medical works, including twelve duplicates, formerly owned by Franklin have been located. Copies of seven other titles known from other evidence to have been in Franklin's library have not been found (marked "nf"). These also appear in the following list. Of the 153 pieces, 98 are in the collec-

11. Williams, 22 October 1807.

12. I have used the standard library symbols: PPAmP and PPL as noted above; PHi, Historical Society of Pennsylvania; PPC, College of Physicians, Philadelphia; and others. It is probable that "Medical Miscellanies (Fr & Lat)" and certain that "Medical Pamphlets" refer to bound volumes of pamphlets which may be any of the surviving ones.

tion of the Historical Society of Pennsylvania (now in the custody of the Library Company of Philadelphia), virtually all bound in pamphlet volumes purchased from the bookseller Dufief in 1801–03 by William Duane, sold by him in 1822 to the Athenaeum of Philadelphia, and sold in turn by it to the Society in 1885. Another 43 are at the American Philosophical Society, the majority of them bought from Dufief and at the auction of Shannon and Poalk in 1803, or acquired with the bulk of the Franklin Papers from Charles P. and Mary Fox in 1840. The Library Company's 7 came variously, as did the single items in 5 other institutions.[13]

Franklin's medical books represented a mixture of an eighteenth-century consensus of what was most useful, specific texts dealing with his interests and his own ailments, and a fluttering of writings given him by friends and admirers. The Bache list of *Medecine & Surgery* included a fair proportion of each of the three categories. The works not on that list were overwhelmingly gifts: medical theses, pamphlets dealing with animal magnetism in which Franklin played an official role, and advertisements for various nostrums in pamphlet or broadside form.

Some of the books would have been familiar to many colonial Americans and Franklin's own Library Company had a good number of them on its shelves.[14] One of the most popular "popular" texts of the seventeenth century was *Via Recta ad Vitam Longam* by Tobias Venner who lived to the age of 83. His self-successful advice urged a moderate diet—in an age of monstrous overeating—and the curative waters of Bath. This kind of regimen appealed to Franklin. Hart also advocated it to some extent in his *Diet of the*

13. Edwin Wolf 2nd, "The Reconstruction of Franklin's Library: An Unorthodox Jigsaw Puzzle," *Papers of the Bibliographical Society of America* 56 (1962): 1–16, tells of the dispersal of Franklin's books.

14. Edwin Wolf 2nd, "Medical Books in Colonial Philadelphia," in John B. Blake, ed., *Centenary of Index Medicus* (Bethesda, Maryland, 1980), 72–92, mentions many of the commonly owned works.

Diseased. Dean Swift's friend Arbuthnot and the French chemist Lémery were more general in their writings on food.

The commonsensical advice in the *Aphorisms* of the classical Greek father of medicine, Hippocrates, would have appealed to the Philadelphia pragmatist, as would the more modern scientific tenets of Boerhaave, dean of the Leyden school of medicine, and their exegesis by his disciple, Freiherr van Swieten. Lester King, writing of eighteenth-century medicine, noted: "Boerhaave was in general fairly inquisitive. Although he had the utmost respect for tradition, he also wanted to know the facts."[15] So did Franklin. *An Essay on Regimen* by Cheyne would have intrigued him. It suggested what is today known as psychosomatic medicine and was written by one of the most frequently read physicians of his day.

From his early years Franklin paid attention to his health. There were satisfactory prescriptions in Culpeper's *English Physician*, an herbal of English plants which remained current long enough to celebrate its sesquicentennial. Another well regarded text was Allen's *Synopsis Universae Medicinae*, a compendium of the esteemed opinions and practice of the most famous physicians from classical times to the compiler's own day. The Library Company considered it essential for laymen and ordered it in its first shipment of 1732. Another commonly used handbook, also requested by the library in 1732, was Quincy's *New Dispensatory* which was joined in Franklin's collection with the Latin works of Lémery and the botanist Tournefort and the unidentified "Medicamenta Simplicia."

Ettmüller's *Opera* and Boerhaave's *Institutiones* were professionally oriented, but Mead's *Precepts and Cautions* and Blankaart's *Physical Dictionary* were satisfactory reference books for literate do-it-yourself doctors. It is surprising that Franklin owned so many works in Latin, such as Willis's classic work on the nervous system, Kerck-

---

15. Lester S. King, *The Medical World of the Eighteenth Century* (Chicago, 1958), 100.

ring's medico-alchemical commentary, and La Framboisière's instructions for medical students. We know he attended Boston Latin School, but was he able to handle the convoluted scientific style of his day, hardly Ciceronian Latin? He could easily comprehend, as did his contemporaries who enthusiastically favored the work, Cheselden's *Anatomy of the Human Body*.

From the time his young son died of smallpox, Franklin was emotionally and intellectually involved in the efficacy of inoculation. In the summer of 1752 he told John Perkins that he had "a French piece printed at Paris 1724, entitled *Observations Sur la Saignee du Pied*," which included *Raisons de doubte contre l'Inoculation*.[16] Dimsdale, who went to Russia to inoculate Catherine the Great and her family, was the foremost exponent of inoculation in Great Britain. Franklin had his major treatise as well as Black's *Observations Medical and Political* that recommended universal inoculation. He also owned Haygarth's *Inquiry How to prevent the Small-Pox* which the author sent him through their mutual friend, Thomas Percival.[17] And, of course, Percival presented copies of his own books. It is strange that none of the writings of William Heberden appear on either list, except for the treatise advocating inoculation to which Franklin added a preface.

One of the aged diplomat's most painful complaints was a stone. In 1785 he asked Jonathan Williams to get for him a medicine that Blackrie boasted would dissolve stones. Franklin tried the recommended solvent that year.[18] The same day he asked Williams to get Blackrie's tract he wrote of it to Caleb Whitefoord: "It treats I

16. Benjamin Franklin to John Perkins, Philadelphia, 13 August 1752, Leonard W. Labaree, et al., eds., *The Papers of Benjamin Franklin* (New Haven, 1959–), 4, 340.

17. John Haygarth to Benjamin Franklin, Chester, England, 15 December 1788, Franklin Papers 36: 108, PPAmP.

18. Benjamin Franklin to Jonathan Williams, Passy, 19 May 1785, Albert Henry Smyth, ed., *The Writings of Benjamin Franklin* (New York, 1906) 9: 329; Benjamin Franklin to Benjamin Vaughan, Passy, April 1785, ibid., 9: 304.

understand of the Sope-Lye, which is recommended in the Pamphlet you were so kind as to send me."[19] Because he continued to suffer, it is doubtful that the solvent helped much, any more than the advice contained in Perry's *Disquisitions* and the appendix to Dobson's *Medical Commentaries* by Falconer, an extoller of the waters of Bath. Copies of the latter had been sent to him by his old friend Richard Price in 1785 and by John Wright a year later.[20]

Franklin talked more about gout, the other ailment that harassed his later years. He does not, however, seem to have had the most widely circulated work on the disease by Cadogan, but he did get lesser treatises by Flower, Jay, and Hardy, and an advertisement of a nostrum by Gondran. Everyone in the age of Enlightenment hoped that cures would be found for all manner of diseases in the burst of scientific development. Electricity promised much. Symptomatic of these hopes were Sans on a cure for paralysis and Gardanne's *Conjectures* on the general subject of electrical medicine. Franklin himself had tried to cure aging James Logan of the effects of a stroke with a mild jolt of static electricity; the weather inhibited the spark on that occasion.

Fever of all kinds was prevalent and all too often fatal. Much medical literature of the period concerned it. One of seventeenth-century Willis's *Diatribae duae* dealt with fevers and Hancocke's *Febrifugium Magnum*, recommending common water as the best cure, enjoyed a temporary vogue. Of much more influence and importance in helping to stem the spread of fevers, one of the major causes of soldiers' mortality, was Pringle's *Observations on the Diseases of the Army*. This was complemented by Bruce's *Inquiry* into army and navy diseases. Gastellier, who in 1781 sent Franklin *"deux ouvrages de ma composition"* on miliary fever, later dedicated to

19. Benjamin Franklin to Caleb Whitefoord, Passy, 19 May 1785, Smyth 9: 330.
20. Edward Bridgen to Benjamin Franklin, London, 7 November 1785, Franklin Papers 33: 239, PPAmP.

him the pamphlet, *Des Specifiques*.[21] More general was Waine-wright's popular *Mechanical Account of the Non-Naturals*, describing the medical effects of external elements such as air, water, and earth.

One of Franklin's closest medical friends had been John Fother-gill. When a posthumous publication of his *Works* was announced in 1783, the diplomat asked the editor, John Coakley Lettsom, to put him down for two copies.[22] One of these, destined for the American Philosophical Society, bears the Franklin shelf-mark and apparently stopped off at Franklin Court on its way to Fifth Street. It is not surprising that Franklin owned Lettsom's *History of the Origin of Medicine*, nor that young Benjamin Rush sent him a copy of his Edinburgh thesis.

The proceedings of the Société Royale de Médecine and the score of tracts for and against the validity of Mesmer's animal magnetism are evidence of the American's official participation in the royal commission appointed to investigate it. With a high pub-lic profile and great popularity in England and France Franklin was the recipient, and sometimes the dedicatee, of doctoral disser-tations and research papers by men who hoped that by presenting their works to the venerable sage some of his fame would rub off on them. It is doubtful that Franklin ever did more than glance at them. A high percentage of the pamphlets in the bound volumes belonging to the Historical Society falls into this category. Cer-tainly, the unexpected clutch of Wittenberg theses must have been given him ceremonially during his trip to Germany in 1766. More personal presentation copies came from acquaintances such as Percival and from Hewson who married Polly Stevenson, Frank-lin's young favorite. A few of the pamphlets, among them Gerbier's

21. René Georges Gastellier to Mme. Defay, Montargis, 30 November 1781, Franklin Papers 23: 83; and Gastellier to Franklin, 9 February 1783, ibid., 27: 93.
22. Benjamin Franklin to John Coakley Lettsom, Passy, 6 March 1783, Smyth 9: 16.

Quatrieme Lettre, belonged to Temple Franklin and—to make a malicious guess—the advertisements of cures for venereal disease, too. We know his books and papers were homogenized with his grandfather's after their return from Europe.

A full analysis of all the medical books would fill more pages than could be justified. Perhaps this paper may bring to the surface some of the still unlocated books of Franklin provenance. We should love to have his copy of Burton's *Anatomy of Melancholy*, and hope that, unlike the books so far found, it would be annotated fully.

Medecine & Surgery

Pharmacopia Lemeriana contracta—1 Vol.

Lémery, Nicolas de. Pharmacopoeia Lemeriana Contracta: Lémery's Universal Pharmacopoeia Abridg'd. *London: for Walter Kettilby, 1700.*

Oeconomia universalis G. Charletoni

*Charleton, Walter. . . . Oeconomia Animalis, Novis in Medicina Hypothesibus superstructa & Mechanice explicata. *London: Ex Officina Johannis Redmayne, prostant venales apud Johannem Creed, 1669.*

Syrop Mercuriale

Common Water, Cure for Fevers

*Hancocke, John. Febrifugium Magnum: or, Common Water the best Cure for Fevers, And probably the Plague. *London: for R. Halsey: And Sold by J. Roberts, 1723.*

Remedes pour la Pierre et la Gravelle

Blackrie, Alexander. Recherches sur les Remedes capable de dissoudre la Pierre et la Gravelle. *London & Paris: P. D. Pierres, 1775.*

Medecines that dissolve the Stone

*Blackrie, Alexander. A Disquisition on Medicines that dissolve the stone. In which Dr. Chittick's Secret is considered and discovered. *London: for the Author, and sold by D. Wilson, 1766.*

Essay on Regimen

*CHEYNE, GEORGE. An Essay on Regimen. Together with Five Discourses, Medical, Moral, and Philosophical. *London: for C. Rivington; And J. Leake, 1740.*

Institutiones Medicae by H. Boerhave

*BOERHAAVE, HERMANN. Institutiones Medicae In usus annuae Exercitationes Domesticos. *Leyden: Apud Johannem vander Linden, 1713.*

Divers Maladies

*HELVETIUS, JEAN ADRIEN. Recueil des Methodes . . . Pour la Guerison de diverses Maladies. *The Hague: Chez Adrian Moetjens, 1710.*

Dissertatio de Arthritide Mantissa Schematica &c

RHIJNE, WILLEM TEN. . . . Dissertatio de Arthritide: Mantissa Schematica: de Acupunctura: et Orationes Tres. *London: Impensis R. Chiswell, 1683.*

Allen's Synopsis Universae Medicinae practicae

*ALLEN, JOHN. Synopsis Universae Medicinae: sive, Doctissimorum Virorum de Morbis Eorumque Causis ac Remediis Judicia. *London: Impensis R. Knaplock, J. Tonson, G. & J. Innys, 1719.*

Philosophical Principles of Medecine by T Morgan

*MORGAN, THOMAS. Philosophical Principles of Medicine, In Three Parts. Containing, I. A Demonstration of the General Laws of Gravity, with their Effects upon Animal Bodys. *London: by J. Darby and T. Browne, and sold by J. Osborne, T. Longman, and J. Batley, F. Clay, E. Symon, S. Billingsley and S. Chandler, 1725.*

L'Art de saigner les Pieds

LAFOREST, NICOLAS LAURENT. L'Art de Soigner Les Pieds. *Paris: Chez l'Auteur, Méquignion, [et] Blaizot, à Versailles, 1782.*

Practica Medicina, Dan Sennerto

*SENNERT, DANIEL. Practicae Medicinae Liber Primus (-Secundus). *Wittenberg: Typis Haeredum Salomonis Auerbach, 1628–29.*

Perry on the Gravel & Stone

PERRY, CHARLES. Disquisitions on the Stone and Gravel. *London: 1777.*

Discours prononcé aux Ecoles de Medecine

De Morbibus cutaneis

[LORRY, ANNE CHARLES.] Tractatus De Morbis Cutaneis. *Paris: Apud P. Guillelum Cavelier, 1777.*

Hist^y. of the Origin of Medecine by J Coakly Lettsom

LETTSOM, JOHN COAKLEY. History of the Origin of Medicine: An Oration, Delivered at the Anniversary Meeting of the Medical Society of London, January 19, 1778. *London: by J. Phillips, for E. and C. Dilly, 1778.*

Sur les Poisons et sur le Corps Animal—2

FONTANA, FELICE. Traité sur le Vénin de la Vipere, sur les Poisons Americains, sur le Laurier-cerise et sur quelques autres Poisons Vége-taux. On y a joint des Observations sur la Structure Primitive du Corps Animal. *Florence: Et se trouve à Paris Chez Nyon l'Ainé [&] A Londres chez Elmsley, 1781.*

MacBride's Practice of Phisic—1

MACBRIDE, DAVID. A Methodical Introduction to the Theory and Practice of Physic. *London: for W. Strahan; T. Cadell; A. Kincaid and W. Creech, and J. Balfour, Edinburgh, 1772.*

Hist^e de la Soc. de Medecine (Work incompl.)—4

PARIS, SOCIÉTÉ ROYALE DE MÉDECINE. Histoire de La Société Royale de Médecine. Année M.DCC.LXXVI (–M.DCC.LXXXI, M.DCC.LXXXIII, Seconde Partie). *Paris: De l'Imprimerie de Philippe-Denys Pierres, Et se trouve Chez Didot le jeune, 1779; De L'Imprimerie de Monsieur, sous la direction de P.-Fr. Didot le jeune, 1782; Chez Théophile Barrois le jeune, 1785; De l'Imprimerie de Philippe-Denys Pierres, Et se trouve Chez Théophile Barrois, 1784.* PPAmP

Cat^e of Rarities in Grasham College, anat of Hum body—1

GREW, NEHEMIAH. Musaeum Regalis Societatis. Or A Catalogue & Description of the Natural and Artificial Rarities Belonging to the Royal Society And preserved at Gresham Colledge. *London: by W. Rawlins, for the Author, 1681.*

Medical Miscellanies (Fr & Lat)

L'Électricité medicale—2

GARDANNE, JOSEPH JACQUES DE. Conjectures sur l'Électricité Medicale, avec des recherches sur la colique métallique. *Paris: Chez la veuve d'Houry, 1768.*

De Circuitione Sanguinis—1

Th¹. Kerckringii Comentarius de Medicina

*KERCKRING, THEODOR. . . . Commentarivs in Cvrrvm Trivmphalem Antimonii Basilii Valentini. *Amsterdam: Sumptibus Andreae Frisii, 1671.*

F. Plazzonius de Partibus generationis

*PLAZZONI, FRANCISCO. . . . De Partibus Generationi inservientibus Libri Duo. *Leyden: Ex Officinâ Felicis Lopez de Haro, 1664.*

Pathologia Cerebri &c.

*WILLIS, THOMAS. Pathologiae Cerebri, et Nervosi Generis Specimen. *London: Typis S. Roycroft; Impensis Jo. Martyn, 1678.*

Hippocratica Praxis

TILEMANN, JOHANN. . . . Hippocratica Praxis in cognitione medica affectuum. *Ulm: Typis Haeredum C. B. Künnii, 1681.*

Willis's Fermentation

WILLIS, THOMAS. Diatribae duae Medico-Philosophicae: Quarum prior agit De Fermentatione, sive De motu intestino particularum in quovis corpore; altera De Febribus. *London: Typis Tho. Roycroft; Impensis Jo. Martin, Ja. Allestry, & Tho. Dicas, 1662.*

Enquiry into the Cause of the Pestilence (Eding–59)

BRUCE, ALEXANDER. An Inquiry concerning the Cause of Pestilence, and the Diseases in Fleets and Armies. *Edinburgh: S. Bladon, 1759.*

Philosophical, Med¹. & Exper¹. Essays by T Percival

PERCIVAL, THOMAS. Philosophical, Medical, and Experimental Essays. *London: for Joseph Johnson, 1776.*

Mélanges de Medecine

Mélanges de Medecine. Seconde Partie. *Paris: 1776.*

Guerison de la Paralysie par l'Électricité—1 Vol

SANS, abbé ——. Guerison de la Paralysie, par l'Électricité . . . lue à la
Société Royale de Medecine le 9 & le 30 Septembre 1777. *Paris: Chez
Cailleau, 1778.*

Dimsdale on Inoculation & epidemical Small-Pox

DIMSDALE, THOMAS, baron. Tracts on Inoculation, Written and published at St. Petersburg in the Year 1768 . . . With Additional Observations On Epidemic Small-Pox. *London: by James Phillips; For W. Owen;
and Carnan and Newbery, 1781.*

Black's Medical & Pol¹. Observations

*BLACK, WILLIAM. Observations Medical and Political, on the Small-
Pox, And the Advantages and Disadvantages of General Inoculation,
especially in Cities. *London: for J. Johnson, 1781.*

Dʳ Fothergill's Works

FOTHERGILL, JOHN. The Works of John Fothergill, M.D. . . . with
Some Account Of His Life. By John Coakley Lettsom. *London: for
Charles Dilly, 1784.*

Medecina Secreta

Medicamenta Simplicia

Elemens de Medecine en forme d'Aphorismes

BARBEU-DUBOURG, JACQUES. Élémens de Médecine, en Forme
D'Aphorismes. *Paris: Chez P. Fr. Didot, 1780.*

The English Phisician enlarged

*CULPEPER, NICHOLAS. The English Physician enlarged. *London: for
S. Ballard, R. Ware, S. Birt, C. Hitch, J. Hodges, J. Wood, and C. Woodward,
1741.*

on the Generation of Heat in Animals by R. Douglass

DOUGLAS, ROBERT. An Essay Concerning the Generation of Heat in
Animals. *London: for R. Dodsley, and Sold by M. Cooper, 1747.*

Mudge's Cure of a Cattarhous Cough

*MUDGE, JOHN. A Radical and Expeditious Cure for a Recent and
Catarrhous Cough. *London: by E. Allen; and sold by J. Walter; B. Thorn, at
Exeter; and M. Haydon, at Plymouth, 1779.*

Pringle's Diseases of the Army (given to J. Williams)

*PRINGLE, JOHN. Observations on the Diseases of the Army. *London: for A. Millar; D. Wilson; and T. Durham; and T. Payne, 1761.*

Medical Pamphlets

Anatomy of the Human Body by W. Cheselden

*CHESELDEN, WILLIAM. The Anatomy of the Human Body. *London: for C: Hitch & R: Dodsley, 1756.*

Small Pox & Inoculation by J. Haygarth

*HAYGARTH, JOHN. An Inquiry How to prevent the Small-Pox. And proceedings of a Society For promoting General Inoculations at stated Periods, and preventing the Natural Small-Pox, in Chester. *Chester: by J. Monk, for J. Johnson, London, and P. Broster, Chester, 1784.*

Schola Medica N. A. Frambesarii

*LA FRAMBOISIÈRE, NICOLAS ABRAHAM DE . . . . Scholae Medicae, Ad Candidatorum examen pro Laurea impetranda subeundum. *Leyden: Ex Officina Ioannis Maire, 1647.*

Use of Baths

BAYNARD, EDWARD. Of the General Use of Hot and Cold Baths. *London: 1706.*

Des Difficultés d'uriner par M d'Arran

DARAN, JACQUES. Composition du remède de M. Daran. Remède qu'il pratique avec succès depuis cinquante ans, pour le guérison des difficultés d'uriner. *Paris: Didot [et] Mequignon, 1780.*

Des Spécifiques en Medecine

GASTELLIER, RENÉ GEORGES. Des Spécifiques En Médecine. *Paris: Chez Didot, 1783.*   PHi

Materia Medica

*TOURNEFORT, JOSEPH PITTON DE. Materia Medica; or, a Description of Simple Medicines Generally Us'd in Physick. *London: by W. H. for Andrew Bell, 1716.*

L'Action de l'Air dans les Maladies contagieuses

MENURET DE CHAMBAUD, JEAN JACQUES. Essai sur l'action de l'air dans les maladies contagieuses, Qui a remporté le Prix proposé par la Société Royale de Médecine. *Paris: Rue et Hôtel Serpente, 1781.*
                                                                    PPL

Agenda de Sante par M. A. Honoré

HONORÉ, ANDRÉ. Agenda de Santé, ou Nouveau recueil portatif des plantes, arbres et arbustes. *Paris: 1777.*

Erreurs populaires sur la Medecine par M. Dharce

IHARCE, JEAN LUC D'. Erreurs Populaires sur La Médecine; Ouvrage composé pour l'instruction de ceux qui ne professent pas cette science. *Paris: Chez l'Auteur, [&] Mequignion l'aîné, 1783.* PPAmP

Boerhave's Aphorismes

*BOERHAAVE, HERMANN. Boerhaave's Aphorisms: Concerning the Knowledge and Cure of Diseases. *London: for W. Innys and J. Richardson, and C. Hitch and L. Hawes, 1755.*

Thesis by Dr Bouch de Hysteria

Traité de la Fièvre miliare des Femmes en couche

GASTELLIER, RENÉ GEORGES. Traité de la Fièvre Miliare des femmes en couche, Ouvrage qui a été couronné par la Faculté de Médecine de Paris, dans sa Séance publique tenue le 5 Novembre 1778. *Montargis: Chez Noel Gilles, 1779.* PPC

Essays de J. Rey D^r en Medecin, avec Notes par Gobet

REY, JEAN. Essais . . . Sur la Recherche de la Cause pour laquelle l'Estain & le Plomb augmentent de poids quand on les calcine. . . . avec des Notes, par M. Gobet. *Paris: Chez Ruault, 1777.*

Antiqua Medico-Philosophia orbi novo adaptanda

Via recta ad Vitam longam

*VENNER, TOBIAS. Via Recta ad Vitam Longam: Or, A plaine Philosophicall demonstration of the Nature, faculties, and effects of all such things as by way of nourishments make for the preseruation of health. *London: by Felix Kyngston, for Richard Moore, 1628.*

Quincy's Dispensatory

*QUINCY, JOHN. The New Dispensatory: containing I. The Theory and Practice of Pharmacy . . . V. A Collection of Cheap Remedies for the Use of the Poor. *London: for J. Nourse, 1753.*

De Puerorum Confectione by Dr Morgan

MORGAN, JOHN. Πυοποιεσις, sive Tentamen Medicum de Puris Confectione. *Edinburgh: Cum Typis Academicis, 1763.*

Cullen's first Lines of the Practice of Phisic—2

*CULLEN, WILLIAM. First Lines of the Practice of Physic, For the Use of Students in the University of Edinburgh. *Edinburgh: for William Creech, 1781.*

De Puerorum Generatione by N Romain—1

ROMAYNE, NICHOLAS. Dissertatio Inauguralis, De Puris Generatione. *Edinburgh: Apud Balfour et Smellie, 1780.*

Aphorisms d'Hypocrate—2

*HIPPOCRATES. Les Aphorismes d'Hippocrate expliquez conformément au sens de l'auteur, a la Pratique Médecinale, et a la Mechanique du Corpe Humain. *Paris: Chez Laurent d'Houry, 1727.*

Essays Medical & Experimental

PERCIVAL, THOMAS. Essays Medical and Experimental on the Following Subjects; 1. On the Columbo Root . . . 9. On Coffee. *London: for Joseph Johnson, 1773.*

Hewson on the Blood

HEWSON, WILLIAM. Experiments on The Blood, with some remarks on its morbid appearance . . . Read before the Royal Society, 1770. *London: by W. Bowyer and J. Nichols, 1771.*   PHi

Gerbier's Medical Observations

GERBIER, HUMBERT. Lettres et Observations . . . au sujet de deux nouveaux remèdes contre les maladies squirrheuses, cancereuses, &c. *Geneva: 1777.*

Medecine nouvelle 1785

Dissertation of Melancholy by C Wistar (Latin)

WISTAR, CASPAR. Dissertatio Medica Inauguralis, De Animo Demisso. *Edinburgh: Apud Balfour et Smellie, 1786.*

De Aere, Locis et Aquis

*HIPPOCRATES. . . . Liber de aere, aquis, et locis, de caelestibus nempe in aerea, terrena, et aquea in toto terrarum orbe influentibus. *Paris: Chez l'aucteur, 1662.*

The Anatomie of Melancholy by Democritus Junr.

*[BURTON, ROBERT.] The Anatomy of Melancholy . . . By Democritus Junior. *London: [by R. W.] for Peter Parker, 1676.*

Diet of the Diseased by Dr Hart

HART, JAMES. Κλινική; or The Diet of the Diseased. *London: by Iohn Beale, for Robert Allot, 1633.*

Mechanical Account of the Non-Naturals by Wainewright

*WAINEWRIGHT, JEREMIAH. A Mechanical Account of the Non-Naturals: Being a Brief Explanation Of the Changes made in Humane Bodies, by Air, Diet, &c. *London: by J. H. for Ralph Smith, 1718.*

Aphorismes de Mesmer

*MESMER, FRANZ ANTON. Aphorismes de M. Mesmer, dictés à l'Assemblée de ses Eleves, & dans lesquels on trouve ses Principes, sa Théorie & les Moyens de Magnétiser. *Paris: Chez Quinquet, 1785.*

Mead's Medical Precepts & Cautions

*MEAD, RICHARD. Medical Precepts and Cautions. *Edinburgh: by A. Donaldson and J. Reid, For Alex. Donaldson, 1763.*

Enquiries into the Lymphatic System

HEWSON, WILLIAM. Experimental Inquiries: Part the Second. Containing A Description of the Lymphatic System In the Human Subject, And in Other Animals. *London: for J. Johnson, 1774.*

Phisical Dict^y.

*BLANKAART, STEVEN. The Physical Dictionary. Wherein The Terms of Anatomy, the Names and Causes of Diseases, Chirurgical Instruments, and their Use, are accurately described. *London: by R. B. for Sam. Crouch and John & Benj. Sprint, 1715.*

Of Diet, Arbuthnot

ARBUTHNOT, JOHN. Practical Rules of Diet In the various Constitutions and Diseases of Human Bodies. *London: for J. Tonson, 1732.*

Van Helmonts Phisick refined

*HELMONT, JAN BAPTISTA VAN. Oriatrike or, Physick Refined. The Common Errors therin Refuted, And the whole Art Reformed & Rectified. *London: for Lodowick Loyd, 1662.*

Celsus de Medicina

*CELSUS, AULUS AURELIUS CORNELIUS. . . . De Medicina Libri Octo. *Rotterdam: Apud Joh. Danielem Beman, 1750.*

Med¹. Essays on the Human Body

Physical Essays on Parts of the Human Body and Animal Oeconomy. *London: for John Clarke, 1734.*

Diaeleticon Poly-historicon Quercitani

[DUCHESNE, JOSEPH.] Ios. Qvercitani ... Diateticon Polyhistoricon. *Leipzig: Impensis Thomas Schureri & Bartholomaei Voigtij, 1607.*

Lemery on Foods, transl. by Hay

*LÉMERY, LOUIS. A Treatise of all Sorts of Foods, Both Animal and Vegetable: also of Drinkables ... Translated by D. Hay, M.D. *London: for T. Osborne, 1745.*

Maladies epidemiques par de a Cloture—3

LÉPECQ DE LA CLÔTURE, LOUIS. Observations sur les maladies epidémiques. *Paris: Vincent, 1776.*

Blackerie on Medecines that disolve the Stone—1

*BLACKRIE, ALEXANDER. A Disquisition on Medicines that dissolve the stone. In which Dr. Chittick's Secret Is Considered and Discovered. *London: for the Author; and sold by D. Wilson and G. Nicol; J. Wilkie; and F. Newbery, 1771.*

Dʳ Rush's Medical Dissertation

RUSH, BENJAMIN. Dissertatio Physica Inauguralis, De Coctione Ciborum In Ventriculo. *Edinburgh: Apud Balfour, Auld, et Smellie, 1768.*
                                                                    PHi

Morgan's philosophical Principles of Medecine

*MORGAN, THOMAS. Philosophical Principles of Medicine, In Three Parts. *London: for J. Osborn and T. Longman, 1730.*

Hancocke on Fevers & Quinsy on Woodward's State of Phisic

*HANCOCKE, JOHN. Febrifugium Magnum: or, Common Water the best Cure for Fevers, And probably for the Plague. *London: for R. Halsey: And Sold by J. Roberts, 1722.*

QUINCY, JOHN. An Examination of Dr. Woodward's State of Physick and Diseases. *London: for William Bell, William Taylor, and John Osborn, 1719.*

Maclurg's Experiments on the Bile

MCCLURG, JAMES. Experiments upon the Human Bile: and Reflections on the Biliary Secretion. *London: for T. Cadell, 1772.*

Rogers Essay on epidemic Diseases

ROGERS, JOSEPH. An Essay on Epidemic Diseases; and More particularly on the Endemial Epidemics of the City of Cork, such as Fevers and Small-Pox. *Dublin: by S. Powell, For W. Smith, 1734.*

Aphorisms of Hypocrates & Sentences of Celsus

*HIPPOCRATES and AULUS AURELIUS CORNELIUS CELSUS. The Aphorisms of Hippocrates, and the Sentences of Celsus. *London: for R. Wilkin, J. and J. Bonwick; S. Birt; and T. Ward and E. Wicksteed, 1735.*

Pringles Diseases of the Army

*PRINGLE, JOHN. Observations on the Diseases of the Army, in Camp and Garrison. *London: for A. Millar; D. Wilson and T. Durham; and T. Payne, 1753.*

Ettmulli Opera—3

*ETTMÜLLER, MICHAEL . . . . Opera Medica Theoretico-practica. *Frankfurt: ex Officina Zunneriana, 1708.*

Vanswietens Commentaries on Boerhave—8

*SWIETEN, GERARD, freiherr VAN. The Commentaries upon the Aphorisms of Dr. Herman Boerhaave . . . concerning The Knowledge and Cure of the several Diseases incident to Human Bodies. *London: for Robert Horsfield; and Thomas Longman, 1765.*

Lectures on the Rationale of Medecine by Strother—2

STROTHER, EDWARD. Praelectiones Pharmaco-mathicae & Medico-practicae: or, Lectures on the Rationale of Medicines. *London: for C. Rivington, 1732.*

Potterfrield's Treatise on the Eye

PORTERFIELD, WILLIAM. A Treatise on the Eye, The Manner and Phaenomena of Vision. *Edinburgh: for A. Miller at London, and for G. Hamilton and J. Balfour at Edinburgh, 1759.*

Nat^l. Cure of the Diseases of the Body & Mind—1

CHEYNE, GEORGE. The Natural Method Of Cureing the Diseases of the Body, and the Disorders of the Mind Depending on the Body. *London: for Geo. Strahan, and John and Paul Knapton, 1742.*

Recherches sur la Nature et Effets du Mephetisme

HALLÉ, JEAN NOËL. Recherches sur la Nature et les effets du Méphitisme des Fosses d'Aisance. *Paris: de l'Imprimerie de P. D. Pierres, 1785.*

Faulconer's medical Commentaries on Fixed-Air

*DOBSON, MATTHEW. A Medical Commentary on Fixed Air . . . With an Appendix on the use of the solution of fixed alkaline salts saturated with fixible air, in the stone and gravel. By William Falconer. *Dublin: for W. Gilbert, 1785.*

L'Anti-Magnetisme

[PAULET, JEAN JACQUES.] L'Antimagnétisme, ou Origine, Progrès, Décadence. Renouvellement et Réfutation du Magnétisme Animal. *London [Paris]: Desenne, 1784.*

Recherches et Doutes sur le Magnetisme Animal

THOURET, MICHEL AUGUSTIN. Recherches Et Doutes sur Le Magnétisme Animal. *Paris: Chez Prault, 1784.*

Precis des Faits relatifs au Magnetisme-Animal

MESMER, FRANZ ANTON. Précis Historique Des Faits Relatifs au Magnétisme-Animal Jusques en Avril 1781. *London [Paris]: 1781.*

Recherches sur le Mephetisme by Hallé

HALLÉ, JEAN NOËL. Recherches sur la Nature et les effets du Méphitisme des Fosse d'Aisance. *Paris: de l'Imprimerie de P. D. Pierres, 1785.*

Percival's Medical Essays

*PERCIVAL, THOMAS. Essays Medical and Experimental, viz. I. On the use and abuse of theory and reasoning in physic . . . VII. On the efficacy of external applications in the ulcerous sore throat. *London: for J. Johnson, 1777.*

Le Colosse aux Pieds d'Argille

Essay sur la Fievre milliare, Par Mr Gastellier

GASTELLIER, RENÉ GEORGES. Avis A Mes Concitoyens, ou Essai sur la fievre miliare. *Paris: Chez Gogué, 1773.*

MEDICAL BOOKS NOT ON BENJAMIN FRANKLIN BACHE'S LIST

ANDRIEU, ANTOINE. Notice Intéressante, Ayant rapport à un Ouvrage récemment publié, où l'on indique, d'après la saine expérience, Des Nouveaux Moyens, également sûrs, agréables & commodes pour guérir la Maladie Anti-Sociale. *Paris: Chez l'Auteur, 1782.* PHi

[BARBEU-DUBOURG, JACQUES.] Lettre D'Un Medecin de la Faculté de Paris, A un de ses Confreres, Au suject de la Société Royale de Médecine. *[Paris: 1779.]* PHi

[——.] Opinion d'un médecin De La Faculté de Paris, sur l'inoculation de la petite vérole. *Paris: l'Imprimerie de Didot; chez Lacombe, [1768].*
(2 copies) PHi

[——.] Recherches sur la durée De La Grossesse et le temps de l'accouchement. *Amsterdam: 1765.* nf

BARD, SAMUEL. Tentamen Medicum Inaugurale, De Viribus Opii. *Edinburgh: Apud A. Donaldson et J. Reid, 1765.* PPAmP

BERDMORE, THOMAS. A Treatise on the Disorders and Deformities of the Teeth and Gums, explaining The most rational Methods of treating their Diseases. *London: for the Author, Sold by Benjamin White; James Dodsley; And Becket and De Hondt, 1770.* PPL

BERGASSE, NICOLAS. Considérations sur Le Magnétisme Animal, ou sur la théorie du monde et des êtres organisés. *The Hague: 1784.* PHi

BERKELEY, GEORGE, bishop of Cloyne. Siris: A Chain of Philosophical Reflexions and Inquiries Concerning the Virtues of Tar Water. *London: For W. Innys, and C. Hitch; and C. Davis, 1744.* PHi

BERNARD, ——. Nouvelles Inventions Du Sr. Bernard, Orfèvre-Méchanicien pour les Instruments de chirurgie. Sondes flexibles pour les Rétentions d'urine & autre maladies de l'Uretre. *[Paris:] De l'Imprimerie de L. Jorry, [1778].* PPAmP

BERTHOLON, PIERRE. De L'Électricité du Corps Humain Dans L'État De Santé et de maladie. *Paris: Chez P. F. Didot le jeune, 1780.* nf

BONNEFOY, JEAN BAPTISTE. Analyse Raisonnée des rapports Des Commissaires chargés par le roi De L'Examen du Magnétisme Animal. *Lyon: Et se trouve à Paris, Chez Prault, 1784.* PHi

BOURZEIS, JACQUES AMABLE DE. Observation très-importante sur les effets Du Magnétisme Animal. *Paris: Chez P. Fr. Gueffier, 1783.* PHi

BOYVEAU-LAFFECTEUR, PIERRE. Observations sur les effets Du Rob Anti-Syphilitique. *Paris: De l'Imprimerie de Ph.-D. Pierres, 1783.*
PPAmP

BROWN, GUSTAVUS RICHARD. Disputatio Physica Inauguralis, De Ortu Animalium Caloris. *Edinburgh: Apud Balfour, Auld, et Smellie, 1768.* PPAmP

BROWNRIGG, WILLIAM. Considerations on the means of preventing The Communication of Pestilential Contagion, and of eradicating it in infected places. *London: for Lockyer Davis, 1771.* PHi

BUCQUET, JEAN BAPTISTE MICHEL. Mémoire Sur la manière dont les Animaux sont affectés par différens Fluides Aériformes, Méphitiques; & sur les moyens de reédier aux effets de ces Fluides. *Paris: De l'Imprimerie Royale, 1778.* PPAmP

BURRINGTON, GEORGE. An Answer to Dr. William Brakenridge's Letter concerning the Number of Inhabitants, within the London Bills of Mortality. *London: for J. Scott, 1757.* (2 copies) PHi

[CABANIS, PIERRE JEAN GEORGES.] Serment D'Un Médecin, Prononcé le jour de sa réception, dans des écoles, en face d'une église, et près d'un hôpital. *[Paris: P. F. Didot le jeune, 1783.]* PPAmP

CADET-DE-VAUX, ANTOINE ALEXIS. Avis Sur les Moyens de diminuer l'Insalubrité des Habitations qui ont été exposées aux Inondations. *Paris: De l'Imprimerie de Ph.-D. Pierres, 1784.* PHi

——. Mémoire sur le méphitisme des puits . . . Lu à l'Académie Royale des Sciences, le 25 Janvier 1783. *[Paris: 1783.]* PPAmP

CHASTENET, ARMAND MARC JACQUES DE, marquis de PUYSÉGUR and JACQUES MAXIME PAUL DE CHASTENET, comte de PUYSÉGUR. Détail Des Cures Opérées A Buzancy, Près Soissons, par Le Magnétisme Animal. *Soissons: 1784.* PHi

CHASTENET, JACQUES MAXIME PAUL DE, comte de PUYSÉGUR. Rapport Des Cures Opérées A Bayonne par le magnétisme animal. *Bayonne: Et se trouve à Paris, Chez Prault, 1784.* PHi

CLIFTON, FRANCIS. Tabular Observations Recommended, as the Plainest and Surest Way of Practising and Improving Physick. *London: for J. Brindley, 1731.* PHi

COSTE, JEAN FRANÇOIS. Oratio Habita In Capitolio Gulielmopolitano In Comitiis Universitatis Virginiae Die XII Junii M.DCC. LXXXII. *Leyden: 1783.* PHi

CRAWFORD, ADAIR. Experiments and Observations on Animal Heat, and the Inflammation of Combustible Bodies. *London: for J. Murray; and J. Sewell, 1779.* PPAmP

[DAMPIERRE, ANTOINE ESMONIN, marquis de.] Réflexions Impartiales sur Le Magnétisme Animal, Faites après la publication du Rapport des Commissaires. *Geneva: Chez Barthelemi Chirot, Et se vend à Paris, Chez Perisse le jeune, 1784.* PHi

DEMOURS, PIERRE. Lettre De M. Demours . . . A M. Petit . . . En Réponse à sa critique d'un rapport sur une maladie de l'oeil. *Paris: Chez P. Fr. Didot, [&] Dessain Junior, 1767.* PPAmP

ELMER, JONATHAN. Dissertatio Medica, Inauguralis, De Sitis in Febribus Causis et Remediis. *Philadelphia: Apud Henricum Miller, 1771.* nf

ESLON, CHARLES D'. Observations sur Les Deux Rapports de MM. Les Commissaires nommés par sa majesté, pour l'examen du magnétisme animal. *Philadelphia [Paris]: Et se trouve a Paris, Chez Clousier, 1784.* PHi

[———.] Supplément aux deux Rapports De MM. les Commissaires de l'Académie & de la Faculté de Médecine, & de la Société Royale de Médecine. *Amsterdam: [Paris]: Et se trouve a Paris, Chez Gueffier, 1784.* PHi

FABRE, PIERRE. Recherches sur la nature De L'Homme consideré dans l'état de santé et dans l'état de maladie. *Paris: Chez Delalain, 1776.* PPL

FALCK, NIKOLAI DETLEF. A Treatise on the Venereal Disease. *London: for the Author, and Sold by B. Law, 1772.* nf

FAYNARD, JACQUES. Copie Fidelle de l'Honorable Liste de la Noblesse, &c. Qui ont eu la Bonté d'encourager la Poudre Vulnéraire & Incomparable. *[London: 1774.]* PPAmP

FLOWER, HENRY. Observations on the Gout and Rheumatism. Exhibiting Instances of Persons who were greatly relieved in the Fit of the Gout. *London: for E. Cooke, 1766.* PHi

[FONTETTE-SOMMERY, ———, comte de.] Lettre A Monsieur D'Eslon, médecin ordinaire de monseigneur Comte D'Artois. *Glasgow [Paris]: Et se trouve à Paris, Chez Prault, 1784.* PHi

FOWLER, THOMAS. Medical Reports, of the Effects of Tobacco, Principally with Regard to its Diuretic Quality, in the Cure of Dropsies and Dysuries. *London: for J. Johnson, and William Brown, 1785.*   PPAmP

FRANCE, COMMISSION CHARGÉ DE L'EXAMEN DU MAGNÉTISME ANIMAL. Exposé Des Expériences qui ont été faites pour l'examen Du Magnétisme Animal. Lû à l'Académie des Sciences, par M. Bailly, en son nom & au nom de M^rs. Franklin, le Roy, de Bory & Lavoisier, le 4 Septembre 1784. *Paris: de l'Imprimerie Royale, 1784.*   PHi

——. Another edition. *Paris: Chez Moutard, 1784.*   PHi

——. Rapport Des Commissaires chargés par le roi De L'Examen du Magnétisme Animal. *Paris: de l'Imprimerie Royale, 1784.*   PHi

——. Another edition. *Paris: Chez Moutard, 1784.*   PHi

——. Report of Dr. Benjamin Franklin, and other Commissioners, charged by the King of France, with the examination of the Animal Magnetism, as now practised at Paris. Translated from the French. *London: for J. Johnson, 1785.*   PHi

[FRANKLIN, BENJAMIN.] Some Account Of the Success of Inoculation for the Small-Pox in England and America. *London: W. Strahan, 1759.*   PHi

GARDANNE, JACQUES JOSEPH DE. Catéchisme sur Les Morts Apparentes, Dites Asphyxies; ou Instruction sur les manieres de combattre les différentes especes de Morts apparentes. *Paris: de l'Imprimerie de Valade, 1781.*   PHi

——. Another copy.   MBAt

GARREAU, ——. Rue De Maurepas, près les bains, A Versailles, Le Sr. Garreau, Chirurgien Herniare, par Brevet du Domaine du Roi & du Gouvernment. *[Paris: 1780.]*   PPAmP

GERBIER, HUMBERT. Quatrieme Lettre de M. Gerbier, Docteur en Médecine, l'un des Médecins de Monsieur, A MM. les Auteurs de la Gazette de Santé. *Geneva and Paris: Chez les Libraires assortis en Livres de Médecine, 1777.*   PPAmP

GLASS, SAMUEL. An Essay on Magnesia Alba. Wherein Its History is attempted, Its Virtues pointed out, and The Use of it recommended. *Oxford: for R. Davis, and J. Fletcher, London, 1764.*   PHi

[GONDRAN, ——.] Elixir Avec lequel on prépare Bain contre les accidents de la Goutte. *[Paris: ca. 1780.]* PPAmP

GRAHAM, JAMES. Dr. Graham is now preparing the largest and most elegant Medico-Electrical-Aërial and Magnetic Apparatus in the world. *[New-castle: 1779.]* PHi

——. The General State of Medical and Chirurgical Practice, Exhibited; Shewing them to be Inadequate, Ineffectual, Absurd, and Ridiculous. *Bath: sold by all the principal Booksellers in Great Britain, 1778.* PHi

[GRAINGER, JAMES.] An Essay On the more common West-India Diseases; and the Remedies which that Country itself produces. *London: for T. Becket and P. A. De Hondt, 1764.* PHi

GREAT BRITAIN, PARLIAMENT, HOUSE OF COMMONS. Resolutions of the House of Commons upon the Report made from the Committee appointed to enquire into the state of health of the Prisoners Confined in the King's House, at Winchester. *[London:] 1780.* PU

HALES, STEPHEN. An Account of some Experiments and Observations on Tar-Water. *London: for R. Manby and H. S. Cox, 1747.* PHi

[HANNAY, SAMUEL.] Directions for using the Antivenereal Specifick . . . To the Public. Philpot Lane, August 15, 1774. *[London: 1774.]* PPAmP

HANWAY, JONAS. The Great Advantage of eating Pure and Genuine Bread, comprehending the Heart of the Wheat, With All its Flour. Shewing how it may contribute to the Health and Profit of the People . . . Second Edition. *London: Sold at Mrs. Woodfall's; [&] J. Brotherton and Sewell, 1773.* PHi

HARDY, JAMES. An Answer to the Letter addressed by Francis Riollay, Physician of Newbury, to Dr. Hardy, on Hints given concerning the Origin of the Gout, in his Publication on the Colic of Devon. *London: for T. Cadell; and Richardson and Urquhart, 1780.* PHi

——. A Candid Examination of what has been advanced on The Colic Of Poitou and Devonshire, with Remarks on the most probable and Experiments intended to ascertain the true Causes of The Gout. *London: W. Mackintosh, for T. Cadell, 1778.* PHi

HAYES, THOMAS. A Serious Address on the Dangerous Consequences of Neglecting Common Colds and Coughs . . . The Second Edition. *London: 1785.* DeU

[HAYGARTH, JOHN.] Bite of a Mad Dog. *[Chester: 1788.]*   PPAmP

[HECQUET, PHILIPPE.] Observations sur La Saignée Du Pied, et sur La Purgation Au commencement de la Petite Vérole, des Fiévres malignes & des grandes maladies. *Paris: Chez Guillaume Cavelier, 1724.*   nf

HEINSIUS, JOHANN AUGUST. Dispvtatio Inavgvralis Medica De Vena Fonte Haemorrhoidvm non satis limpido. *Wittenberg: Litteris Caroli Christiani Dürrii, [1768].*   PHi

[HELMONT, FRANCISCUS MERCURIUS VAN.] One Hundred Fifty Three Chymical Aphorisms. Briefly containing Whatsoever belongs to the Chymical Science. *London: for the Author, and are to be Sold by W. Cooper; and D. Newman, 1688.*   PPL

HENRY, THOMAS. A Letter from Mr. Henry to Mr. Delamotte. *[London: for Joseph Johnson, 1774.]*   PHi

[——.] A Letter to Dr. Glass, containing a reply to his examination of Mr. Henry's Strictures on the magnesia sold under the name of the late Mr. Glass. *London: for Joseph Johnson, 1774.*   PHi

[——.] A Short View of the State of the Controversy between Mr. Delamotte and Mr. Henry. *[London: for Joseph Johnson, 1774.]*   PHi

HEWSON, WILLIAM. . . . Descriptio Systematis Lymphatici, iconibus illustrata. *Utrecht: Typis H. van Otterloo, 1783.*   nf

HILDBRANDT, GEORG FRIEDRICH. Avis sur l'utilité d'une nouvelle Machine Fumigatoire. *[Paris: 1781.]*   PPAmP

HILL, sir JOHN. Cautions Against the immoderate Use of Snuff. Founded on the known Qualities of the Tobacco Plant; And the Effects it must produce when this Way taken into the Body . . . The Second Edition. *London: for R. Baldwin, and J. Jackson, 1761.*   PHi

INGENHOUSZ, JAN. Experiments upon Vegetables, discovering Their great Power of purifying the Common Air in the Sun-shine, and of Injuring it in the Shade and at Night. *London: for P. Elmsly; and H. Payne, 1779.*   PPAmP

JAY, sir JAMES. Reflections and Observations on the Gout. *London: for G. Kearsley; H. Parker, and J. Ridley, 1772.*   PPAmP

——. The Second Edition. *London: for G. Kearsley; H. Parker, and J. Ridley, 1772.*   PHi

[JUSSIEU, ANTOINE LAURENT.] Rapport de l'un Des Commissaires chargés par le roi De L'Examen du Magnétisme Animal. *Paris: Chez La Veuve Herissant; [&] Théophile Barrois, le Jeune, 1784.* PHi

LANGGUTH, GEORG AUGUST. ... de Morbi Bovm Contagiosi Cavsa et sanatione probabili praefatvs. *Wittenberg: Prelo Ephraim Gottlob Eichsfeldi, [1753].* PHi

———. ... de Vtilitate Atqve Dignitate Artis Veterinariae Praefatvs. *Wittenberg: Prelo Ephraim Gottlob Eichsfeldi, [1753].* PHi

———. ... De Haemorrhoidvm Venosarvm Vindicatione pavca praefatvs. *Wittenberg: Litteris Caroli Christiani Dürrii, [1768].* PHi

———. ... Ad Loc. Hippocr. Praedict. II. XXVII. pavca praefatvs. *Wittenberg: Literis Caroli Christiani Dürrii, [1766].* PHi

LA PLANCHE, LAURENT CHARLES DE. Lettre de monsieur De La Planche D.M.P. Aux auteurs du journal de médecine, sur l'origine de la section du pubis. *[Paris: 1781.]* PHi

LASSONE, JOSEPH MARIE FRANÇOIS. Rapport Des Inoculations faites Dans La Famille Royale, au château de Marli. Lû à l'Académie Royale des Sciences, le 20 Juillet 1774. *[Paris: de l'Imprimerie Royale, 1774.]* PHi

[LAUGIER, ESAÏE MICHEL.] Parallele entre Le Magnétisme Animal, L'Électricité Et Les Bains Médicinaux par distillation, &c. appliqués aux Maladies rebelles. *Paris: Chez Morin, 1785.* PHi

LIND, JAMES. Dissertatio Medica, Inauguralis, De Febre remittente putrida paludum quae grassabatur in Bengalia A.D. 1762. *Edinburgh: Apud Balfour, Auld, et Smellie, 1768.* PHi

LOGAN, GEORGE. Tentamen Medicum Inaugurale, De Venenis. *Edinburgh: Apud Balfour et Smellie, 1779.* PHi

LONDON, GENERAL DISPENSARY FOR RELIEF OF THE POOR. An Account of the General Dispensary for relief of the poor. Instituted 1770. *London: 1772.* (2 copies) PHi

LONDON HOSPITAL. Charter of Incorporation of the London-Hospital. *[London: 1758.]* PHi

LONDON, PARISH CLERKS. A General Bill of all the Christnings and Burials, from the 14. of December, 1686. to the 13. of December, 1687. *[London: 1687.]* PPL

——. A General Bill of all the Christnings and Burials, from the 16. of December, 1712. to the 15. of December, 1713. *[London: 1713.]*   PPL

Lyon, École Vétérinaire. Procés Verbal De l'Expérience Magnétique fait à l'Ecole Vétérinaire de Lyon, le lundi 9 Août 1784, en présence de Monsieur le Comte d'Oels. *Lyon: de l'Imprimerie de la Ville, 1784.*   PHi

MacBride, David. Experimental Essays on Medical and Philosophical Subjects: particularly, I. On the Fermentation of Alimentary Mixtures, and Digestion of Food . . . The Second Edition. *Dublin: for Thomas Ewing, 1767.*   PPL

——. An Historical Account of a New Method Of Treating the Scurvy, At Sea. *Dublin: Re-printed by W. G. Jones, for Thomas Ewing, 1767.*
                                                                                                                             PPAmP

McClurg, James. Tentamen Medicum Inaugurale, De Calore. *Edinburgh: Apud Balfour, Auld, et Smellie, 1770.*   PHi

[Macquer, Pierre Joseph.] Dictionnaire de Chymie, contenant La Théorie & la Pratique de cette Science, son application à la Physique, à l'Histoire Naturelle, à la Médecine & à la Economie animale. *Paris: Chez Lacombe, 1766.*   PPAmP

Majault, ——. Réflexions sur quelques Préparations Chymiques, apliquées à l'usage De La Médecine; Lues à la Séance publique de la Faculté de Médecine de Paris, le 5 Novembre 1778. *Paris: Chez Quillau, 1779.*   PPAmP

Mardon, L—. The English Malady Removed; or, a New Treatise on the Method of curing the Land Scurvy, Leprosy, Elephantiasis, and Evil, with other cutaneous Eruptions. *London: for the Author, and sold at Mr. Pearch's; and Mr. Spilsbury's, 1771.*   PHi

Martin, Hugh. A Narrative of a Discovery of a Sovereign Specific, for the Cure of Cancers. *Philadelphia: Robert Aitken, 1782.*   PHi

[Meyer, Johann Friedrich.] Alchymistische Briefe. *Hanover: H. E. C. Schlüter, 1767.*   PHi

Mittié, Jean Stanislas. Lettres De M. Mittié . . . envoyant le Recueil des pieces qu'il a publiées sur la Maladie Vénérienne, sur les inconvéniens du Mercure, & sur l'efficacité des Végétaux de l'Europe, pour la guérison de cette Maladie. *Brussels: 1784.*   PHi

——. Observation Sommaires, Sur tous les Traitemens des Maladies Vénériennes, particulierement avec les Végétaux. *Montpellier: Et se trouve a Paris, Chez Didot, le jeune, 1779.* PHi

MORAND, JEAN FRANÇOIS CLÉMENT. Catalogue des Piéces D'Anatomie, Instrumens, Machines, &c. Qui composent l'Arsenal de Chirurgie formé à Paris pour la Chancellerie de Médecine de Pétersbourg. *Paris: de l'Imprimerie Royale, 1759.* PHi

MORRIS, BENJAMIN. Dissertatio Medica Inauguralis, De Angina Vera Seu Inflammatoria. *Leyden: Apud Phil: Bonk, [1750].* PHi

PARIS, FACULTÉ DE MÉDECINE. De Mandato J. Caroli, H. Sallin, Saluberrimae Facultatis Medicae Parisiensis Decani, Et MM. Doctorum Regentium ejusdem Facultatis Ob Serenissimi Principis Ludovici-Caroli Ducis Neustriae Natalia. *[Paris:] Typis Quillau, 1785.* PHi

——. Extrait Des Registres De la Faculté de Médecine de Paris. *[Paris:] De l'Imprimerie de Quillau, 1784.* (2 copies) PHi

——. Extractum É Commentariis Saluberrimae Facultatis Parisiensis. *[Paris:] Typis Quillau, 1784.* PHi

——. Rapport De MM. Cosnier, Maloet, Darcet, Philip, le Preux, Desessartz, & Paulet, Docteurs-Régens de la Faculté de Médecine de Paris; Sur les avantages reconnus de la Nouvelle Méthode d'administrer l'Électricité dans les Maladies Nerveuses. *Paris: de l'Imprimerie de Philippe-Denys Pierres, 1783.* PHi

——. Rapport Fait par Messieurs les Commissaires nommés par la Faculté de Médecine. Pour L'Examen des Eaux D'Enghien, Au dessous de l'Etang de Saint-Gratien. *[Paris: 1785.]* PHi

PARIS, MAISON DE BICÉTRE. Puits De La Maison De Bicétre, dependante de l'hopital général De Paris. *[Paris: ca. 1760.]* PPAmP

PARIS, SOCIÉTÉ ROYALE DE MÉDECINE. Extrait Des Registres De La Société Royale De Médecine. Rapport sur les Aimans présentés par M. l'Abbé le Noble; Lu dans la Séance tenue au Louvre, le Mardi premier Avril 1783. *Paris: de l'Imprimerie de Ph.-D. Pierres, [1783].* PPAmP

——. Prix Distribués & annoncés par la Société Royale de Médecine, dans sa Séance publique, tenue au Louvre le Mardi 27 Août 1782. *Paris: De l'Imprimerie de Ph.-D. Pierres, [1782].* PPAmP

——. Rapport Des Commissaires de la Société Royale de Médecine, nommés par le roi pour faire l'examen du Magnétisme Animal. *Paris: De l'Imprimerie Royale, 1784.*   PHi

——. Another edition. *Paris: Chez Mouard, 1784.*   (2 copies) PHi

——. Rapport Des Commissaires de la Société Royale de Médecine, sur le Mal Rouge De Cayenne ou Éléphantiasis. *Paris: De l'Imprimerie Royale, 1785.*   PPAmP

——. Rapport sur Plusieurs Questions Proposées à la Société Royale de Médecine, par M. l'Ambassadeur de la Religion, de la part de Son Altesse Eminentissime Monseigneur le Grand Maitre. *Malta: 1781.*   PHi

——. Another copy.   PPAmP

——. Réflexions sur la nature et le traitement De La Maladie qui règne Dans Le Haut Languedoc; Lues dans la Séance tenue au Louvre par la Société Royale de Médecine, le 4 Juin 1782. *Paris: De l'Imprimerie de P. Fr. Didot le jeune, 1782.*   PPAmP

PASCHE, GOTTLOB ANDREAS. ... Dissertatio Inavgvralis Medica De Ocvlorvm Integritate Improvidae Pverorvm Aetati Solicite Cvstodienda. *Wittenberg: Prelo Ephraim Gottlob Eichsfeldi, [1754].*   PHi

PAUL DE LAMANON, ROBERT DE. Mémoires sur Différentes Parties D'Histoire Naturelle Et De Physique. *Paris: de l'Imprimerie de Demonville, [1782].*   PPAmP

PERCIVAL, THOMAS. Dissertatio Medica Inauguralis De Frigore. *Leyden: Apud Theodorum Haak, 1765.*   PPAmP

[——]. Further Observations on the State of Population in Manchester, and other adjacent Places. *[Manchester: 1774.]*   PHi

——. Observations and Experiments on the Poison of Lead. *London: for J. Johnson, 1774.*   PPAmP

——. Observations on the Medicinal Uses of the Oelum Jecoris Aselli, or Cod Liver Oil, In the Chronic Rheumatism, and other painful Disorders. *[Manchester: 1782.]*   PHi

——. Another copy.   PPAmP

[——.] Observations on the State of Population in Manchester, and other adjacent Places. *[Manchester: 1773.]*   PHi

———. Tables shewing the Number of Deaths occasioned by the Small-Pox in the several Periods of Life, and different Seasons of the Year, together with its comparative Fatality to Males and Females. *[Manchester: 1775.]* PHi

———. Another copy. PPAmP

[PETIAU, ———, abbé.] Lettre De M. L'Abbé P*** De l'Académie de la Rochelle, A M*** de la même Académie. Sur le Magnétisme Animal. *[n. p.: 1784.]* PHi

PETZSCH, ERNST HEINRICH. Dispvtatio Inavgvralis Medica De Scabie Viva. *Wittenberg: Litteris Caroli Christiani Dürrii, [1767].* PHi

POTTS, JONATHAN. Dissertatio Medica Inauguralis, De Febribus Intermittentibus, Potentissimum Tertianis. *Philadelphia: Typis Johannis Dunlap, 1771.* PPAmP

PRICE, RICHARD. Observations on the Expectations of Lives, The Increase of Mankind, The Influence of great Towns on Population, and particularly The State of London, with respect to Healthfulness and Number of Inhabitants. *London: W. Bowyer and J. Nichols, 1769.* PHi

[RABIQUEAU, CHARLES.] Lettre Electrique Sur La Mort De M. Richmann. *[N.p.: 1753.]* MB

REDMAN, JOHN. Dissertatio Medica Inauguralis De Abortu. *Leyden: Apud Conradum Wishoff, [1748].* PHi

Reflections On Antient and Modern Musick, with the Application to the Cure of Diseases. *London: for M. Cooper, 1749.* PHi

REGIMEN SANITATIS SALERNITANUM. L'École de Salerne, ou l'art De Conserver La Santé, En vers latins & françois. *Montecassino: Et se trouve à Paris: Chez Segaud, 1779.* nf

RICHTER, FRIEDRICH TRAUGOTT. Dispvtatio Inavgvralis Medica De Haemorrhoidibvs Morbo Caeco. *Wittenberg: Litteris Caroli Christiani Dürrii, [1766].* PHi

ROWLEY, WILLIAM. Medical Advice, For the Use of the Army and Navy, In the present American Expedition. *[London:] for F. Newbery, 1776.* PPAmP

———. Seventy four Select Cases, with the Manner of Cure, and the Preparation of the Remedies, in the Following Diseases. I. The Schirrus, Cancer, and Ulcers of the Breast and Womb . . . The Second Edition. *London: for F. Newbery, 1779.* PPAmP

RUSTON, THOMAS. Dissertatio Medica, De Febribus Biliosis Putridis. *Edinburgh: Apud A. Donaldson et J. Reid, 1765.* PPAmP

———. Proposals for Inoculation. *London: T. and J. W. Pasham, 1768.* PPAmP

[SERVAN, JOSEPH MICHEL ANTOINE.] Doutes D'Un Provincial, proposés A Messieurs Les Médecins-Commissaires chargés par le roi De L'Examen Du Magnétisme Animal. *Lyon: Et se trouve à Paris, Chez Prault, 1784.* PHi

SHIPPEN, WILLIAM. Dissertatio Anatomico-Medica, De Placentae cum Utero Nexu. *Edinburgh: Apud Hamilton, Balfour, et Neill, 1761.* PPAmP

SMALL, ———. Extrait Du Journal De Médecine, du mois de septembre 1780. *[Paris: 1780.]* (3 copies) PHi

———. Another copy. PPAmP

SMITH, ADDISON. Visus Illustratus; or, the Sight rendered clear and distinct; being an Enquiry or Examination into the cause of the Inefficacy or Defect of the present mode of constructing Spectacles for the relief of Presbytes, or weak-sighted Eyes. *London: for the Author, and sold by Egerton; and Walker, 1783.* PHi

[SMITH, D. A.] Notes On Mr. William Bromfeild's Two Volumes Of Chirurgical Observations And Cases. *London: for T. Longman, 1773.* PPAmP

SMITH, HUGH. Essays Physiological and Practical, on the Nature and Circulation of the Blood, and the Effects and Uses of Blood-Letting. *London: for W. Johnston, 1761.* PPAmP

[STARKEY, GEORGE.] Opus Tripartitum de Philosophorum Arcanis. videlicet, I. Ennaratio Methodica trium Gebri Medicinrum . . . *London: Apud Guilielmum Cooper, 1678.* PPL

SUE, JEAN BAPTISTE. Programma de Oesophagotomia. *[Paris:] Typis Michaelis Lambert, 1781.* PPAmP

TERRAS, JOHANN JUSTUS. . . . De Valetvdine Sexvs Elegantioris A Coma Calamistrata. *Wittenberg: Prelo Ephraim Gottlob Eichsfeldi, [1749].*
PHi

TILTON, JAMES. Dissertatio Medica, Inauguralis. *Philadelphia: Typis Gulielmi & Thomae Bradford, 1771.* PHi

[TISSART DE ROUVRES, JACQUES LOUIS NOËL, marquis DE.] Nouvelles Cures opérées par le magnétisme animal. *Paris: 1784.*
PHi

TRILLER, DANIEL WILHELM. . . . De Scarificatione Et Vstione Ocvlorvm ab Hippocrate Descripta. *[Wittenberg:] Prelo Ephraim Gottlob Eichsfeldi, [1754].* PHi

ULMANN, JOHANN GOTTFRIED. De Morbo Bovm Adhuc Epidemice Grassante. *Wittenberg: Literis Caroli Christiani Dürrii, [1765].* PHi

VATER, ABRAHAM. . . . De Valetudine Sexus Elegantioris A Coma Calamistrata. *Wittenberg: Prelo Ephraim Gottlob Eichsfeldi, [1749].* PHi

VITET, LOUIS and JEAN HENRI DESIRE PETETIN. Extrait du n°. 16, 1781, des observationes sur les maladies régnantes à Lyon . . . De la section de la symphyse des os pubis. *[Paris: 1781.]* PHi

WILKINSON, JOHN. The Case of Mr. Winder who was cured of a paralysis by a flash of lightning . . . communicated to the Society of Gottinghen by Dr. Wickmann. *Göttingen: bey Pockwitz und Barmeier, 1765.*
PHi

William Byrd Reports on His Mission to the Cherokee in 1758

W. W. ABBOT

AFTER General Edward Braddock's defeat in the summer of
1755 in his expedition against the French and their Indian
allies at the forks of the Ohio, things went from bad to worse for
British forces in North America and, beginning in 1756, in Europe
and on the high seas as well. The tide was turning in favor of
Britain in 1758 when William Pitt sent an army to Philadelphia to
mount a second expedition against the French and Indians at Fort
Duquesne on the Ohio.

In the early stages of this second British expedition in Pennsylvania, the young colonel of the First Virginia Regiment, George
Washington, held it as an article of faith that the expedition which
his regiment was joining would be "extremely arduous, perhaps
impractical . . . unless assisted by a considerable Body of [southern]
Indians."[1] It was not Washington, however, but another young
Virginian who took the initiative in recruiting these Indians. William Byrd III, of Westover, had been with John Campbell, earl of
Loudoun, for about a year at the time Loudoun was laying plans
for the expedition; he had no trouble in persuading Loudoun in
February 1758 to send him to Carolina to "Collect a large body of

---

1. Washington to Gen. John Forbes, 19 June 1758, in John C. Fitzpatrick, ed.,
The Writings of George Washington (Washington, D.C.: U.S. Government Printing
Office, 1931), 2: 215–18.

93

Cherokee Indians" and bring them for "Their rendesvous" to Winchester in Virginia.[2]

Not yet thirty years old, William Byrd had already frittered away much of his "very large and opulent fortune in Virginia."[3] In the fall of 1756, after returning from an earlier mission to the Cherokee for the colony of Virginia,[4] he put the three older of his five little children aboard one of the ships of the tobacco fleet bound for England to begin their schooling. This done, he abandoned forever his wife Elizabeth Hill Carter Byrd, leaving her with two infants at Belvedere, the house he had built for himself and his wife at the falls of the James, upriver from Westover.[5] In early 1757 Byrd was in New York serving as a gentleman volunteer in the

2. The quotation is in Loudoun to John Blair, 13 February 1758, Forbes Papers, Alderman Library, University of Virginia, Charlottesville. Forbes wrote Loudoun on 4 February 1758: "Mr Bird assures me that he could easily bring 500 Cherokees"; and on 12 February Forbes drafted a letter for Loudoun to Gov. William Henry Lyttelton of South Carolina declaring that Byrd "proposes to go into their Country, & to Engage a body of their nation to join those Troops that I intend to employ upon the Ohio this ensuing Campaign" (James, ed., *Writings of John Forbes*, 37–39, 41–42).

3. The quotation is from John Forbes to William Pitt, 10 July 1758, ibid., 141–43. Peter Randolph, of Chatsworth, Byrd's neighbor and friend, wrote him on 20 September 1757 of his wish that a "measure cou'd be fallen on to prevent the sale of your lands" and suggesting that he might "sell the young Negroes," of whom Byrd owned more than two hundred too young to work (Tinling, ed., *Three William Byrds* 2: 627–29).

4. William Byrd and Peter Randolph went to the Cherokee country in February 1756 with a commission from Lt. Gov. Robert Dinwiddie of Virginia, dated 23 December 1755, to "settle a firm Treaty of Peace and Friendship" (Robert Dinwiddie Papers, Virginia Historical Society, Richmond), and with Dinwiddie's instructions "to prevail with them to send a Number of their Warriors to our Assistance in the Spring" (Robert Dinwiddie to George Washington, 14 December 1755, in W. W. Abbot, ed., *The Papers of George Washington*, Colonial Series [Charlottesville: University Press of Virginia, 1983], 2: 213–16).

5. The sequence of Byrd's actions in the fall of 1756 may be pieced together from the letters from his mother Maria Taylor Byrd, 15 March 1757, 21 September 1757, 6 November 1757, 24 December 1757, and 20 February 1758, and from his wife Elizabeth Hill Carter Byrd, 13 May 1757, and 12 May 1758, in Tinling, ed., *Three William Byrds* 2: 622–24, 625–26, 629–31, 631–33, 633–35, 635–37, 624–25, 653–54.

military family of the earl of Loudoun, who himself had arrived in New York on 23 July 1756 with several aides and seventeen servants, including a maître d'hotel, a valet, a cook, a groom, a coachman, a postilion, footmen, and a mistress, to assume command of the British forces in North America. Byrd accompanied Loudoun when the earl sailed with an army to Halifax in May 1757 to mount an attack on the French fortress at Louisburg. After the expedition failed, Byrd and Loudoun returned in September to New York where they both remained until Byrd set out by ship for South Carolina on 18 February 1758.[6]

Landing in Charleston on 10–11 March, after what he described as "the most dismal passage that ever was,"[7] Byrd met with Gov. William Henry Lyttelton, paid a call on Lyttelton's predecessor, James Glen, and drank wine with the British army officers stationed in the town. He also spent nearly two weeks preparing for his meeting with the Cherokee. The plan was that he would meet as soon as possible with as many of the Cherokee headmen and warriors as could be assembled at the lower Cherokee town of Keowee. Keowee was near a frontier fort, Fort Prince George, in the northwest corner of South Carolina, some three hundred miles from Charleston.

Byrd left Charleston on 24 March, and a few days later met on the road a Cherokee chieftain headed for Charleston "with seventy four Warriors & eight squaws" to obtain presents. Byrd persuaded the Cherokee chieftain, the Little Carpenter, to let him take fifty-six of the warriors and seven of the squaws back to Keowee with him.[8] When he got to Keowee, Byrd found no warriors or headmen

6. Stanley M. Pargellis, *Loudoun in North America* (New Haven: Yale University Press, 1933), 81, 228–52.

7. Byrd to John Forbes, 21 March 1758, printed below.

8. Byrd to William Henry Lyttelton, 31 March 1758, in Tinling, ed., *Three William Byrds* 2: 644–45. On 21 January 1758 the Little Carpenter (Atta-kullaculla) and "the great Warrior of Chotee [Chota]" came into Fort Loudoun, the recently built fort near Chota manned by Capt. Paul Demeré and men of his

there.[9] He soon learned that "upwards of four hundred Cherokees were already gone to Winchester in different partis," mostly from the lower towns, "& that all the inhabitants of the middle settlements [in the North Carolina mountains] were at home, but refused to come to me."[10] In early April Byrd visited all of the lower

---

independent company, with "two Frenchmen, and Twictwee [Twightwee] Indian women Prisoners, Six Frenchmen's and six Twectwee Indians Scalps" (Paul Demeré to Henry Bouquet, 21 February 1758, in S. K. Stevens et al., eds., *The Papers of Henry Bouquet*, [Harrisburg, Pa.: Pennsylvania Historical and Museum Commission, 1972], 1: 306–8). The Little Carpenter was rewarded in Charleston and was back at Keowee before Byrd left for Winchester at the beginning of May.

9. Loudoun and Gov. William Henry Lyttelton of South Carolina had done what they could to get the Indians to Keowee. Loudoun wrote to Pres. John Blair of Virginia on 13 February 1758 instructing him to "immediately send off Interpreters to the Upper Cherokee Towns to warn the Chiefs to go to the meeting to be held . . . at Keave, and that they should go prepared for an Expedition." Two days later he ordered Capt. Abraham Bosomworth, an old Cherokee hand, to go to Williamsburg and from there to "proceed with the Interpreter and Guides" to South Carolina (ViU: Forbes Papers). On the advice of the provincial council, Blair sent to the frontier for the Indian interpreter Abraham Smith, the brother of Richard Smith, the interpreter at Keowee. When Smith had not shown up by 8 April, and having received reports of the parties of Cherokee pouring into Winchester, Blair and Bosomworth agreed that Bosomworth should go to Winchester to deal with the Cherokee there rather than go on to South Carolina (7 March 1758, *Executive Journals of the Council of Colonial Virginia* [Richmond: Virginia State Library, 1966], 6: 82–83; Bosomworth to Loudoun, Blair to Bosomworth, both 8 April 1758, ViU: Forbes Papers). After Byrd got to Charleston, Gov. William Henry Lyttelton sent Lt. Henry Howarth, "who is well acquainted with the Cherokee," to the Cherokee Nation with instructions to "collect as large a Number of the Indians as possible to bring them to Keowee" (extract of a letter from William Henry Lyttelton to Loudoun, 21 March 1758, ViU: Forbes Papers).

10. The quotation is from Byrd to Loudoun, 30 April 1758, Tinling, ed., *Three William Byrds* 2: 649–50. Various reports of the number of southern Indians who came to Winchester in the spring of 1758 range from Byrd's "upward of four hundred" to Forbes's "above 700 Men" (Forbes to William Pitt, 19 May 1758, ViU: Forbes Papers). In the Forbes papers there is a return of the southern Indians, dated at Winchester on 21 April 1758, which gives not only the number of arrivals but also the date of arrival and departure with destination of each party, the name of commander of each party, and the town from which each came. Between 28 February and 21 April, five hundred and ninety-four southern Indians came to Virginia, all but about seventy were Cherokee. There is no mention in the correspondence of George Washington, the commander of Fort Loudoun at

Cherokee towns, and on 10 April he reported that about one hundred "fine young fellows" had agreed to join him. He then visited the Cherokee's middle settlements and sent messages to the Overhills Cherokee towns in what is now southeastern Tennessee.[11]

It was 1 May before Byrd left Keowee for Winchester. He set out with about sixty warriors from the lower towns and with a promise from the Little Carpenter that he and two hundred warriors from the upper towns would soon follow. When Byrd got to Winchester on 27 May with fifty-seven Indians, he found that many of the Cherokee, some of whom he met on his way up, had already gone home, and that still others were preparing to go.[12] Byrd's arrival slowed the departures for only a time, but as colonel of the newly-formed Second Virginia Regiment, Byrd was able to hold his own party of Cherokee in Virginia, and took them with his regiment to Fort Cumberland, Maryland, at the beginning of July.[13] The other Cherokee continued to leave, however, and finally, on 3 August, Byrd confessed to Gen. John Forbes that "everyone of my cussed Indians has left me."[14]

Winchester, of the arrival of any more Cherokee before Byrd got there at the end of May. In fact by 11 May those who had come earlier were already beginning to return home (see note 12).

11. Byrd to William Henry Lyttelton, 10 April 1758, in Tinling, ed., *Three William Byrds* 2: 647–48.

12. About the time Byrd left Keowee for Winchester, a party of Cherokee on their way home from that place clashed with white settlers in Bedford and Halifax counties in southern Virginia. On 11 May Colonel Washington pursued a party of twenty-five Cherokee heading south and persuaded them to return to Winchester. John St. Clair wrote to Forbes from Winchester on 19 May to report that "Seven Cherokees left this day and many others are bent on going home"; to John Blair on 23 May, that the Cherokee were "returning in great numbers"; and to Forbes again on 25 May, that "Fifty Cherokees left this yesterday morning" (ViU: Forbes Papers).

13. General Forbes reported to his superior, Gen. James Abercromby, on 9 July 1758: "The Cherokees are all gone except about 100 with Col. Bouquet at Raestown [Pa.] and 60 with Colo. Byrd at Fort Cumberland" (James, ed., *Writings of Forbes*, 135–40).

14. ViU: Forbes Papers.

The Little Carpenter did not keep his promise to follow Byrd to Virginia in the early summer of 1758; he did come to Pennsylvania in the fall with a party of Cherokee as Forbes's army moved in slow stages toward Fort Duquesne. There was also a party of Catawba Indians in Pennsylvania in the summer and fall of 1758 serving as scouts for the army. But by the time Forbes and his forces reached the forks of the Ohio on 17 November 1758 and looked down upon the smoking remains of Fort Duquesne on 11 November, all of the southern Indians, Cherokee and Catawba, had gone home.[15]

Between 21 March and 3 June 1758 William Byrd wrote four letters about his Cherokee mission to Col. John Forbes. Forbes was Loudoun's adjutant general until Loudoun's recall to London shortly after Byrd's departure for Carolina, at which point Forbes was made the general in command of the proposed expedition against Fort Duquesne. These letters came to light when the University of Virginia acquired a collection of John Forbes's papers in 1972. The letters appear neither in Alfred Proctor James, ed., *The Writings of General John Forbes* (Menasha, Wis.: The Collegiate Press, 1938) or in Marion Tinling, ed., *The Correspondence of the Three William Byrds of Westover, Virginia, 1684–1776* (Charlottesville: University Press of Virginia, 1977), but Marion Tinling printed substantially without comment, the Byrd correspondence found in the Forbes Papers (ViU) in the *Virginia Magazine of History and Biography* 88 (July 1980): 277–300.

To John Forbes
Charles Town S. Carolina March 21st 1758

Dear Sir

I set down with great great Pleasure to obey your Commands, & to return you my thanks for your Letter of Recommendation to Colo. Mont-

---

15. The best printed source for following the progress of Forbes's army in the summer and fall of 1758 is in S. K. Stevens, Donald H. Kent, and Autumn L. Leonard, eds., *The Papers of Henry Bouquet: The Forbes Expedition* (Harrisburg: Pennsylvania Historical and Museum Commission, 1951), vol. 2.

gomery, who has shewn me very great Civilities.[16] That Gentleman & all his Officers are heartily 'tir'd of this Place & sincerely wish to be removed to the Northward. I think 'tis pitty so fine a Battalion shoud lay the Summer in Garrison here, for it is realy the finest I ever saw, but the hot weather disagrees very much with them.[17] Colo. Montgomery will give you an Account of his Disputes with the Town's People about Quarters, I fear he will have but an uncumfortable time with them.[18]

I had the most dismal Passage that ever was, I was one & twenty Days at Sea in a Conti[n]ual Storm, & was a good deal allarm'd by the violent Gales, Lightening, Waterspouts &c. however I arrived here safe & sound on the 11th of March. Mr Atkins is detain'd at Cape Fear by Sickness, so no Difficulties will arrise from him.[19] Mr Littleton, who is rather slow,

16. Forbes wrote to Loudoun on 15 February 1758: "I send you a Letter inclosed for Colonel Montgomery, to go by Colonel Byrd, if you approve of it" (James, ed., *Writings of Forbes*, 44–45). Forbes's letter to Lt. Col. Archibald Montgomery, commander of the First Highland Battalion, has not been found.

17. Archibald Montgomery landed at Charleston on 3 September 1758 with his ten companies of Highlanders to join Lt. Col. Henry Bouquet and his five companies of the second battalion of the Royal American (sixtieth) Regiment and the two companies of the provincial first Virginia Regiment under Lt. Col. Adam Stephen, for the defense of South Carolina. On 10 September Bouquet wrote Loudoun: "They [i.e., The Highlanders] were very healthy when they arrived, but grow very Sickly." He and Montgomery later reported "above 500 Sick," six of whom died (Stevens, ed., *Bouquet Papers* 1: 197, 248).

18. Henry Bouquet's letters in the fall and winter of 1757–58 have frequent allusions to the failure of the South Carolina Assembly to provide the British troops in Charleston with proper quarters and supplies as required by law. See particularly Bouquet to Loudoun, 25 August 1757, 16 and 21 October 1757, and Bouquet to Forbes, 1 February 1758, in Stevens, ed., *Bouquet Papers* 1: 172–77, 212–20, 223–25, 288–89. Bouquet did write to John Hunter, 16 February 1758, a month before he and his Royal Americans sailed for New York on 26 March, that the Assembly had at last provided sufficient quarters, which "will be ready next month" (ibid., 303–4). The main point at issue was not housing but the unwillingness of the Assembly to provide the supplies for the troops that the local inhabitants by British law were required to provide. For a discussion of the quartering of troops at this time in America in general and in South Carolina in particular, see Pargellis, *Loudoun in America*, 187–210.

19. Loudoun and Byrd shared George Washington's lack of confidence in the Indian superintendent Edmond Atkin. See Loudoun to William Henry Lyttelton, 13 February 1758, in Stevens, ed., *Bouquet Papers* 1: 299–301; and Washington to Robert Dinwiddie, 5 October 1757, in Fitzpatrick, ed., *Writings of Washington* 2: 138–43. Atkin, a South Carolina merchant, returned in 1756 from a six-year stay in England with the king's commission as the superintendent of Indian affairs in

has at last been of great Service, & obtain'd of the Assembly £20,000 Currency, which is 700 pr Ct worse than Sterling, to fit out the Indians for the Expedition.[20] I have been detain'd by them much longer than I expected, which I am very uneasy at; but hope to get away the Day after tomorrow, & without an unforeseen Accident I shall get to Keawee on the 4th of Apriel, which will be sooner than the upper Indians can get there. I certainly will loose no time & hope by the last of May to be at Winchester, where I flatter myself I shall have the Honor to hear from you. I will not give you the trouble of reading the Particulers of my transactions here, 'tho, if 'tis worth your while, you will undoubtedly see my Letter to his Lordship.[21] I deliver'd your letter to Mr Glen who proposes to go to New York purely to make you a Visit,[22] I fear it will not be in his Power to

---

the southern colonies. Atkin went to Williamsburg from Winchester in October 1757 and seems to have spent most of the winter in New Bern, N.C., or in Wilmington, on the Cape Fear River. He arrived in Charleston the night before Byrd left for Keowee on 24 March; and, after allowing Byrd and Governor Lyttelton to persuade him he need not go with Byrd, Atkin gave the Virginian what "Hints" he thought necessary and then "set up the whole . . . Night" to write further instructions as well as a formal message for delivery to the Cherokee headmen (Atkin to Loudoun, 25 March 1758, Amherst Papers, P.R.O., W.O. 34/47, fols. 197–98; Atkin to Byrd, and enclosures, 24 March 1758, ibid., fols. 194–95).

20. In July 1757 the South Carolina Assembly voted £140,000 in colonial currency for seven new companies of provincial soldiers, but by mid-October it was clear to Henry Bouquet at Charleston that the companies would never be raised (Bouquet to Loudoun, 16 October 1757, in Stevens, ed., *Bouquet Papers* 1: 212–20). Governor Lyttelton was able to persuade the Assembly to allow him to use for outfitting the Cherokee £20,000 of the unexpended balance appropriated for the new companies. Byrd is saying here that the par of exchange is £700 in South Carolina currency for £100 in sterling, or 7 to 1.

21. See Byrd to Loudoun, 21 March 1758, in Tinling, ed., *Three William Byrds* 2: 640–42.

22. James Glen, governor of South Carolina from 1738 to 1756, was a "near relation," a "cousin," of Forbes (Forbes to William Pitt, 10 July 1758, in James, ed., *Writings of Forbes*, 140–42; Last Will and Testament, ibid., 299–300). Henry Bouquet wrote Forbes from Charleston on 1 February 1758: "I have had Several times the Pleasure of Speaking of you wth Govr Glen and his family. . . . I shall not fail to give him your Compliments he lives in a very agreable Country seat at 3 miles from Town, where he enjoys more happiness than he ever knew in the hurry of business" (Stevens, ed., *Bouquet Papers* 1: 288–89). Glen visited Forbes in Philadelphia in early June 1758. Not long after he arrived, Forbes sent Glen to the frontier to visit Colonel Bouquet at Raystown, Pennsylvania, and colonels Wash-

comply with your Order for there is no good Wine of any kind here, except a Little Colo. Montgomery has, & that we have lower'd considerably by drinking Bumpers to your Health. I have not seen Mrs Draton, but Mrs Glen has ask'd me ten thousand Questions about you, & amongst others if 'twas likely you woud Marry in America, I told her I imagin'd you woud when the warm Weather set in.[23]

Some Cherokees are gone to the Fronteers of Virginia, in all about a hundred & thirty, they appear by all accounts sincere in our Interest. The little Carpenter is just return'd from a Scout to Fort Loudoun, where he has been I am not able to learn, but Capt Demoree writes he has brought in two French Men 12 French Women thirty Indian Women & six & twenty Scalps.[24] I propose to do myself the Pleasure to write to you again from Keawee, & will inform you fully of every thing that Passes there. That the Begger's Benison may ever attend you my Dear Collonel is the Prayer of Your Most Obedient & Obliged Humble Servant

W. Byrd

To John Forbes
Keawee in Cherokee Apriel the 30th 1758

Dear Sir

Altho' I am honor'd with your Commands to omitt no Opportunity to let you hear what I am about, yet I fear I shall be troublesome in paying my Respects too often. By Colonel Boquet I had the Pleasure to inform you of my Proceedings in Charles Town & the Success I met with at last in my Application to the Assembly there.[25] I left that Place on the 24th of

---

ington and Byrd at Fort Cumberland, Maryland, in order to use his influence to persuade the remaining Cherokee to stay with the army. Glen had little luck with the Indians, but he wrote entertaining and informative letters about the military posts and their commanders (see James Glen to Forbes, 26 July, 8 August 1758, ViU: Forbes Papers).

23. Mrs. Drayton was probably Governor Glen's sister Margaret, third wife of John Drayton. Forbes was ill throughout the campaign and died a bachelor in Philadelphia on 11 March 1759.

24. Byrd was probably referring to Capt. Paul Demeré's letter to Col. Henry Bouquet, 21 February 1758, cited in note 8.

25. Byrd is referring to his letter of 21 March. He also wrote brief letters to Governor Lyttelton on 31 March and 10 April 1758 (Tinling, ed., *Three William Byrds* 2: 644–45, 647–48).

March, I got here on the 7th of Apriel, in the way I join'd a Party of fifty six Indians & brought them with me. I found on my arrival many of the Cherokees had gone out of the Lower Towns to Virginia in small Parties to the amount of two hundred & fifty, & I am credibly informd one hundred & fifty more are gone there from Over the Hills. The Messengers who went to bring the Warriors down here to a meeting with me return'd without a single Man, they reported all the over Hills Indians were gone either down the Tanasy or to Winchester, & that all the Inhabitants of the Mountain Settlements were at home & refused to come. I went immediately there, & prevail'd on them at last to take up the Hatchet, & they have faithfully promis'd to send off one hundred Men in a very few Days who are to meet me on the Path. There are above one hundred more who are intirely at the Dissposal of the Little Carpenter; who have given me their Word to be at Winchester by the first of June. I march tomorrow with near a hundred more who I have pick'd up about the Nation, & hope to join the Army by the last of May; I flatter myself that will be time enough. You may be assur'd Sir I have been as expeditious as possible, when you recollect my Situation since the 28th day of March, as well as my Inclination to serve his Lordship & my Country, but I was two Months too late. I have six hundred Miles to Winchester, I am three hundred & twenty from Charles Town, & ninety six above any house but little Forts & Whigwhams, & am just return'd from a Tour of a hundred & forty Miles among the Savages settled on the Waters of the Messessepy. These Gentry have made me promiss, before they will ster one Step, that I will not quit them 'till the Campain is over; so I can not tell when I shall have the Pleasure to drink Mrs Cuylers Health in a glass of your Clarret.[26] The Squaws are the only good things to be met with here, & I can not break them of anointing themselves with Bears-Grease, & depriving themselves of the greatest Ornament of Nature. As I am now intirely out of the Christian World, you must expect me to converse about nothing but Savages, therefore lest the Topict shoud be as dissagreable to you as their Society is to me, I will conclude with wishing you all Health & Happiness. Believe me my Dear Colonel no Man is more sincere in his

26. Mrs. Cuyler has not been identified, but it is perhaps worth noting that Cornelius Cuyler and several of his relatives were engaged in trade in New York and that in the summer of 1789 Alexander lived in a boardinghouse in New York run by a Mrs. Cuyler.

Professions than I, when I assure you I am Your Most Obedient & Most Obliged Humble Servant

W. Byrd

P.S. Several of my Warriors choose to go to Virginia with the over-Hills People, so that my Party will not exceed sixty Men. This change proceeds from a Dream. I send Mr Turner over the Mountains tomorrow to conduct them to Virginia.[27] Pray Sir present my Compliments to my Partner & Captain Cunninghame, I am so haunted by the Indians I can not find time to write to them.[28]

Dear Sr I had this Moment the Pleasure to recieve your Letter by Express, & heartily wish you joy of your Preferment. I am sincerely concern'd at Lord Loudoun's being recall'd.[29] Depend I will use the utmost

---

27. George Turner wrote to Byrd from Fort Loudoun in the upper Cherokee country, 23 June 1758, in great detail about waiting at Keowee until 1 June for wagons from Charleston with presents for Little Carpenter and of his subsequent dealings with Little Carpenter at Fort Loudoun between 13 and 21 June. Turner summed it all up in a letter to Forbes of the same date: "They trump'd a Story of their Conjuror foretelling them a great deal of Sickness & Death that wou'd attend them in case they went [to Virginia] & they positively refus'd to go till the Fall." He concluded in his letter to Byrd: "But I sincerely believe that They never intended to go" (ViU: Forbes Papers). The Little Carpenter (Attakullaculla) wrote Byrd from Keowee on 27 May indicating that he would soon be leaving for Winchester (Tinling, ed., *Three William Byrds* 2: 656–57).

28. James Cuninghame, a captain in the Forty-fifth British Regiment, was Loudoun's aide-de-camp. Byrd's "Partner" in Loudoun's entourage may have been Capt. Francis Halkett. See Byrd to Forbes, 3 June, printed below, and note 37.

29. Forbes's letter has not been found, but on 21 March 1758 Forbes wrote both Gov. Horatio Sharpe of Maryland and Gov. Arthur Dobbs of North Carolina about his appointment "to the Command of the Kings Regular Forces and provincial Troops, who are to be employ'd jointly in the opperations to be carryed on this ensuing Campaign to the Southard of Pensilvania included" (James, ed., *Writings of Forbes*, 59–63). Byrd received at Keowee on 30 April two letters from Governor Lyttelton, dated 14 and 20 April (also not found), as well as the missing letter from Forbes. Presumably, Lyttelton received in Charleston on or before 20 April a letter from Forbes similar to those written to the other governors and also the letter for Byrd which he sent up to Keowee with his own letters (Tinling, ed., *Three William Byrds* 2: 650–52). William Pitt's order recalling Loudoun reached New York on 4 March 1758 but was not read by Loudoun until he returned to the city on 11 March. Pitt's instructions were for Gen. James Abercromby to take

dispatch to get to you with my Party, which 'tho small are very good Men. There is no alteration since I wrote the Letter. I am quite overjoy'd I am to serve the Campain under you my Dear Colonel, & make no doubt but I shall get to Winchester as soon as the Highland Battalion.[30] Mr Turner goes off tomorrow to conduct in the upland Warriors, he will be a week or ten Days behind me. God bless you & send you Success.

To John Forbes

Bedford May 21st 1758

Dear Sir

'Tis with very great Pleasure I tell you I am at Bedford Court House with fifty seven Warriors, & without an Accident hope on Sunday next to have the Honor to kiss your Hands. I left Mr Turner to conduct the Little-Carpenter & two hundred Men into Winchester, who I expect in a fortnight. I gave you my Reasons by an Express from Keawee for coming with the smallest Party. I make no doubt Sir but you have heard of the two Skirmishes that happend in this County between some Cherokees & our People on the Fronteer, which had near proved Fatal to my Affair. In case the Indians with you shoud be allarm'd at the Story, I think Proper to send an Express with a Message from my Savages to prevent their Coming away in Numbers; the Consequence of which woud be the utter Distruction of this part of the Country, & an unavoidable Warr with those People.[31]

---

temporary command of the forces in America and for John Forbes, colonel of the Seventeenth Regiment of Foot, to hold the rank of brigadier general while he commanded the army marching on Fort Duquesne.

30. Col. Archibald Montgomery and his Highlanders did not leave Charleston until 20 May and landed in Philadelphia on 12 June 1758. By then it was clear that Winchester was not to be the rendezvous for Forbes's army. Three additional companies of Highlanders had already come directly from Britain to Philadelphia.

31. Byrd and his Cherokee warriors crossed from North Carolina into the frontier county of Halifax in Virginia and traveled from there northward through the frontier counties of Bedford and Augusta into Frederick County, which borders on Maryland and is the site of Winchester. For contemporary accounts of how settlers in Halifax and Bedford in early May twice attacked and killed a number of Cherokee passing through the counties on their way home, see "Depositions

I must once more express my Satisfaction in the Thoughts of serving the Campain under you my Dear General for I sincerely am with the highest Regard Your Most Obedient & Obliged Humble Servant.

W. Byrd

To John Forbes

Winchester, June 3d 1758

My Dear Sir

Your Favor of the 30th of May gave me vast Pleasure, 'tho I must acknowledge myself greatly Dissapointed at the distant Prospect of paying my Duty to you, especially as Sr John St Clair will not agree to my accepting of your kind Invitation.[32] Indeed it woud not be very convenient at this time, as most of the Savages insist upon going home, & I am sorry to tell you we can prevail on very few to join those I brought, who are realy in earnest; however I am daily employing my Party to use their Influence to stop the rest. I sent the Cawtawbaws to join Colo. Boquet.[33] I have heard nothing lately of Mr Turner, but flatter myself he will be here in time with the Little Carpenter & his Company, notwithstanding the Dissputes in the South. It woud give me the greatest Satisfaction to have my Dear General approve of my Conduct in the Cherokee

---

Concerning Indian Disturbances in Virginia," in *Colonial Records of South Carolina: Documents Relating to Indian Affairs 1754–1765*, ed. by William L. McDowell, Jr. (Columbia: University of South Carolina Press, 1970), 463–70.

32. John St. Clair, deputy quartermaster general under Forbes, as he had been in 1755 for Gen. Edward Braddock and his army, arrived in Winchester on 16 May to see to the equipping of the Virginia forces assembled there to join Forbes's army. Before the campaign was over, the eccentric quartermaster general was much at odds with General Forbes and many of Forbes's senior officers.

33. The return of the southern Indians dated 21 April 1758 (see note 10) indicates that a party of twenty-seven Catawba Indians under Captain Bullen, a favorite of George Washington's, came to Winchester on 1 April. Col. Henry Bouquet wrote Forbes on 3 June from Carlisle, Pa., that he had "Captain Bullen, chief of the Catawbas, here" (Stevens, ed., *Bouquet Papers* 2: 15–21). Bullen was killed on 22 or 23 August 1759 near Fort Cumberland, Maryland, where he was to join colonels Byrd and Washington.

Nation, for I assure you Sir I did my utmost endeavour (under every Dissadvantage) to execute my Commission. The Regiment the Country has sent to my care is a Body of fine stout Fellows; Officers & Men intirely raw & undisciplined, & I myself unexperienced.[34] I can not by any Means procure Serjeants who know their Duty to instruct them, but we will do our best. We are at present in want of every Conveniency to enable us to take the Field, I hope we shall be soon supply'd. If you have no objection I propose to dress my Soldiers after the Indian Fashion.[35] I wish I coud get a few Drummers & Fifers.

As to my Horse I believe he will suit you Sir, therefore I shall think myself Honor'd if you will accept of him, as I have enough for the Campain.[36]

My Room, according to custome, is crouded with Savages, some drunk some sober, which confuses me a good deal, as you will Judge from my Letter, & obliges me to conclude sooner than I intended; but first give me Leave to ask a few impertinent Questions. Pray Sir who is your Aid de Camp, & Majr of Bregade? if tis Majr Halket I beg my Compliments to him in a particular Manner; & to be short, who all are on your Staff?[37] What is become of Capt Cuninghame & the rest of my Acquaintance who composed the Family of the good Earl of Loudoun? & when shall I have

34. Byrd was ill during much of the campaign, but he stayed with his 2d Virginia Regiment until it was disbanded after Fort Duquesne was taken. In the spring of 1759, he succeeded Washington as colonel of the 1st Virginia Regiment. Washington gave up his command after more than three years in December 1758, and in January 1759 he took his bride to Mount Vernon.

35. Apparently unaware that it was at this earlier suggestion by Byrd that Forbes chose the informal dress for his army, George Washington took credit for the adoption of "Indian dress" by British forces. See particularly Washington to Henry Bouquet, 3, 13 July, Washington to Adam Stephen, 16 July 1758 (Fitzpatrick, ed., *Writings of Washington* 2: 226–29, 235–36, 240); and Forbes to Bouquet, 27 June 1758 (James, ed., *Writings of Forbes*, 124–26).

36. William Byrd II bought, bred, raced, and bet on fine horses. He had earlier promised to get a Virginia horse for Loudoun (Peter Randolph to Byrd, 20 September 1757, in Tinling, ed., *Three William Byrds* 2: 627–29).

37. Capt. Francis Halkett of the 44th Regiment was Forbes's brigade major during the campaign. He also had been both brigade major for Gen. Edward Braddock and Lord Loudoun.

the much wish'd for Opportunity of telling you in Person[38] how sincerely I am Dear Sir Your Most Obedient & Most Obliged Humble Sert.

W. Byrd

38. Colonels Byrd and Washington saw Forbes—Washington for the first time—only after the ailing general finally reached Raystown on the night of 15 September 1758. Shortly after his arrival Forbes summoned the two Virginia officers from Fort Cumberland to reprimand them for opposing his decision not to use the old Braddock Road for his army's march to Fort Duquesne.

The Men of '68: Graduates of America's First Medical School

RANDOLPH SHIPLEY KLEIN

THE young man rose and faced the distinguished audience; although he had rehearsed carefully, the importance of the occasion created anxiety in him which he could not master. He took a deep breath, made eye-contact with several in the audience, and spoke: "Illustres academiae Curatores—Qui his praeses es dignissimus vir gravissime."[1] As he continued in Latin, explaining that the importance of the occasion demanded that he speak of honors which through the ages had been bestowed on the real champions of medicine, the words began to flow. After touching on the earliest beginnings—the Egyptians, Greeks, and Romans—he considered the Arabs and then the Renaissance when "the pursuit of scientific inquiry was all the rage with some." He no longer needed the prompts such as "Slow, loud, etc." when he looked at Drs. John Morgan and William Shippen, Jr., and stated "The fame of the School of Medicine at Edinburgh steals away from all the rest the limelight of our time; and no where has medicine stood more respected than it is by the British." Finally, he looked to his fellow students—the first to graduate from an American medical school—and urged "strive through your efforts to raise to the highest the reputation of this Academy. Accordingly, there shall

1. John Lawrence, Oration, 1768, College of Physicians. This Latin oration is reproduced with a translation and introductory remarks by Herbert J. Dietrich, Jr., in *Transactions and Studies* of the College of Physicians of Philadelphia, ser. 4, 25 (1957–58): 41–49.

never be lacking a succession of intelligent and skilled physicians who in their concern for the health of mankind will have been a great credit to themselves and their country."

Although 21 June 1768 and the names John Archer, David Cowell, Jonathan Elmer, James Tilton, and the six other graduates may not loom as large in the annals of medical history as John Lawrence's epic sketch suggests, the event and the individuals honored formed a notable step in the rise of Philadelphia as the medical capital of the New World and in the progress of medical science in early America.

A portion of the public had its first look at the new medical men at the graduation ceremony. David Crowell and Humphrey Fullerton argued "ingeniously" and "with great acuteness" about the "immediate Seat of Vision." "With great learning" Samuel Duffield and Nicholas Way debated the question "Num detur Fluidum Nervosum?" Tilton delivered an essay "On Respiration." Jonathan Potts, the valedictorian, spoke "On the Advantages derived in the Study of Physics, from a previous liberal Education in the other Sciences." After the awarding of degrees Provost Smith gave a brief account of the college and its progress and observed that because they were "first to receive medical honors in America on a regular Collegiate plan, it depended much on them . . . to place such honors in estimation among their countrymen . . . never neglect the opportunity the profession would give them." Dr. Shippen delivered the remainder of the charge, urging that they "support the dignity of their Profession by a laudable perseverance in their studies and by a Practice becoming the character of a gentleman." After a prayer by the vice president the Royal North British Fusileers band played as the ceremony concluded.[2]

2. Trustees of the College of Philadelphia, Minutes, 21 June 1768. See also *Pennsylvania Gazette*, 30 June 1768. Archer's diploma is displayed in Baltimore at the Library of the Medical and Chirurgical Faculty of Maryland. His and Jonathan Elmer's notes on Morgan's courses are in the University of Pennsylvania

Analysis of this group of ten medical graduates and their activities can provide insights into the development of professional careers during the last third of the eighteenth century and the first decades of the nineteenth century. It is important to understand the background of the group and how it diverged from or continued past practices. It is worthwhile exploring how it was chosen, its training, and the ways in which its members established themselves in the emerging medical community and the larger public world. The American Revolution, especially its military facets, created an exciting and often grisly arena in which to develop and demonstrate talents and of course played a crucial role in the formative period of the graduates' lives. For some the war and Revolution reinforced their chosen path, while others rethought their priorities—enthusiasm for medicine diminished as other endeavors beckoned. After the war the yellow fever epidemics created a dramatic and frightening contrast with the more mundane activities which generally occupy the attention of busy practitioners. Although not the first professionally trained "doctors" in North America, the first medical graduates of the College of Philadelphia began their adult lives with a particular identity and a special sense of mission.[3] The stages of life tested their ability and determination to make a difference in the world and benefit themselves, their country, and even mankind.

*

Through much of the colonial period, most who practiced medicine and the healing arts gained their knowledge and training informally. Apprenticeship to a practitioner of physick or druggist

---

Archives. See Whitfield J. Bell, Jr.'s "Medical Students and Their Examiners in Eighteenth Century America," *Transactions and Studies* of the College of Physicians of Philadelphia, ser. 4, 21 (1953): 14–24.

3. Their mentors clearly intended to inculcate a special sense of identity and mission; their efforts and being first evidently had the desired results. The Haw-

was common, and others acquired accepted knowledge and folk wisdom related to health in yet more informal ways or taught themselves. Rare was the individual with an actual M.D. degree; few colonists could afford the expense of living abroad in order to earn one, and few Europeans with sound medical degrees ventured to the British colonies.

By about the mid-eighteenth century, however, especially in the middle and southern colonies, leaders of the medical profession had degrees from Edinburgh, Leyden, or Rheims, and the apprentice system (still the most common training ground) was much improved. Philadelphia, the most populous and cosmopolitan town in British North America, with its coterie of talented physicians and practitioners such as Thomas and Phineas Bond, John Redman, William Shippen, Sr., Thomas Graeme, and Thomas Cadwalader; Pennsylvania Hospital; and host of people interested or engaged in scientific inquiry, provided a natural breeding ground for rising standards.[4]

A major step in the movement to promote professionalization occurred soon after John Morgan and William Shippen, Jr., returned to Philadelphia after receiving extensive university education including the M.D. degree and practical experience in Edinburgh and London. While abroad Morgan and Shippen had spoken with John Fothergill about the need and usefulness of establishing a medical school in America. Although Shippen returned first and offered medical lectures for several years, it was Morgan who approached the trustees of the College of Philadelphia with a pro-

---

thorne effect which modern sociologists perceive when groups become aware they have been singled out was at work, for a quick survey suggests that the first graduates outachieved their prewar successors. In 1775 fewer than 400 of the 3,500 medical practitioners had attended any medical school in America or abroad.

4. Whitfield J. Bell, Jr., *The Colonial Physician* (New York: Science History Publications, 1975), 5–40 and Bell's "The Scientific Environment of Philadelphia, 1775–1790," American Philosophical Society *Proceedings* 92 (1948): 6–14.

posal to establish a medical school. His initial failure to involve Shippen in the venture created bitter feelings which troubled both men and the medical community in general; Shippen's appointment to the medical faculty as Professor of Anatomy and Surgery failed to heal the wounds. Morgan's creation of a medical society which excluded the Shippens further infected the relationship which festered until Morgan's death almost two decades later. In some ways the atmosphere surrounding the creation of the medical school resembled the tumultuous political climate that was concurrently launching the new nation.[5]

America's first medical school began on a formal basis in 1765 when the trustees of the College (which included five physicians) accepted Morgan's proposal and on 3 May appointed him Professor of the Theory and Practice of Physick. In 1767 regulations enunciated requirements for the M.B. and M.D. degrees. Requirements for the former were: knowledge of Latin and those branches of mathematics and natural philosophy deemed necessary for medical study; attending at least one course of lectures on anatomy, materia medica, chemistry, and theory and practice, and one course of clinical lectures and having attended practice at the Pennsyl-

---

5. Chapter 8 of Whitfield J. Bell, Jr.'s *John Morgan, Continental Doctor* (Philadelphia: University of Pennsylvania Press, 1965), is an excellent modern treatment. The founding of the medical school is recorded in a number of works including William Pepper's "The History and Progress of Medical Education in the United States," in *University of Pennsylvania Lectures Delivered by Members of the Faculty* (Philadelphia, 1916); Thomas Harrison Montgomery's *A History of the University of Pennsylvania from Its Foundation to A.D. 1770* (Philadelphia: George Jacobs & Co., 1900), 304–309, 479–486; Francis R. Packard, *History of Medicine in the United States* (2 vols., New York: Hafner Publishing Co., 1963), 1: 339–394; and George Bacon Wood, *Medical Department of the University of Pennsylvania*, APS pamphlet collection; Joseph Carson, *History of the Medical Department of the University of Pennsylvania* (Philadelphia: Lindsay, 1869). Essential to an understanding of the topic is Richard H. Shryock's *Medicine and Society in America, 1660–1860* (New York: New York University Press, 1960), and "A Century of Medical Progress in Philadelphia, 1750–1850," *Pennsylvania History* 8 (1941): 7–28. Also useful is George Washington Norris's *Early History of Medicine in Philadelphia* (Philadelphia, 1886).

vania Hospital for one year; and knowledge of pharmacy gained through apprenticeship to a reputable physician. After meeting those requirements, a private examination by medical professors and trustees preceded a public examination. The doctor's degree required that the candidate be over twenty-one years old, practice for three years after receiving the M.B., and write and defend a thesis (in Latin) on a medical topic.[6] Although neither Morgan nor Shippen were modest about their medical knowledge and training, it is fair to state that they were eminently well qualified to launch a successful medical school. Added to extensive theoretical and practical experience gained at home and abroad, both burned with a sense of mission in promoting the professionalization of medical practice; their ambitious natures and the infancy of the professions in the Delaware Valley enhanced the drive for excellence. Provost William Smith offered lectures for medical students in natural and experimental philosophy, coupled with lectures on electricity by Ebenezer Kinnersley. In 1766 Thomas Bond initiated a course of clinical lectures, and in January 1768 Adam Kuhn was appointed professor of botany. The faculty was completed in 1769 with the appointment of Benjamin Rush as professor of chemistry; however, he joined the faculty after the first class graduated.

Students enrolling in courses soon learned, if they did not already know, that "medicine is a science as important in its object as it is difficult in the acquisition."[7] Because theirs was a new program in the far reaches of the empire, it generated excitement and self-consciousness. The example of Edinburgh and the great Scottish and English hospitals was emulated as well as possible. The Americans would strive to be like the best. Students gathered at Morgan's house to discuss natural curiosities, cases, and careers.

6. The requirements were published in *The Pennsylvania Gazette* that June and appear in full in Packard's *History of Medicine* 1: 355–356.

7. John Morgan, *A Discourse Upon the Institution of Medical Schools in America* (Philadelphia, 1765).

Like counterparts abroad they also created organizations known as the Hospital Medical Society and the Junior Medical Society.

Graduation records identify each of the candidates by name and geographical origin. Not surprisingly, all came from the Delaware Valley. Five came from Pennsylvania, three from the Lower Counties (now Delaware), and two from New Jersey. Samuel Duffield and Jonathan Potts were identified with Philadelphia, although the former was originally from Lancaster County and the latter originally from Berks County. Humphrey Fullerton came from the west, Lancaster County; David Jackson hailed from Chester County, south of the city on the way to the Lower Counties from whence came John Archer (New Castle—originally from Harford County, Maryland), James Tilton (Kent County) and Nicholas Way (Wilmington). David Cowell was listed as from Bucks County; his family owned land there and across the river in New Jersey where his father lived in Trenton. The young men from New Jersey were Jonathan Elmer (Cumberland County) and John Lawrence (East Jersey). In short all came from within a radius of about 60 miles.[8]

8. Trustees of the College of Philadelphia, Minutes, 21 June 1768. Information about the graduates is scattered in many primary and secondary sources. Details beyond those mentioned in the notes are available from the author. Only John Archer, Jonathan Elmer, David Jackson, Jonathan Potts, and James Tilton appear in the *Dictionary of American Biography*, edited by Allen Johnson and Dumas Malone (20 vols., New York: Charles Scribner's Sons, 1928–1936). The sketch of each covers but one half to a full page. In part this reflects the relative appreciation of the history of science and medicine earlier in this century. Only Tilton appears in James Thacher's *American Medical Bibliography* (2 vols., 1828, edition with introduction by Whitfield J. Bell, Jr., New York: Da Capo Press, 1967). See also Mrs. Thomas Potts James, *Memorial of Thomas Potts, Junior* (Cambridge, 1874); Stephen Wickes, *History of Medicine in New Jersey* (Newark, 1879). Archer, David Cowell, and John Lawrence appear in *The Princetonians*, edited by James McLachlan (3 vols. to date, Princeton: Princeton University Press, 1976–). Also very useful are the files on alumni and the annotated copies of the 1877, 1887, and 1897 *Medical School Alumni Catalogs* at the University of Pennsylvania Archives. According to descendants, Wickes, and McLachlan, David Cowell is erroneously listed as "Benjamin" in the trustees minutes (21 June 1768), *Pennsylvania Gazette* (30 June 1768), and early alumni catalogs. "David" Cowell's diploma and an explanation

All but one were born in the Delaware Valley where their families had lived since the first third of the eighteenth century or earlier. Cowell, Elmer, and Tilton had family roots in New England. The first was probably born in Dorchester, Massachusetts, the second's ancestors settled in Hartford, Connecticut in 1636 with Thomas Hooker, and the Tilton family supposedly settled in Lynn, Massachusetts during the Great Puritan Migration of the 1630s. In the late seventeenth century the Lawrence, Potts, and Way families settled in the Delaware Valley. The Archers, Duffields and Fullertons were part of the eighteenth-century migration of Scotch-Irish to Pennsylvania. The Duffields, however, were originally French Huguenots who fled to England before migrating to the North of Ireland and then Pennsylvania. As a group the ten were a mixture, although almost all were of English or Scotch-Irish ancestry and seven were Presbyterian. Potts and probably Way were Quakers. Lawrence was buried in a Baptist cemetery although the family had been Anglican.

Of course financial considerations played a significant role in the decision to study in Philadelphia rather than abroad; nevertheless more than finances and geographical proximity comes into play. The reputation of Philadelphia's physicians and the scientific and cultural life proved a potent magnet. Beyond that, personal contacts counted heavily as well. For instance, David Jackson's brother Paul was Professor of Languages at the College of Philadelphia and had been a physician before that. The fact that three students were graduates of the College of New Jersey which William Shippen, Sr. helped found and served as trustee is no mere coincidence. In fact his son also trained the brother of one of the graduates, upon

---

are in the University Archives and the correction appears in the 1900 *Alumni Register*. Cowell is an uncommon name in eighteenth-century Pennsylvania and New Jersey. I found no reference to "Benjamin" Cowell at the Historical Society of Pennsylvania, Genealogical Society of Pennsylvania, or records dealing with New Jersey.

occasion remarking "Johnnie is well."[9] Lancaster to the west on a well traveled road to Philadelphia was the home of William's influential brother Edward, with whom he kept up a steady correspondence. Similarly, Jonathan Potts's father was a good friend and political ally of Benjamin Franklin, and it appears likely that Jonathan Elmer's family knew the Franklins as well. Those from the Lower Counties were part of a tradition one historian has dubbed "the Philadelawareans."[10] John Archer and James Tilton, originally from Maryland, doubtless felt a kinship with the Bonds. Both Archer and Tilton had attended the Rev. Finley's Nottingham Academy in Chester County near the Maryland border. Like Morgan and Shippen, David Jackson had also attended the academy. All this is not to neglect the obvious attraction Philadelphia held to those with intellectual and scientific interests, nor is it to slight the individual's inclination to become a doctor. Still, young men are often influenced by personal contact, all the more so in the highly personal world of the eighteenth century. It is obvious that several of these graduates might easily have chosen a career in law, the ministry, or business. John Archer, for example, studied theology in Princeton and went so far as preaching a trial sermon before turning to medicine. Jonathan Elmer, as another example, later demonstrated extraordinary ability in both theology and law, as well as in medicine. Personal contacts helped them select the pathway to medicine.

So far as age is concerned, the group which ranged from twenty-one to thirty-six years of age at graduation was young by twentieth-century standards; six were twenty-one to twenty-three. In the eighteenth century, however, college students were on average

9. William Shippen, Jr. to Ebenezer Cowell, quoted in Cowell Family Portraits, DuBois Collection, Historical Society of Pennsylvania (hereinafter HSP).
10. John A. Munroe, "The Philadelaweareans," *Pennsylvania Magazine of History and Biography* 69 (1945): 128–149. His *Federalist Delaware* (New Brunswick: Rutgers University Press, 1954) is useful for the later period.

considerably younger than at present; hence Jackson, Lawrence, and Way, at twenty-one were not extraordinarily young. Their professors were about thirty. Samuel Duffield was the "old man," aged thirty-six upon graduation, while Cowell at twenty-eight and Archer at twenty-seven were also several years older than the majority.

Ascertaining socioeconomic background presents some problems, for no standard ranks all individuals or families in the Delaware Valley. After acknowledging a degree of imprecision, it seems safe to assert that most of the graduates came from more humble backgrounds than their professors and American physicians who had earned degrees in Scotland, England, or on the Continent. No overwhelming tendency to send first sons into medicine appears. Archer, Cowell, and Duffield were first sons (the last was a twin). Although Jackson was not first-born, his older brother had originally trained as a physician before changing careers. The ordinal position of most of the others remains unclear. Potts, the seventh child in a family of thirteen, is clearly an exception.

Almost all the students came from families which had enough money to provide them with an education through college, academy, or tutor in order to know enough Latin and Greek to satisfy the medical department. Archer, Lawrence, and Cowell graduated from the College of New Jersey. Archer, Jackson, and Tilton attended Nottingham Academy, while Fullerton and Potts received an education in Lancaster and Ephrata, respectively. Elmer was tutored privately and also by his brother. It is unclear how Duffield and Way acquired their education before arrival in Philadelphia. Samuel Duffield and his brother George were both born in 1732; George went to an academy and then the College of New Jersey, but no evidence indicates Samuel received such formal training. Because he was considerably older than the others in his class, it seems likely that his family could not afford to provide an equal education for both and that George worked for a time while possibly preparing himself.

The standing of the Potts family certainly equaled that of many with European degrees, and the local college was not the original choice. Jonathan's grandfather Thomas was an early Quaker settler of Germantown, and his son John owned many acres in Berks County, the location of the family iron works and Pottsgrove. In the summer of 1766 Jonathan Potts, after serving an apprenticeship with Dr. Phineas Bond and taking Morgan's course, joined his cousin Benjamin Rush and sailed for England "for an Improvement in the Duties of his Profession."[11] They carried letters of introduction to Benjamin Franklin from Samuel Wharton, Joseph Galloway, and William Franklin. Because of limitations on time, the two went directly to Edinburgh, where Dr. William Cullen welcomed them saying "that his attention to his lectures and practice was so great and constant that he had not time to treat his pupils with hospitality he would wish to do. 'But however close my attention' said he, 'may be to these necessary avocations, young gentlemen recommended to me from Dr. Morgan may always depend upon my immediate patronage and friendship.' He then made us welcome to his house and commanded us to visit him very often—and to show us how desirous he was to treat us like his friends, he introduced us to his family, with whom we have since had the honor of spending an evening."[12] Benjamin Franklin's introduction of Rush and Potts to Cullen and Sir Alexander Dick was written several months later. Franklin noted that "they are at Edinburgh to improve themselves in the Study of Physic, and from

11. Samuel Wharton to Benjamin Franklin, Philadelphia, 30 August 1766, Franklin Papers, American Philosophical Society (hereinafter APS).

12. Benjamin Rush to John Morgan, Edinburgh, 16 November 1766, *Letters of Benjamin Rush*, edited by Lyman H. Butterfield (2 vols., Princeton: Princeton University Press, 1951), 1: 28.

the character they bear of Ingenuity, Industry and good Morals, I am persuaded they will improve greatly under your learned Lectures and do Honour to your medical school."[13] To the young men Franklin wrote that he was happy to recommend them, but that they should

> apply diligently to your Studies, refraining from all idle useless Amusements that are apt to lessen or withdraw the Attention from your main Business . . . be very circumspect and regular in your Behaviour at Edinburgh, (where the People are very shrewd and observing) that you may bring from thence as good a Character as you carry thither, and in that respect not be inferior to any American that has been there before you.

After praising the great men who would teach them, Franklin urged "that besides the Study of Medecine, you endeavour to obtain a thorough Knowledge of Natural Philosophy in general . . . because I have observed that a number of Physicians, here as well as in America, are miserably deficient in it." The great scientist wished both young men "all Happiness and Success in your Undertakings."[14]

Jonathan's studies ended after only a few months, for he had to return because of what writers have usually described as his fiancée's illness. The "illness" was a pregnancy and the following year Rush wrote "I cannot help smiling every time I think of Jonathan being so early the father of two children."[15] Hopefully he shared Rush's enthusiasm for study in Philadelphia, which Rush described

13. Benjamin Franklin to William Cullen, [20 December 1766], *The Papers of Benjamin Franklin*, edited by Leonard W. Labaree and others (23 vols. to date, New Haven: Yale University Press, 1959–   ), 13: 531.

14. Benjamin Franklin to Benjamin Rush and Jonathan Potts, London, 20 December 1766, *Papers of Franklin* 13: 530.

15. Benjamin Rush to Samuel Fisher, Edinburgh, 28 July 1768, *Letters of Rush* 1: 64.

as "becoming the *Edinburgh of America*."[16] There Potts joined other
students, who were glad that an aspiring medical man "no longer
tears himself from every tender engagement and braves the danger
of the sea in pursuit of knowledge in a foreign country."[17] He would
walk the streets of Philadelphia which were "crowded with sons of
science."[18] Perhaps Potts's disappointment at the change in his
plans lessened because Dr. Cullen had also spoken so highly of
Morgan and the medical school which he had founded. The professor imagined Morgan's fame in America "will be more durable
than his own in Europe." John Archer's early career plans altered
significantly, but for very different reasons. Archer taught school in
Baltimore and studied for the ministry. Although he passed a pre-
liminary examination by the New Castle Presbytery, his subsequent examiners regretfully informed him that "though we would
gladly encourage youths who offer themselves for the sacred ministry . . . through the whole course of his tryals [Mr. Archer] discovers such a want of knowledge in divinity & the other particulars
he has been examined on, as well as an incapacity to communicate
his ideas on any subject, yet we cannot encourage him" to continue.[19]

Although the Revolution was brewing when the graduates left
the apparatus room when the ceremonies concluded, the immediate
task at hand was establishing oneself in the community. This process had already begun for some. It appears that several were
already practicing, although surviving evidence rarely indicates
whether they did so wholly on their own or as an assistant or
apprentice to an established practitioner. Samuel Duffield had
bought out Dr. Samuel Ormes and three months before graduation
advertised that he "carries on the business of a DRUGGIST, whole-

16. Rush to Morgan, 16 November 1766.
17. Ibid.
18. Ibid.
19. Quoted in *The Princetonians* 1: 301.

sale and retail" on Second Street near Walnut at the sign of Boerhaave's Head.[20] There he sold the usual patent medicines such as
Godfrey's cordial and Daffy's elixir, and also "a few setts of very
neat surgeon's capital and pocket instruments, and a variety of
shop furniture."[21] He also put up medicine chests for sea captains
and owners of remote estates and enterprises. Within a year he had
acquired a partner in Sharp Delany and the firm of Duffield and
Delany advertised regularly for almost the next decade, although
the partnership dissolved in 1775 when Duffield left. Selling drugs
fell short of the lofty ideals Dr. Morgan advocated for a physician;
however, it proved a necessity for virtually all medical men in the
colonies. Duffield was not simply an apothecary and purveyor of
medical supplies. He practiced medicine, being appointed a physician to the Almshouse in 1772 and a physician to the Society for
Inoculating the Poor in 1774. The latter was formed in response to
the smallpox epidemic of 1773 and Duffield, Bond, Rush, and
Shippen were among those who served. Jonathan Potts already
had a wife and two children; both Archer and Jackson were married before graduation. The former returned to the family's strong-
hold in Berks County and practiced medicine in Reading. The
latter married his sister-in-law shortly after his brother Paul Jackson, a professor at the college, died. David practiced medicine in
Chester County for several years and then moved to Philadelphia
prior to the Revolution where his training and acquaintances gained
him ready admission to the scientific and social life of the city.

In general graduates returned to their hometown where personal ties would make it easier to establish a practice. John Archer,
who practiced medicine in Newcastle County before receiving his
degree, hesitated in Philadelphia less than a year before returning

---

20. *Pennsylvania Gazette*, 17 March 1768. A biographical sketch of Duffield by
Randolph S. Klein appears in the *Transactions and Studies* of the College of Physicians of Philadelphia, ser. 4, 35 (1967–68): 119–125.
    21. Ibid.

to the area where he practiced for the next four decades. Some accounts suggest that in doing so he turned down Dr. Morgan's offer of a partnership. Neither John Lawrence nor James Tilton married; both remained lifelong bachelors. Upon graduation each returned to the place from whence he came. Tilton established his practice in Dover, while Lawrence returned to Monmouth County, New Jersey where he built up a huge practice throughout the county and even into adjoining Middlesex County. Success again greeted him when in 1775 he moved to Perth Amboy.

Presumably Humphrey Fullerton and Nicholas Way commenced their careers at about the same time as the others. Both are rather obscure figures at this time. Fullerton may have stayed in Philadelphia, though it is possible that he returned to Lancaster County. Way practiced in Delaware as before. He joined with other Delawareans and also Philadelphians such as Provost Smith, the Rev. William White, and Chief Justice Allen in promoting a successful scheme to establish a grammar school, based on the European model, in Wilmington. His inclusion in this group of movers and shakers suggests that he was gaining ground as a successful practitioner and man of the community. He remained a trustee for many years and also served as secretary of the board.

Jonathan Elmer also married shortly after graduation. He settled in Bridgeton, New Jersey where his wife Mary was near her father, Col. Ephraim Seeley. Dr. Elmer extended his practice into neighboring counties, but poor health forced him to forego long horseback rides. In 1772 Governor William Franklin appointed him sheriff of the county, and soon political activities filled the void created by a contracted practice.

Upon his graduation David Cowell received two houses in Trenton and some acreage to add to the more than 500 acres in Sussex County, New Jersey which his father had given him the year before. This induced young Cowell to leave his budding practice in Northampton, Bucks County, Pennsylvania, and build his life in Trenton. David lived with his younger brother Ebenezer.

The faculty hoped that most recipients of a M.B. degree would write a dissertation and return for an M.D. Evidently most found the additional degree unnecessary or not worth the trouble. The disappointment expressed by later historians may reflect more the importance of an M.D. degree in the twentieth century than the sentiments of the faculty or professional organizations in the eighteenth century.[22] At any rate, in 1771 Jonathan Elmer, Jonathan Potts, James Tilton, and Nicholas Way defended, in Latin, their dissertations. Elmer's "De Causis et Remediis in Febribus," Potts's "De Febribus intermittentibus potissimum tertianis," Tilton's "De Hydrope," and Way's "De Variolarum Insitione" were questioned by Drs. Kuhn, Morgan, Shippen, and Rush, respectively.[23] The trustees' minutes indicate that "each of the candidates having judiciously answered the objections made to some parts of their Dissertations, the Provost conferred upon them the Degree of Doctor of Physick, with particular solemnity, as the highest mark of honour which they could receive in the Profession."[24] Dr. Morgan then urged the four to continue pursuit of diligent study, and indicated that rather than prescribing the Hippocratic oath as did many universities, "laying aside the form of oaths the College, which is of a free spirit, wished only to bind its Sons and Graduates by the ties of Honour and Gratitude."[25] He impressed upon them that they had received the most distinguished degree and as "the foremost sons of the Institution, and as the Birth Day of its Medical Honours had arisen upon them with auspicious lustre," they might serve well their patients, the community, and their profession.[26] The dissertations were more exercises in literature than medical

22. For example, see Packard, *History of Medicine* 1: 367.
23. The dissertations are at the American Philosophical Society.
24. Trustees of the College of Philadelphia, Minutes, 28 June 1771.
25. Ibid.
26. Ibid.

research. The glowing phrases for public consumption and the trustees's minutes were not always echoed precisely in private. Thomas Bond, for example, sent Franklin a copy of Elmer's dissertation observing that "the Author is really a Man of Merit."[27] Although Tilton mentioned Bond's great success in curing dropsy, "the Facts are there so badly chosen, and the Principles so much mistaken that I cannot patronise it." Overall, however, Dr. Bond thought the commencement, or as he called it "the Farce was prettily played off."[28]

Franklin replied with thanks to Bond for the dissertation and to "the young gentleman who has done me the Honour to inscribe his Performance to me." The great scientist wished Elmer "the Success which his Ingenuity seems to Promise." Franklin also expressed pleasure that "our School of Physic begin to make a Figure. I know not why it should not soon be equal to that in Edinburgh."[29] Elmer gained election to the country's first permanent medical society the following year. Those forming the New Jersey Medical Society in 1766 observed that "medicine, comprehending properly Physic and Surgery, is one of the most useful sciences to mankind, and at the same time the most difficult to be fully attained"; they gathered together to promote that end through "friendly correspondence and communication of sentiment" in a "well-regulated society." Members planned to judge candidates, establish standard fees, and maintain the dignity of the profession. They justified their fee schedule by explaining that being aroused from bed at night, being exposed "often to great dangers from contagious diseases, &c.; besides the great expense of education, and the many painful years

27. Thomas Bond to Benjamin Franklin, Philadelphia, 6 July 1771, *Papers of Franklin* 18: 166.
28. Ibid.
29. Benjamin Franklin to Thomas Bond, London, 5 February 1772, *Papers of Franklin* 19: 64.

to be employed in preparatory studies . . . entitled [them] to a just and equitable reward for their services."[30]

The four who received the M.D. had moved ahead of their fellow graduates of 1768 and continued their growth to become medical leaders. That they continued for the degree probably reflects as much their relative economic standing as their drive. Nevertheless, socioeconomic standing, talent, and ambition continued to enable them to make the most of opportunities presented to them and remain the outstanding members of the medical class of 1768.

*   *

The date of the founding of the medical school proved fortuitous. Although the Stamp Act crisis created great disruptions, elation followed its repeal and the controversy over the Townshend duties lacked the intensity of its predecessor and soon dissipated into a rather misleading period of calm during the early 1770s. The attraction of politics which captured several graduates a number of years later did not divert them from their chosen path. Relative prosperity followed the economic hard times which came after the French and Indian War. Thus despite the disruptive undercurrents of the times, the political and economic climate helped foster success for diligent, young medical graduates. For about seven years they could go about the work essential to building a foundation for a successful career. When the war broke out, they were no longer fresh out of school. Their training, coupled with significant experience placed them in a position to play meaningful and sometimes influential roles in the Revolution. The war provided a severe test of medical knowledge, organizational ability, and stamina, as well as a real and present threat to life. Congress lacked an understanding of the importance of the medical department and that

30. Constitution of the New Jersey Medical Society, 23 July 1766, quoted almost in full in Wickes, *Medicine in New Jersey*, 44–48. Elmer's accounts with patients for 1785–1795 appear in Wickes, 110–118.

made the situation more difficult. The performance of doctors in the Revolution is not one of unblemished heroism; yet without question the war and Revolution were pivotal events in the ten graduates' lives. Most were about thirty as the war commenced. All were affected by it; three did not survive it.

The Men of '68 fall into three groups. Although not recalled as national heroes, Archer, Elmer, Potts, Jackson, and Tilton were prominent in the American cause. Cowell, Duffield, Fullerton, and Way served the American cause less conspicuously. John Lawrence, by contrast, strenuously opposed the Revolution and became a well-known Loyalist.

Wars, as Dr. Morgan indicated in his proposal for establishing a medical school, provide ample opportunity to acquire medical knowledge and experience. Professional concerns, however, only partially explain involvement in the Revolution. Like most people, the physicians had complex motivations. At the outset of the war, Archer, Elmer, and Tilton were officers in the militia, while Potts served on the Berks County Committee of Safety and as a delegate to the Provincial Congress, and Jackson managed a lottery to raise money "for defraying the expenses of the next campaign."[31] Archer and Elmer became increasingly involved in politics, whereas Jackson, Potts, and Tilton took on more responsibility for medical problems. Archer and Elmer both excelled in politics and helped write the state constitutions of Maryland and Delaware. Elmer also served in the U.S. Congress, as did Jackson shortly after the war. Potts reached the highest medical position of the group, in time he held successively positions as director of the General Hospital of the Northern Department and then of the Middle Department. Tilton achieved well-earned praise for his reforms which significantly reduced mortality rates in hospitals.

John Archer served on local committees from November 1774, at

31. *Journals of the Continental Congress*, 26 November 1776, 6: 982.

which time he also accepted a commission as captain in the militia. In 1776 he became a major in the militia and during the summer helped frame the Maryland state constitution and bill of rights. He enjoyed politics; the following year he was appointed commissioner of the peace. He served in that Harford County court for thirteen years. In 1779 while aide-de-camp to General Anthony Wayne in New York, Dr. Archer participated in the victory at Stony Point. Washington commended his "zeal, activity and spirit [which] are conspicuous upon every occasion."[32] By the end of the year, however, ill health caused him to retire to his 450-acre plantation.

Although a withdrawal from medicine was new to Archer, it simply continued Jonathan Elmer's prewar decision. Elmer's family helped spearhead revolutionary activities in their neighborhood. His brother Ebenezer was among those arrested for burning tea taken from the *Greyhound* at the Greenwich tea party. With brother Daniel as foreman of the jury and brother Jonathan (a member of the committee of observation) and like-minded men among the jurors, no chance of conviction existed. Jonathan served in the Provincial Congress of New Jersey for several months during 1775 and after his uncle Theophilus filled in for a few months, returned the next year to serve on the committee which drafted the new state constitution which was adopted 2 July 1776. In the meantime Dr. Elmer helped organize militia units and headed the Bridgeton association which published the *Plain Dealer*. For five years during the war he was clerk of Cumberland County and for six years he was a member of the Continental Congress. That tenure was interrupted for a year or so while he attended family affairs, a necessary step he argued because of the inadequacy of congressional salaries. Naturally he received an appointment to the medical committee

32. George Washington to President of Congress, 21 July 1779, *Writings of George Washington*, edited by John C. Fitzpatrick (39 vols., Washington, D.C.: U.S. Government Printing Office, 1931–1944), 15: 452.

and visited several army hospitals. He also served on the treasury board.

Jackson, Potts, and Tilton practiced medicine throughout the war and witnessed its horrors and frustrations first-hand. All expressed strong emotional reactions to what they saw and experienced. Potts and Tilton took major steps to improve conditions which won accolades and all three demonstrated medical and administrative abilities that were often in short supply during the struggle. On occasion their careers intertwined and their paths crossed. Potts came to hold authority over broad groups, whereas Jackson and Tilton remained associated with particular state units. On the whole all three excelled at what they did; all three emerged with reputations greater than their former major professors.

Although two of Jonathan Potts's brothers were Loyalists, he and four others were rebels. In addition to political concerns, Jonathan also dealt with medical emergencies by caring for Pennsylvania troops and prisoners of war at Reading, Pennsylvania. In April 1776 he petitioned for appointment as director general of a hospital in Canada. That summer he headed north with a commission as physician and surgeon from the Continental Congress and a letter commending him as "a gentleman of Character in every respect and most indisputable zeal in the publick cause."[33] Toward the end of June he arrived at Crown Point and was assigned by Samuel Stringer, head of the Northern Department, to Fort George where Potts had sheds built on the shore to house the "very numerous" sick.[34] He complained that summer to John Morgan that "without Clothing, without bedding or a shelter sufficient to Screen them from the weather" over a thousand suffered "under the vari-

33. Pass for Dr. Potts from Joseph Reed to Brig. Gen. John Sullivan, 25 June 1776, Jonathan Potts Papers, HSP.
34. Jonathan Potts to Owen Biddle, Field of Action, near Princeton, 5 January 1777, Jonathan Potts Papers, HSP.

ous & cruel disorders of Dysentaries, Bilious Putrid Fevers & the effects of a confluent small pox."[35] Morgan, already upset with Stringer's defiance of his authority, informed a congressional committee that "from all I am to learn, everything in the Medical department, in Canada, displays one scene of confusion and anarchy."[36]

Dr. Potts returned to Philadelphia by the end of the year and officers in charge of sick soldiers were to make returns to him. He accompanied American troops when they attacked the British at Princeton. He saw friends die of wounds and regretted that when maneuvers began he "was obliged to fly before the Rascals, or fall into their hands, and leave behind my wounded Bretheren" including Gen. Hugh Mercer. Potts expressed disbelief at "the inhuman Monsters" who robbed and insulted the general while he lay incapacitated in a hospital bed. Of course young Potts delighted in reports that "the Hero Washington [had] so shamefully Drubb'd and outgeneral'd [the enemy] in every Respect."[37]

Conflicts among medical authorities and charges of incompetence were common during the war. In January 1777 Dr. Samuel Stringer was dismissed as head of the Northern Department and Potts was soon appointed in his place and headed for Albany. William Shippen, Jr.'s recommendation of Potts to replace Stringer came just after Washington wrote to Congress that "I am so well assured, that you would not recommend Doctr. Potts to succeed Doctr. Stringer in the northern Department except you had sufficient proof of his Abilities in the medical line, that I readily concur with you in the Appointment."[38] The large number of sick and

35. Jonathan Potts to John Morgan, 10 August 1776, Jonathan Potts Papers, 77, HSP.

36. *Journals of the Continental Congress* 5: 460.

37. Ibid.

38. George Washington to President of Congress, Morristown, 20 January 1777, *Writings of Washington* 7: 40. George Washington to Phillip Schuyler, 27

dearth of medical supplies caused great consternation. Dr. Potts triumphed despite adversity. Through a system of inoculation his efforts eradicated smallpox as a major threat. In the fall Gen. Gates informed Congress of "the great Care, and Attention with which Dr. Potts and ye gentlemen of the General Hospital have conducted the business of their Department."[39] Just after the British surrender at Saratoga, the American general recommended that Congress honor Potts and his subordinates, and thanks from Congress soon arrived.

Some months before while establishing a hospital on Mt. Independence, Potts had indicated "I am determined to do my duty at every risk."[40] The danger proved greater than anticipated; by the end of the year ill health caused Potts to return to Reading, from whence he directed his hospitals in the Northern and Middle Departments. His former professor, William Shippen, Jr., who was now Director General of Hospitals of the Middle Department welcomed him back to Pennsylvania and expressed appreciation of the offer to help. Dr. Shippen informed the younger physician that "if your inclination is equal to your abilities the business will be well done."[41] In February 1778 Potts was appointed deputy director general of the Middle Department. His performance again earned high praise. General Washington stayed at the house of Isaac Potts (Jonathan's brother) while at Valley Forge and again praised the physician.

While Director General John Morgan suffered unrelenting abuse

---

January 1777, George Washington to William Shippen, Jr., 27 January 1777, *Writings of Washington* 7: 70, 71.

39. Gen. Horatio Gates to John Hancock, 20 October 1777, Collections of the Genealogical Society of Pennsylvania, Genealogical Notes 19: 183.

40. Jonathan Potts to Gen. Gates, Albany, 3 April 1777, Jonathan Potts Papers, HSP.

41. William Shippen, Jr., to Jonathan Potts, Bethlehem, 14 December 1777, Collections of the Genealogical Sociey of Pennsylvania, Genealogical Notes 19: 183.

and finally removal from office largely because of soldiers' suffering during that winter, other doctors fared better. William Shippen, Jr.'s self-congratulatory reports to Congress were promoted by his cousins the Lees of Virginia, who soon rejoiced at Shippen's appointment as Director General, replacing his old rival Morgan. Dr. Potts's reputation was also becoming well known. Dr. John Warren, brother of the more famous Dr. Joseph Warren, wrote from Boston that "the Honour and reputation you have gained in the Medical Department over which you presided would have rendered you a most welcome guest in this part of the continent."[42] Dr. Potts was also elected surgeon of the First City Troop of Philadelphia.

Clearly the young doctor had solid abilities and a promising career. Unfortunately by October 1780 declining health forced him to retire. Within a year he died at his home in Reading at the age of thirty six, leaving a widow and five children.

Quite possibly James Tilton encountered his classmates Potts and Fullerton several times during the war. For example, Tilton served as a regimental surgeon with the Delaware Regiment during the campaigns around New York City during which Fullerton was captured by the British. While Potts lamented the loss of particular friends at Princeton and recounted individual wounds inflicted by musket ball, bayonet, and sabre, Tilton saw his regiment decimated. After that he was in charge of hospitals at Princeton, Trenton, and New Windsor, Maryland. Physical suffering seemed inescapable during war; however, unnecessary suffering remained intolerable in Tilton's mind. He could reconcile himself to broken bodies and endure a "severe and tedious" fever which almost killed him and laid him up for nine months before he sought rest in Bethlehem, Pennsylvania for a few days. There he found "all manner of excrementitious matter was scattered indiscriminately through the camp, insomuch that you were offended by a disagree-

---

42. John Warren to Jonathan Potts, Boston, 30 April 1778, Collections of the Genealogical Society of Pennsylvania, Genealogical Notes 19: 183.

able smell almost everywhere within the lines."[43] Additional reports about the hospitals bothered him greatly. Affection for his former professor, William Shippen, Jr., retreated in the face of mounting evidence of atrocious conditions and the irresponsible behavior of the Director General. One of the earliest reports was that hospital staffs "were very deficient in even the commonest necessaries; that when the wounded arrived they immediately became affected with fever; and that the commissary, matron, nurses, and waiters, and all but one of the surgeons had the infection."[44] Crowded conditions made matters worse. The hospitals were a veritable death camp. One Virginia regiment reported that of forty admitted, only three emerged alive; "all the rest had to be buried."[45] Dr. Samuel Finley explained to Tilton that "we lost from ten to twenty of camp diseases for one by weapons of the enemy."[46] Dr. Benjamin Rush had been gathering a huge compendium of damning facts regarding Shippen. He complained to Washington of hospital deaths at Reading, Lancaster, and Princeton: "eight-tenths of them died with putrid fevers caught in the hospitals." He continued, "this extraordinary mortality among our soldiers is not necessarily entailed upon military hospitals. Dr. Potts lost only 203 men between the 1st of March and the 10th of December last, inclusive of all those who died of wounds. He suffered his patients (who were at one time very numerous) to want for *nothing*. The putrid fever never made its appearance in any one of his hospitals." In outrage Rush wrote, "What satisfaction can be made to the United States? What consolation can be offered to the friends of those unfortunate men who have perished—or rather

43. Quoted in Wickes, *Medicine in New Jersey*, 67.
44. Quoted in John W. Jordan, "The Military Hospitals at Bethlehem and Lititz During the Revolution," *Pennsylvania Magazine of History and Biography* 20 (1896): 149.
45. Ibid., 149–150.
46. Quoted in Benjamin Rush to George Washington, Princetown, 25 February 1778, *Letters of Rush* 1: 201. Tilton repeated the comments *verbatim*, quoted in Wickes, *Medicine in New Jersey*, 68.

who have been *murdered* in our hospitals—for the injustice and injuries that have been done to them?"[47]

Tilton's reaction was two fold. On the one hand he joined the effort to oust Shippen; on the other he experimented with public health techniques to prevent similar conditions from developing in camps within his jurisdiction. William Shippen had powerful political connections in Pennsylvania and in Congress. He had successfully thwarted Morgan's efforts to expose his shortcomings. He forced the resignation of Rush; but ultimately matters came to a head in a courtmartial. Evidence by Rush and others clearly demonstrated the Director General's shortcomings which included speculating in medical supplies while denying them to wounded troops, misuse of power, and not using his own medical abilities to personally treat wounded soldiers. Tilton's solid reputation and graphic descriptions persuaded many who heard of his sworn testimony and outraged reports. Tilton observed that Shippen was "interested in the increase of sickness and the consequent increase of expense, as far at least as he would be profited by a greater quantity of money passing through his hands." The General Hospital "swallowed up at least one half of our army, owing to a fatal tendency to throw all the sick of the army into the general hospital."[48] Rush also complained that

> Dr. Potts' whole expenses in the northward department, with an army larger than Washington's and with 3000 sick, amounted only to 151,000 dollars from January 1, 1777 to February 1, 1778. Dr. Sh----n's in only eight months of that time have amounted to about 400,000 dollars. In the northward department no man ever suffered from the want of *anything*. In ours, hundreds have died from the want of *everything*.

47. Ibid. 1: 202.
48. Quoted in Thacher, *American Medical Biography* 2: 132–133. He personally presented his well thought-out ideas for reform to Congress.

Notwithstanding this, I expect that Dr. Sh----n will be acquitted honorably.[49]

Shippen escaped conviction by a single vote; however, his usefulness as a leader ended and shortly thereafter he resigned.[50]

The major reforms Rush championed soon began to occur, particularly the separation of administration and purveying authorities, improved administrative procedures, and more authority to officers directly in charge of hospitals. Dr. Tilton turned down a faculty appointment in the Medical Department of his reorganized and politicized alma mater and returned to troops in New Jersey. There he took effective action to prevent atrocities from occurring. At the Morristown encampment and Princeton, for example, he successfully promoted the use of groups of small, airy log cabins to replace larger hospital buildings. With only six occupants and enforcement of sanitary regulations, mortality rates decreased dramatically. His success earned not only verbal praise, but also an appointment as senior physician and surgeon. He accompanied the American armies to Yorktown and operated a hospital at Williamsburg at the time of the last major campaign. After the British surrender, he returned to private practice in Dover, Delaware.

Perhaps Tilton encountered his classmate David Jackson, who may have served in a medical capacity at Yorktown. At first Jackson's wartime career combined managing a Congressional lottery and working as a surgeon and physician with the General Hospital in Philadelphia. Medical responsibilities increased in 1777 when he became senior medical officer; hence he resigned his duties with

49. Benjamin Rush to John Adams, Yorktown, 22 January 1778, *Letters of Rush* 1: 191.

50. For details see Whitfield J. Bell, Jr., "The Court-Martial of William Shippen, Jr., 1780," *Journal of the History of Medicine* 19 (1964): 218–238, and *John Morgan*, chapter 13, and also Butterfield's *Letters of Rush* 2: 1204; Packard, *History of Medicine* 1: 513–618.

the lottery. Financial and medical talents remained in demand; Jackson continued exercising both. He served the Pennsylvania militia as a surgeon, and in October 1779 became quartermaster general of the Pennsylvania militia. The following year he became surgeon general of Pennsylvania troops.

At about the time Philadelphians heard the Declaration of Independence read from "that awfull Stage," American troops arrested John Lawrence along with a number of other prominent local citizens of Amboy. Although the Lawrence family divided over the Revolution, the doctor sided with his father and brother who were very successful in recruiting people to fight for the king or at least take an oath of allegiance to the Crown. Dr. Lawrence remained on parole for a year and then was removed to Trenton for about a week before being paroled in Morristown. In July 1777 a petition "from sundry Ladies" including Governor William Franklin's wife expressing fear of "fatal and melancholy consequences . . . if they [and others] should be deprived of the assistance of Dr. Lawrence's skill in his profession, as his attendance is hourly necessary to several patients now much indisposed, who will be left helpless if he is removed, as no other practitioner resides in the place" failed.[51] The Provincial Congress of New Jersey indicated that "motives of consideration to individuals must give place to the safety of the publick."[52] Soon the doctor was in New York City practicing medicine behind British lines. His patients included Governor Franklin, and other prominent Loyalists including Livingstons and DeLanceys. Lawrence continued his practice there until 1783 and also commanded volunteers defending the city against the Americans. Unlike many Loyalists, including his brother, John Lawrence did not go into exile after the war. He retired to Upper Freehold, New Jersey, a wealthy man. Although the government confiscated some

51. New Jersey Historical Society *Collections* 9: 158.
52. Ibid.

of his property, he evidently had more than adequate means. His fortune increased significantly in 1790 upon the death of his father. Dr. Lawrence enjoyed being a country squire; he loved fox hunting, games, and excelled as a horseman. The happy bachelor lived with his two spinster sisters to whom he eventually left his estate when he died in 1830 while playing chess in Trenton.

The American Revolution was not necessarily a call to greatness. Like many of their countrymen, several of the Men of '68 served, but in ways that left no indelible mark upon the event. Samuel Duffield offered his services in October 1775 as "Surgeon and Physician of the People employed on Board Armed Boats."[53] In this he assisted Benjamin Rush. They shared the combined salary of a physician and mate ($16 per day), because Rush could not find a qualified surgeon's mate and the Council of Safety refused to pay for two surgeons. Subsequently the two doctors often worked together. By April 1776 they were both appointed to superintend and direct the hospital and pest house. In time they were responsible for all medicine, surgical instruments, and bandages belonging to Pennsylvania. The responsibility sounded larger than it was, for the Rush-Duffield report of June 1776 indicated that the state had but seven chests with few supplies, and no amputary instruments. Several of the chests "are also wanting of lint and Bandages, and all Kinds of Instruments." The drugs and bandages they managed to collect remained "by no Means a sufficient supply." Duffield continued to serve as physician in the general hospital until the war's end.[54]

Neither Nicholas Way, nor Humphrey Fullerton, nor David Cowell left records of a notable medical career during the war.

53. Council of Safety, Minutes, 10 October 1775, *Minutes of the Provincial Council of Pennsylvania, Council of Safety, and Supreme Executive Council* [usually referred to as *Colonial Records*] (Philadelphia and Harrisburg, 1851–1852), 10: 362.

54. Samuel Duffield and Benjamin Rush, Report, 25 June 1776, Gratz Collection, American Physicians 7: 28, HSP; *Colonial Records* 10: 409, 545, 614.

Way's activities remain mostly unrecorded. In 1776 he was consulting physician to Anthony Wayne. The officer, suffering from a fever after an arduous march to Detroit, wrote to Way "pray ought I to bleed—consult Doctr Rush."[55] Way remained in touch with Wayne, and about two years later asked his help "to recover a hackney of mine pressed by Col. Riddle . . . about the time I attended the wounded prisoners at the fights of Brandywine."[56] That activity had caused the Hessians and Scots garrisoned in Wilmington to sweep away his hay and other property. In 1777 he established an "inoculation hospital" in a barn on Henry Holly-day's Ratcliffe Manor near Easton, Maryland, where he inoculated about one hundred people per day against smallpox. In his practice he favored mild measures such as fresh air, light clothing, and simple diet, rather than the heroic measures others endorsed. Later he encouraged the establishment of another health facility known as "Dr. Way's bath" on the Christiana Creek near Wilmington.

Humphrey Fullerton's loss was more severe. When hostilities commenced, Fullerton went to Cambridge, Massachusetts, and served as one of the hospital surgeons. After the British evacuated Boston, Fullerton went to New York with the hospital department until he received an appointment to the Flying Camp. Unfortunately the British captured him at Ft. Washington in November 1776; then he spent almost one and a half years as a prisoner of war. He returned to York Town on parole and "in a very lingering and ill state of health."[57] Fullerton died in 1781 while still a captive; he

55. Anthony Wayne to Nicholas Way, Detroit, 5 September 1776, Society Collections, HSP.

56. Nicholas Way to Anthony Wayne, Middletown, 4 April 1778, Wayne Papers 5: 3, HSP.

57. Petition of Joseph Chambers . . . for his Sister Ann McKnight (later Ann Fullerton), 10 February 1788, *Pennsylvania Magazine of History and Biography* 42 (1918): 274–276.

left a wife and child. His widow later petitioned for his pension from 1781 until 1786 when she remarried.

David Cowell also died during the war, although no evidence suggests that his death was war-related. In July 1776 those with old sheets and linen for use as bandages were requested to give them to Dr. Cowell in Trenton, who served for the following two years as a physician and surgeon in the military hospitals. At the end of his life he became involved in a public controversy involving his slave and found himself estranged from his family and defending his military service. He died in 1781 and left his medical shop to his brother John, who had studied with William Shippen and served as a hospital mate in the General Hospital which Shippen directed. "A very large concourse of respectable inhabitants of Trenton," as well as the trustees, tutors, and students of the College of New Jersey attended his funeral. Although David's estate included £1,162 in personal property and he bequeathed £100 each to the College of New Jersey and the Trenton grammar school, his life ended in frustration. He left his "unfaithful disobedient negro" to the Presbyterian church, £100 to the U.S. Congress should it permanently locate in Lamberton, New Jersey, and a pack of squabbling heirs.[58]

*   *   *

As the War for Independence came to a successful conclusion, the Men of '68 had been reduced in number from ten to seven. Cowell, Fullerton, and Potts were dead, all by about the age of forty. By and large, the remainder of the group had many years ahead of them. Despite the wartime experiences which had demonstrated the limitations of doctors' ability to deal with pain and suffering, most of the class remained firmly attached to their medical careers. John Lawrence, the exception, evidently retired at a

58. David Cowell's will and related papers are at the State Library in Trenton.

young age and enjoyed life as a self-styled country gentleman. Nothing suggests his disillusionment with the profession. Jonathan Elmer continued his prewar concentration on politics which he found more to his liking, yet retained an interest in scientific inquiries. The others continued with medicine and now had time to develop other interests which starting a career and the war had caused them to forego.

This was a period of challenges, building, inquiry, and achievement. Four of the seven found that medicine alone could not fulfill their ambitions and combined notable activity in politics with their other duties. Clearly, the inadequate training of many who practiced medicine during the war, especially the outrageous incompetence, stimulated a desire to enhance standards for those who practiced after the war. The effort to promote professionalization and more rigorous training came naturally to those "educated upon the first collegiate plan." The establishment of medical societies and training students seemed logical. Well-qualified physicians could set standards and encourage sensible steps to upgrade the quality of health care. Americans studying abroad could still obtain high-quality training. In America, the situation proved problematic. There were only two medical schools in America before the Revolution, one in Philadelphia and the other in New York. Both suffered from the war. The British occupied New York City throughout the war; hence severe interruptions occurred there. From 1774 to 1779 the College of Philadelphia awarded no medical degrees. Philadelphia was occupied by the British for about nine months, and confusion and damaging antipathies remained in the wake. The College of Philadelphia lost its charter in 1779 for a time as the revolutionary government fostered the University of the State of Pennsylvania. Perhaps that and his clash with Shippen explains why James Tilton turned down an appointment as Professor of Materia Medica. Even after the flow of graduates resumed, rivalries continued. Eventually, Philadelphia realized it could not support two colleges and accommodations occurred. In 1780 the medical school functioned again. In 1783 Harvard founded a medical school and in 1798 Dartmouth's became the fourth established in the United States during the eighteenth century. Of course, relying on formally chartered medical schools remained but one option. John Archer was convinced that he could provide excellent training through his own efforts. By the mid-1780s his home on a 450-acre plantation in Harford County, Maryland, was referred to as "Medical Hall" and before his death some 25 years later, he trained at least 50 physicians. They in turn organized a medical society before which they read original papers. Archer himself joined with others in 1799 to found the Medical and Chirurgical Faculty of Maryland which became the state's medical society, and examined potential candidates. Nicholas Way also trained many students in "physick"; a large number of them came from South Carolina.

Meanwhile in Philadelphia, all was not confusion, despite the changes in institutions of higher learning. In 1786 twelve of the city's most distinguished medical men, including Samuel Duffield, John Redman, William Shippen, Jr., Benjamin Rush, and John Morgan and thirteen junior fellows formed the College of Physicians of Philadelphia. It existed "to advance the science of medicine and thereby lessen human misery by investigating the diseases and remedies which are peculiar to our country."[59] Duffield served for several years as an officer, including vice president from 1805 to 1813. The college's first *Transactions* enunciated a prevailing note of optimism concerning science, America, and the future. It announced that

> many disorders, once deemed incurable, now yield to medicine . . . I am fully persuaded there does not exist a disease in

59. Charter of the College of Physicians, available at the College and quoted in its *Transactions and Studies* 35 (1967–68): 70.

nature that has not a antidote to it. And when I consider the influence of liberty and republican forms of government upon science and the vigour which the American mind has acquired in the events of the late revolution, I am led to hope that a great portion of the honor . . . of discovering . . . these antidotes may be reserved for the physician in America.[60]

The group was on a solid footing when, in 1793, yellow fever, the most appalling collective disaster that has ever struck an American city, presented a fierce challenge to the nation's capital. "Fear seemed to absorb all the finer feelings of the heart," observed one contemporary.[61] Amid conflicting theories about causes and cures, the mayor of Philadelphia, a member of the American Philosophical Society, turned to the College of Physicians for advice and leadership. The college conferred in Philosophical Hall in August "in consequence of the prevalence of the fever of a very alarming nature in some parts of the city" and considered "what steps should be taken."[62] They decided upon thirteen specific recommendations including establishing a large and airy hospital for the poor, generally avoiding infected people or protecting oneself with camphor or vinegar when encountering them, and burying of the dead in closed carriages. Some of the other steps such as exploding gun powder in the air, do not sound so sensible two centuries later. Eighteenth-century doctors divided over the cause and cure; some argued it was contagious and isolation the cure; others suggested domestic origins in putrid air—hence the gun powder. No one knew that mosquitoes carried the disease, a fact not discovered until a century later.

60. *Transactions* of the College of Physicians of Philadelphia 1 (1793): xxxi. W. S. W. Ruschenberger's *An Account of the Institution and Progress of the College of Physicians of Philadelphia* (Philadelphia: Wm. J. Dornan, 1887) is useful.

61. Quoted in John Powell, *Bring Out Your Dead* (Philadelphia: University of Pennsylvania Press, 1949), 90.

62. Ibid., 30–32. In *College of Physicians*, 55–56, Ruschenberger quotes the minutes in full. The originals are at the College.

Dr. Duffield was appointed physician of the port after the death of his friend, James Hutchinson. Duffield went about his duties with vigor. Soon Benjamin Rush recorded "Dr. Harris and Dr. Duffield are confined. The former uses the new, the latter Dr. Kuhn's remedies."[63] Unlike over four thousand victims whose names are recorded in Mathew Carey's *A Short Account of the Malignant Fever*, Duffield recovered. Perhaps his personal triumph and relief explains why, back at work, he prematurely reported to Governor Mifflin that "the disease has so rapidly declined . . . that there is the greatest reason to conclude it will not be known . . . in two weeks."[64]

Yellow fever repeatedly visited Philadelphia. It came in 1794, 1796, 1797, and 1798, and often in the nineteenth century. Duffield remained an active combatant as physician to the port. Although unaware of its cause, his efforts to detect sailors and passengers with fevers and to improve sanitary conditions on board ships in the harbor served a useful purpose. Another of the class of '68 who battled yellow fever's invasion of Philadelphia was Nicholas Way. Way practiced in Wilmington, but sometimes traveled to Philadelphia to treat patients. Many contemporaries respected him for his work in Wilmington with refugees during Philadelphia's yellow fever epidemics. From 1794 to 1797 he was in Philadelphia as treasurer of the United States Mint. The physical hazards were great and doctors enjoyed no immunity. Nicholas Way fell victim to the mosquito, and some charged Dr. Rush. When strickened Nicholas Way "lost at six bleedings between 40 and 45 ounces of blood, a quantity by far too little to kill any person," according to Rush.[65] Not so, argued Drs. Hugh Hodge and Caspar Wistar,

---

63. Benjamin Rush to Julia Rush, Philadelphia, 29 September 1793, *Letters of Rush* 2: 687.

64. Samuel Duffield and James Mease to Governor Thomas Mifflin, Philadelphia, 4 November 1793, Dreer Collection, Continental Congress, HSP.

65. Benjamin Rush to Ashbel Green, 10 September 1797, *Letters of Rush* 2: 789.

whose charge Rush dismissed as "rash and cruel in the highest degree."[66] Controversy over the case raged on for a time. Rush's appointment by President John Adams to succeed Way as treasurer of the Mint was perhaps not the best means of calming tempers in the medical community, which seemed to thrive on personal rivalries and innuendo. David Jackson lived in Philadelphia from the end of the war until his death in 1801. He practiced medicine and ran an apothecary shop; however, his role in the yellow fever epidemic remains obscure. James Tilton, on the other hand, was concerned with the fever and wrote physicians in Philadelphia and New York to both expand and share his knowledge. His thoughts were published in William Currie's *Memoir of the Yellow Fever* (Philadelphia, 1798) and *Yellow Fever 1799*.

Among the many lofty ideals held up to the Men of '68 was that of continuing their studies and contributing to the community and mankind. John Morgan, for example, urged "place before your eyes the illustrious examples of great men, who, by pushing their researches into the bosom of nature, have extended the bounds of useful knowledge."[67] Of course a medical practice in itself could accomplish that, and political service could be viewed in a similar light. Another way to benefit mankind was through the promotion of useful knowledge, and the American Philosophical Society was only one of several specific avenues to accomplish this. Publishing scientific papers and engaging in discussions of improvements found much reinforcement. To varying degrees, with the exception of John Lawrence, those who survived the war participated in the scientific circle whose hub was Philadelphia.

The American Philosophical Society, originated by Benjamin Franklin in 1743, was reorganized and revitalized by the merger of two organizations in 1768 to form The American Philosophical

66. Ibid.
67. Morgan, *Discourse*, 55.

Society Held at Philadelphia for Promoting Useful Knowledge. It quickly achieved an international reputation through its impressive participation in the observations of the Transit of Venus, which were conducted in Europe, America, and also the Pacific Ocean. Samuel Duffield gained election to the Society in 1768, the year he earned his M.B. degree. Jonathan Potts was a member of the American Society at the time of the merger. Shortly thereafter he published a detailed defense of inoculation against smallpox in the Philadelphia press. Within four or five years Jonathan Elmer, James Tilton, and Nicholas Way were elected. Not until 1792 did David Jackson gain election. The other four original graduates never received the honor. Cowell and Lawrence lived in New Jersey and Archer resided in Delaware and later Maryland. That did not preclude election, but may explain why it did not occur. Fullerton was from Lancaster County and died during the war. Only the omission of John Archer appears a bit unusual.

At the same meeting at which Elmer (who upon graduation had communicated a letter "on the Constitution of Air, its effects on the human body &c" which intrigued the Society) was elected, the Society considered a " 'Dissertation of an amphibious animal discovered in . . .' Baltimore County" by John Archer, who sent along "a stuffed skin of that animal."[68] By contrast, only a month after James Tilton's paper on "an extraordinary case of hydrophobia" came before the Society, he and Nicholas Way, both of Delaware, were elected.[69] The fact that Archer had turned down a partnership with Morgan, an influential member of the Society, seems at least suggestive.

Duffield proved most active; in fact the others' involvement in the Society seems negligible. After election he signed for the Committee on Medicine and in 1773 he became part of a committee

---

68. APS, Minutes, 21 January 1774. Ibid., 30 June 1769.
69. APS, Minutes, 18 December 1772.

which conferred "with such people in this city as are concerned in the *Paper Manufactory* on the most probable means of establishing that branch of Business among us."[70] Just prior to the British invasion in 1777, Duffield, a curator of the Society, retrieved copper plates used in the *Transactions*, which illustrated "the Canal plan, as it is the Theater of War at present and has been made use of . . . without the knowledge of the Society, and in a way that may give offense."[71] About eight months after the British evacuation Duffield joined others in requesting that members convene "to take into consideration the present State of the Society, with such Matters as may be necessary to promote the Design of their Institution."[72] He supported the revival with his vote and subsequent attendance. He continued as curator from 1776 to 1791, and restored the collections. When in his seventies, Duffield, who "kept a register for many years" of temperature, barometric readings, winds, and weather, communicated some of his records to the Society.[73]

The American Philosophical Society was the first organization of its type in America and remained the only one until emulated after the war by Massachusetts. Other avenues to promote useful knowledge existed. For example, David Jackson served as a trustee of the University of the State of Pennsylvania and subsequently the University of Pennsylvania from 1789 until his death a dozen years later. John Archer contributed several papers to the *Medical Repository* of New York; he introduced seneka snakeroot as a treatment for

70. APS, Minutes, 5 March 1773.

71. William Smith and Thomas Bond to David Rittenhouse, Dr. Samuel Duffield, and Mr. Du Simitière, 7 September 1777, APS Archives.

72. APS, Minutes, 16 January 1779.

73. Samuel Duffield, Meteorological Observations for May 1803, Manuscript Communications to the APS, Natural Philosophy 1: 37, APS. The history of the Society and the larger context receive skillful treatment in Brooke Hindle's *The Pursuit of Science in Revolutionary America, 1735–1789* (Chapel Hill: Institute for Early American History and Culture, 1956).

the croup. Jonathan Elmer, whose medical knowledge exceeded that of all others in the United States according to Benjamin Rush, published articles and exercised leadership in the New Jersey Medical Society. He joined with others in 1781 to resurrect that Society which noted that "the war (which has been productive of the happy Revolution in America) having claimed the attention of all ranks of Freemen, most of the members of this Society took an early decided part in the opposition to British tyranny and oppression, and were soon engaged either in the civil or military duties of the State. Added to this, the local situation of the war (the scene of action being chiefly in this and the adjoining States), rendered an attendance on the usual stated meetings, not only unsafe but in a great measure impracticable."[74] In 1787 he served as its President, and was named in the incorporation act of 1790 which continued its life. The charter explained that the organization's purposes included

> uninterrupted intercourse and communication of sentiments with one another, to cultivate liberality and harmony among themselves, to promote uniformity in the practice of physic on the most modern and approved systems, to correspond with and receive intelligence from the like societies abroad, and generally to improve the science of medicine and to alleviate human misery.[75]

At about the same time, 1789, James Tilton, Nicholas Way, and others gained a state charter for the Delaware Medical Society. It was the third such state society, and Tilton served as its first president. He also contributed a lengthy article in William Currie's *An Historical Account of the Climate and Diseases of the United States* (Philadelphia, 1792). Tilton enjoyed cultivating his orchards, joined the

---

74. Dr. Beatty's report, "State of the Society since 1775," accepted by the Medical Society in May 1782, is quoted in Wickes, *Medicine in New Jersey*, 49.
75. An Act For Incorporating . . . the Medical Society of New Jersey, 2 June 1790, reproduced in Wickes, *Medicine in New Jersey*, 105–108.

Philadelphia Society for Promoting Agriculture, and contributed "On Peach Trees" and "On the Fruit Curculio" to the first volume of its Memoirs. In response to forty-four questions on American agriculture submitted by the French consul, Barbe-Marbois, to the Society, Tilton submitted an excellent series of answers which elaborated upon crops, techniques, pests, and other concerns of a member of the Académie des Sciences who ran an experimental farm.[76] In 1804 Dr. Tilton and others founded an agricultural association in Delaware which sought "to rouse a spirit of rational enquiry; to fertilize the soil; to improve, increase, and preserve the produce of the earth and livestock . . . , to favor . . . measures of economy . . . , and . . . to record and make known such useful hints, accidental discoveries, and successful experiments in husbandry as may be deserving of public regard."[77] Like Way, he was also interested in promoting the Delaware-Chesapeake canal.

Although Tilton in time gave up the practice of medicine, he did not forget his medical experience. During the War of 1812, which was mostly a series of embarrassments and defeats for the Americans, Tilton published *Economical Observations on Military Hospitals: and the Prevention and Cure of Diseases Incident to an Army.*[78] It reiterated his recommendations concerning the construction and administration of hospitals which brought him recognition some thirty to thirty-five years before. History repeated itself, or at least partially so. The sixty-eight-year old Tilton was appointed physician and surgeon-general of the army. His tour of the northern frontier brought old memories to life, for unsanitary conditions faced him

76. See "James Tilton's Notes on Agriculture of Delaware in 1788," edited by R. O. Bausman and John A. Munroe, *Agricultural History* 20 (1946): 176–187, and *Delaware Register and Farmer's Magazine* (1838–1839), 433–440. "On the Chemical Principles of Bodies," *Columbian Magazine* 2 (1788): 493–497; Hindle, *Pursuit*, 294–395, 343.

77. *Mirror of the Times*, 4 April 1804.

78. James Tilton, *Economical Observations on Military Hospitals* (Wilmington, Del.: J. Wilson, 1813).

everywhere and challenged his abilities. His reform efforts led to *Regulations for the Medical Department*. This important work "defined clearly for the first time the duties of medical officers and other sanitary personnel."[79]

Although the publication record of the group is not overwhelming, in the context of the times it is respectable. All the more so when the other demands on time are taken into consideration. These men had significant commitments which often ranged beyond treating patients and scientific inquiry. In the postwar years five devoted considerable energy to political activities. As a group they held important positions on the national, state, and local levels.

On the national level, John Archer served in the United States Congress from 1801 to 1807 and as a presidential elector in 1797. Jonathan Elmer attended the convention which nominated DeWitt Clinton for president in 1812. He served five years in Congress between 1778 and 1788, and in 1789 as a United States senator. His term expired in two years and voters refused to renew it because "this little doctor" supported moving the capital south.[80] Duffield, Tilton, and Jackson also served in Congress (1778, repeatedly from 1783, and 1785 respectively).

As treasurer of the United States Mint from 1794 to 1797, important people in the capital considered Nicholas Way among "certain characters highly respectable."[81] Archer, Jackson, and Tilton were ardent Democratic Republicans, while Elmer and probably Way were Federalists. No question about Jackson's outlook existed, for he along with David Rittenhouse, James Hutchinson, and other Philadelphians founded the nation's first Democratic Society. Similarly, in 1794, "at a respectable Meeting of

79. *Dictionary of American Biography* 18: 550.
80. His colleague Sen. William Maclay's comment of 3 September 1789 is quoted in Wickes, *Medicine in New Jersey*, 244.
81. APS, Minutes, 16 December 1796.

Citizens of New Castle County in the town of Newcastle . . . it was unanimously agreed to form themselves into a political society."[82] James Tilton was conspicuous among those who called themselves "the Patriot Society of Newcastle County, Delaware." He was also the first president of the Delaware Society of the Cincinnati and a powerful leader in state politics who was instrumental in leading the Democratic Republicans to success. He proved a fiery champion of prison reform, the abolition of slavery, and public education. In quieter moments, he served as a trustee of the Wilmington Academy and on the "select committee" of the Lyceum of Delaware. On the state and local level all these men were also active, at times fulfilling duties as officeholders. Elmer served a term in the New Jersey legislature and for nine years as clerk and eighteen as surrogate of Cumberland County. In 1814 he finally retired from public life after forty-two years of involvement. He taught himself extensively in law and mastered it well enough to earn considerable respect of his colleagues. Although he contemplated revising the state statutes, the task was accomplished by another. David Jackson was an alderman of Philadelphia, the nation's largest city, at the time of his death.

Another arena for leadership lay in religion. Duffield and Elmer involved themselves in the Presbyterian church. Dr. Duffield helped found the Third Presbyterian Church in Philadelphia and served many years as clerk, ruling elder, and trustee. He and William Shippen were among the original four trustees. The church was established in controversy with the Market Street church which tried to control it and prevent the election of Samuel's evangelical brother George as minister. George, a graduate of the College of New Jersey, was used to conflict, for he came from a ministry on the Pennsylvania frontier and had survived Old-Side Presbyterian opposition to his ordination. On one occasion after a favorable vote

82. *Pennsylvania Magazine of History and Biography* 69 (1945): 138.

for George, the Market Street group locked the Pine Street doors. Undeterred, the people got in and the Reverend Duffield commenced a service. In the midst of it, the king's messenger arrived, ordered them out, and read the Riot Act. The interruption ended when someone picked up the intruder and tossed him out. The minister was arrested for instigating a riot. His brother and the mayor of Philadelphia (Shippen's kinsman) offered bail. When the minister refused, the mayor dismissed the charges. Samuel remained a constant support to his politically and culturally active brother, and a great strength in the church for almost twenty-five years after his brother's death. Jonathan Elmer's role proved more tranquil. Although he associated with the church from the time of his marriage, it was not until 1798 that he became a member. His political and judicial skills served him well when he became a ruling elder. In time he attended the Presbytery and the General Assembly as a delegate. Clearly of an intellectual bent, he enjoyed probing religion, and contemporaries regarded him as the peer of most theologians.

Being a doctor was not the safest occupation in the eighteenth century. Although the average lifespan of the Men of '68 was about sixty, several shortened their lives because of their work. Jonathan Potts, Humphrey Fullerton, and possibly David Cowell died as a result of illnesses contracted while serving during the war. Nicholas Way died of yellow fever, a killer which many Philadelphians avoided by leaving town when it struck. Ironically, Jonathan Elmer, whose frail health encouraged him to take up medicine and the rigors of which practice caused him to leave it, lived until age seventy-three. James Tilton lost a leg at age seventy, but lived on another seven years. Duffield and Lawrence survived into their early eighties.

A legacy can be measured in many ways. In financial terms, several of this group succeeded well and left rather significant estates. John Archer, for example, left a 750-acre plantation, 8 slaves, and considerable personal property. Jonathan Elmer's estate was described as "a very handsome fortune." Jonathan Potts was born into a rather wealthy family, as was John Lawrence; both died holding considerable estates. Lawrence's personal property alone was inventoried at over $2,100. Nicholas Way seemed well off too. Cowell's estate was valued at over £1,160 in 1783. In Duffield's 1791 will he recorded bequests ranging from $300 to $3,000 to his five children, in addition to other belongings, though he did not die for another 20 years. Of more moderate means was David Jackson, and it appears Humphrey Fullerton's early death placed his widow in financial straits, although her remarriage suggests she was not left destitute.[83]

A financial legacy is gratifying to heirs; more intriguing to historians is a professional legacy. Clearly certain families took steps to continue the tradition in medicine. Leading the way was John Archer, who persuaded (coerced?) five of six sons to follow in his footsteps. David Jackson's eldest son and namesake continued his father's drug business, and his second son was a member of the medical faculty at the University of Pennsylvania for thirty-six years. Although three of Jonathan Potts's sons died young (the twins fell to yellow fever in Philadelphia), Francis Richard Potts became a physician and practiced in Pottstown. His older brother, named after Benjamin Rush, carried that famous medical name west when he left the area. David Cowell bequeathed his medical shop and books to his brother Dr. John Cowell. Samuel Duffield's son was Bryant Duffield (1770–1841) who received an M.D. degree from the University of Pennsylvania in 1790 and later inherited from his father "all my shop furniture and Medicine together with my Medical Books including those on Surgery and Anatomy." Jonathan Elmer trained his brother Ebenezer who became a very

83. Samuel Duffield's will, like Jackson's and Way's, is recorded in Philadelphia. Probate papers for the others were consulted when available.

successful physician in Cumberland County, New Jersey, as did Jonathan's only son and several of his nephews. Not everyone could pass on professional standing along bloodlines. Lawrence and Tilton were bachelors and Fullerton's only child died young. Of course there existed other ways to leave a medical legacy. The more than fifty physicians whom Archer trained is one. The example of high standards and achievement, so clearly associated with Tilton, Potts, and others, is another. Work to establish medical societies in Pennsylvania, Delaware, New Jersey, and Maryland also prompted professionalism and high standards. For the most part the Men of '68 died optimistic about the future of medicine in America. The burgeoning of proprietary medical schools, which greatly damaged the standing of doctors in nineteenth-century America, was just getting under way as these men reached the latter years of their careers. By and large they could look back upon satisfying lives and feel confident that they had effectively pursued the goals defined by Morgan, Shippen, and Provost Smith.[84]

* * * *

The performance of the Men of '68 presents a positive reflection on their institution and teachers as Provost Smith and the faculty hoped it would. Although most came from humbler backgrounds than Americans who earned degrees abroad, those trained in Philadelphia fared at least equally well. In addition to ambition, an interest in science, and the powerful magnet of the scientific and cultural capital of America, personal contacts help explain why they chose medicine and the College of Philadelphia. Rigorous training prepared them well for participation in their communities. Fortuitous circumstances provided several years of experience before the War for Independence burst upon the scene. All but one

84. Lisabeth H. Holloway, compiler, *Medical Obituaries, American Physicians' Biographical Notices Before 1907* (New York, 1981).

graduate sided with the Americans; at least half of them played significant roles during the contest. Their medical and administrative skills enabled them to defeat smallpox, hospital fever, and tend other physical needs. Courage and righteous anger caused several to attack mismanagement of the Medical Department even when that meant testifying against their former professor at his court-martial. Others played significant, although less dramatic roles. During the Revolution, and especially after it, these intelligent, highly trained, and motivated men took on increasing civic responsibilities. Most held political office—in the Senate, House, state legislature, or in local government. They supported a variety of public-spirited projects. An interest in science extended beyond medicine to agriculture, the American Philosophical Society, and publishing papers. Especially important was their contribution to the medical profession. They helped establish the country's first successful medical societies in Pennsylvania, New Jersey, Delaware, and Maryland, trained students, and passed on a significant legacy not only to their relatives, but also to the nation at large. Only two of the graduates can be characterized as obscure—and death by about age forty may explain that. Although the precise standing of two or three of the others may be debated, they and their distinguished classmates certainly fulfilled Morgan's dream that

> Perhaps this Medical Institution, the first of its kind in America . . . may collect a number of young persons, of more than ordinary abilities, and so improve knowledge as to spread its reputation to distant parts. By sending these abroad duly qualified, or by exciting an emulation amongst men of parts and literature, it may give birth to other useful Institutions of a similar nature, or occasional rise, by its example, to numerous societies of different kinds, calculated to spread the light of knowledge through the whole American Continent.[85]

85. Morgan, *Discourse*, 58–59.

"That Awfull Stage"
(The Search for the
State House Yard Observatory)

SILVIO A. BEDINI

THE transit of Venus, scheduled to occur on 3 June 1769, galvanized men of science throughout the Western world. The British in particular set an example with elaborate plans for observing it in a number of locations. The urgency to observe this celestial phenomenon stemmed from several factors. Edmund Halley had demonstrated that the transits of Venus could be utilized to determine the solar parallax or mean distance of the earth from the sun, which is a fundamental unit of astronomical measurement. Observations made from various points throughout the world would result in differences in the duration of time that Venus required to transverse the face of the sun at the locations selected, and from this data the solar parallax could be accurately determined. The observations made of the transit of 1761 were disappointing, and the phenomenon would not recur for another 105 years.[1]

In Philadelphia, members of the recently founded American Philosophical Society awaited the transit with great anticipation

1. Harry Woolf, *The Transits of Venus. A Study of Eighteenth Century Science* (Princeton: Princeton University Press, 1959), 16–22, 161–75; A. Pannekoek, *A History of Astronomy* (New York: Interscience Publishers, Inc., 1961), 284–87; Silvio A. Bedini, *Thinkers and Tinkers. Early American Men of Science* (New York: Charles Scribner's Sons, 1975), 172–76; Brooke Hindle, *The Pursuit of Science in Revolutionary America 1735–1789* (Chapel Hill: University of North Carolina Press, 1956), 134–65.

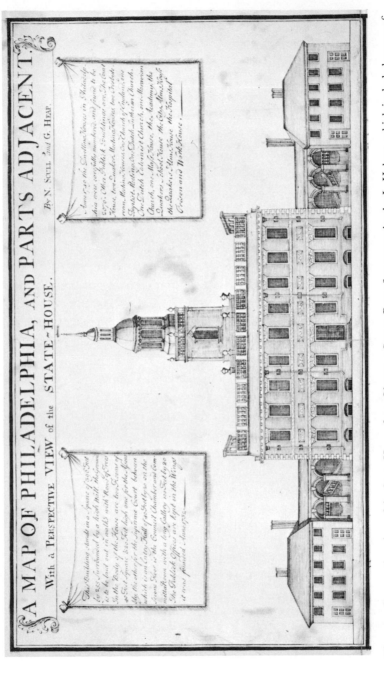

Figure 1. Perspective view of the State House from Chestnut Street. Part of an engraving by L. Hebert, which included a map of Philadelphia drawn by Nicholas Scull and George Hemp, c. 1750. Courtesy of the John Carter Brown Library.

for these and other reasons. First of all, the event encouraged the Society to take steps to establish accurately the longitude of the city. Secondly, the transit of 1761 had not been visible in the region, and no observations had been made in the American colonies. The forthcoming occurrence would provide an opportunity for the Society to attain recognition among scientific societies overseas, and it could in fact be considered a matter of colonial prestige.

Among the Society's leading enthusiasts eager to make the observations were its vice president, the Reverend John Ewing of the faculty of the College of Philadelphia; the college's provost, the Reverend William Smith; the mathematical instrument maker, Thomas Pryor; the statesman, Dr. Hugh Williamson; the instrument maker, David Rittenhouse; surveyor-general of Pennsylvania, John Lukens; John Sellers; and several other local men of science.

Ewing first raised the subject at a Society meeting on 19 April 1768 in a proposal noting that part of the transit would be visible in Philadelphia:

> After having gone thro the Calculation & Projection of the next Transit of Venus over the Sun, on the 3d of June 1769, I find that the Beginning and a great Part of it will be visible at Philadelphia if the weather will be favorable. As much depends on this important Phenomenon & as Astronomy may be brought to a much greater Perfection than it has yet arrived at by a multiplicity of accurate observations made of the Transit in different parts of the World, & compared together; I would hereby propose to this Society that effectual Provision be made for taking the said observation in this City. This is the more necessary as such another Opportunity will not be presented for more than a Century to come.[2]

2. [Henry Phillips, Jr.], "Early Proceedings of the American Philosophical Society for the Promotion of Useful Knowledge Compiled . . . from Manuscript Minutes of Its Meetings From 1744 to 1838," *Proceedings* of the American Philosophical Society, vol. 22, no. 119, pt. 3 (1885) [hereafter *Early Proceedings*], 13–14;

After the committee reviewed the proposal and estimated the probable costs, the plan was approved at the meeting of 21 June. At the same meeting the competition between Ewing and Smith which played a significant role in the formation of the Society's observing teams, became evident. Both were ambitious and interested in establishing their own reputations as well as that of the Society. Smith presented a projection of the anticpted transit which Rittenhouse had prepared. Ewing rendered the results of his own calculations in addition to a projection of the Transit "on a very large scale, containing all the Elements of the Projection, & the Effect of the Parallaxes in Longitude & Latitude; in altering Times of Beginning and End of the Transit at Philad.ᵃ." which the Society ordered to be published.[3]

At first the Society decided to observe the transit at three sites: Rittenhouse's farm in Norriton, Pennsylvania; a site near Cape Henlopen, Delaware; and Philadelphia, where they planned to ascertain the longitude of the city before the transit occurred. Then James Dickinson proposed enlarging the project by making observations also at James Bay in Nova Scotia. This would serve a double purpose by providing an opportunity to reconnoiter and make a map of the country from the south end of Hudson Bay which extended towards the head of the Mississippi. Because the full span of the transit would not be visible in Philadelphia the method of durations could not be used, and observations made at James Bay would be advantageous in providing data for comparison with those of the initial contact of Venus with the sun made at Philadelphia, Norriton, and Delaware. The Society approved the suggestion and requested financial assistance from the Assembly.

". . . Paper by the Revᵈ. Mr. Ewing. . . ," *Transactions of the American Philosophical Society*, 1st ser., 1 (1771): 5–8.

3. *Early Proceedings*, 14–15; *Pennsylvania Gazette*, 15 September 1768, p. 4, col. 1; Hindle, *Pursuit*, 150–51.

The Society agreed to defray all expenses at the sites first selected.[4]

The *Pennsylvania Gazette* reported these activities and the city awaited developments with interest. The Assembly responded favorably to the Society's request and voted "a sum not exceeding *One Hundred Pounds Sterling* for purchasing a reflecting telescope" and to defray the expenses of the observations. Speaker Joseph Galloway advised that he would order the instrument at "the first opportunity, agreeable to an order of the House. He would, therefore, be glad to be furnished with any information relative to the construction thereof that may be thought necessary, as soon as may be." The information was promptly provided. Thomas Penn and Benjamin Franklin, who were in England at the time, established communication for the Society with the Astronomer Royal Sir Nevil Maskelyne and Franklin diverted telescope lenses intended for Harvard College to Philadelphia.[5]

Galloway ordered a reflecting telescope made by Edward Nairne and equipped with a Dollond micrometer from England through Colonel George S. Eddy. Some estimate of the interest generated worldwide in the forthcoming astronomical event may be gained from Eddy's response in early January 1769: "I have bespoke the Telescope they [the Pennsylvania Assembly] have ordered, and hope it will be done in time. The Workmen have promised it, but it should have been thought of sooner; for they have so much upon their Hands by Orders from different Parts on the same Occasion [transit of Venus], that I think it rather doubtful. . . ."[6]

4. *Early Proceedings*, 18, 37; *Gazette*, 10 November 1768, p. 3, col. 1; "Votes and Proceedings of the House of Representatives of the Province of Pennsylvania," *Pennsylvania Archives*, 8th ser., 7 (1935): 6288–89; Hindle, *Pursuit*, 146–51.

5. *Gazette*, 10 October 1768, p. 2, col. 3, 26 October 1769, p. 3, col. 3; *New-York Gazette and Weekly Mercury*, 24 October 1768, p. 2, col. 3; *Early Proceedings*, 19; Hindle, *Pursuit*, 147–50; Bedini, *Thinkers*, 173–74.

6. Folder: Franklin to G, letters from Col. George S. Eddy to Joseph Galloway, 9 January and 21 March 1769, Box of letters to and from the Franklins, Franklin Papers, AM 15946, Princeton University. Entries relating to purchase of telescope

BACK of the STATE HOUSE, PHILADELPHIA.

Figure 2. The State House and yard, with Philosophical Hall in the background, 1800. Engraving by William Birch. The Observatory is believed to have been situated in the area between the trees and Philosophical Hall. Courtesy of the American Philosophical Society.

The Society acknowledged the Assembly's generosity, recognizing that it "was actuated by a laudable desire to promote useful knowledge and the reputation of their country." It then used the opportunity to present a further request "that leave may be given for erecting an Observatory in the State house ground; and that

you would grant such public assistance, as you may think convenient, for erecting the same and also for making an observation of
the transit, at least as far westward as Fort Pit [*sic*], which will be of
great use compared with Observations, in this and other places
more to the eastward."[7]

Eager to proceed, the Society plunged ahead with its plans to
build an observatory without awaiting approval from the Assembly.
The Transit Committee, reconstituted to include Samuel Rhoads,
Jr., J. M. D. Pennington, and Robert Smith, was directed to employ a workman to erect the observatory for "a sum not exceeding
£60 . . . if they cannot find among the houses belonging to the
Assembly, one fit for the purpose. . . ." No such suitable building
was available, and the committee thereupon "agreed with Jas.
Pearson to erect the Observatory to a plan delivered him. . . ." An
order was drawn on the treasurer for an advance of £10 to pay for
building materials. The plan with specifications provided to Pearson had probably been prepared by Rittenhouse at the committee's
request. Pearson was a house carpenter later employed to make
repairs of the State House. He began work on the project after 11
February, completing it by the end of March 1769, and received
£40 in payment. By the middle of May the observing party was
already testing its instruments in the structure, and the Society was
informed "The honorable House of Representatives had generously
granted the privilege of erecting the Observatory in the State house
yard," and provided £100 for the purpose.

Meanwhile in early November 1768 Rittenhouse had begun the
construction of an observatory at Norriton but "through various
disappointments from the workmen and weather," was unable to
complete it until the middle of April 1769. There is reason to
believe that Rittenhouse's observatory structure was similar to an
artist's conception published in 1896. (See Figure 3.) It had a

7. *Early Proceedings*, 29–31; *Pennsylvania Archives*, 8th ser., 7: 6356–57.

Figure 3. Artist's conception of the observatory building constructed by David Rittenhouse in 1768–69 at his farm in Norriton, Pennsylvania. Reproduced from an article on "David Rittenhouse," by Herman S. Davis, in *Popular Astronomy* 4 (July 1896): 5.

shutter in the roof which could be opened for observing and closed in inclement weather. Presumably the same facility was featured in the observatory in the State House Yard, which may have consisted of only an enclosed and roofed observatory room upon the platform.[8]

The Transit Committee next organized itself into groups, one for each site from which observations were to be made. Then when John Lukens claimed engagements that would prevent him from

8. *Early Proceedings*, 31–32, 36–39, 44–45; *Gazette*, 30 March 1769, p. 1, col. 1; *Pennsylvania Chronicle*, 20 March 1769; David Rittenhouse, "Mr. Rittenhouse's Observations at Norriton. . . ," *Transactions*, 1st ser., 1 (1771): 13; Brooke Hindle, *David Rittenhouse* (Princeton: Princeton University Press, 1964), 48.

observing on the western line, plans for a station at that location were cancelled. Ewing was to direct observations at the State House Yard Observatory with Shippen, Williamson, and Pryor; another member of that group was James Pearson. Smith was to direct observations made at Rittenhouse's farm in Norriton, with Rittenhouse, Lukens, and Sellers. Owen Biddle, with Joel Bailey and Charles Thomson to assist him, was to set out in a day or two for Lewes near Cape Henlopen. He received an advance of £20 for his anticipated expenses.[9]

The reconstituted Transit Committee then reviewed the instruments available and arranged to borrow others as required. Shippen planned to use "a small reflecting Telescope belonging to the hon.ble Prop.s" which was in fact the property of former Governor James Hamilton; Pryor owned a fine but relatively small reflecting telescope which he would use; and the committee applied to a Miss Polly Norris for a large 24-inch refractor in her possession, which she agreed to lend. Ewing would use the telescope purchased by the Assembly for the Society. Two instruments made by John Bird, a zenith sector and a transit and equal altitude instrument both owned by the Proprietors, were loaned by Thomas Penn and installed in the State House Yard Observatory "for ascertaining the true Meridian, adjusting the Clock, &c." These instruments had been used previously by Mason and Dixon to establish the boundary between Pennsylvania and Maryland and left in the care of Joseph Shippen when they returned to England. Another instrument forming part of the Mason and Dixon equipment which the State House Yard observers utilized was an astronomical transit also made by John Bird. The clock used was not described.

The best equipment was assembled at Norriton. It included an accurate astronomical regulator, an astronomical transit or meridian instrument, a transit and equal altitude instrument, and a

9. *Early Proceedings*, 37–8; *Gazette*, 8 June 1769, p. 3, col. 2.

refracting telescope, all made by Rittenhouse. He assembled another refractor with the lenses intended for Harvard. Also on hand
was the astronomical quadrant loaned by William Alexander, and
Rittenhouse also was given the use of the new reflecting telescope
which Thomas Penn had purchased for the observations and which
was later given to the College of Philadelphia. Biddle owned a clock
and equal altitude instrument of his own which he planned to use
and he also had a loan of the Library Company's telescope.[10]

The two major instruments to be used were described by Joseph
Shippen as "Two very large fine Telescopes have arrived here
within this Month from London for viewing the Transit." He elaborated that "one of them was sent for by the Assembly for the use of
the Philosophical Society, and the other was sent by Mr. Thomas
Penn as a present to the College here. This last cost about 70
Guineas, and is 3½ feet long the other must have cost much more
as it is near a foot longer. . . ."[11]

The fourteen-man committee met frequently during the week
preceding the transit to eliminate any local obstructions and to
adjust their instruments. Repeated observations were made each
day to adjust the clocks and on the day of the transit the group
assembled at the observatory in the morning for the same purpose,
to "discover the error of the clock & the time of apparent noon."[12]
After anxious waiting during several days of cloudy weather, 3
June proved to be clear and bright, a condition which favored the
observers at all three stations. The transit began at two o'clock in
the afternoon, and observations were made successfully at the
three sites.

10. *Early Proceedings*, 38–39; William Smith, "An Account of the Transit of
Venus. . . ," *Transactions*, 1st ser., 1 (1771): 9–12.
11. Edward Burd to Edward Shippen, 4 June 1769, L. B. Walker, The Burd
Papers, 3, Independence National Historical Park, History Division (hereafter
INHP); Joseph Shippen to Edward Shippen, 29 May 1769, Papers of Edward
Shippen, 1707–1783, The Library of Congress.
12. *Early Proceedings*, 40.

The State House was established to be in the north latitude 39° 56′ 53″, and its longitude from the Royal Observatory at Greenwich was computed west 75° 8′ 45″, or five hours and 35 seconds in time. Williamson further noted that Mason and Dixon had established the latitude of Philadelphia to be 39° 56′ 29.4″ and its southernmost point to be 39° 56′ 29.2″. Because the observatory was found to be 26.2 seconds north of that point, its position was established to be 39° 56′ 55.4″.

Numerous observations were made at Norriton, Philadelphia, and Lewes and eventually the results were calculated and rendered into reports. It required more than a year to compile the results and arrange for their publication. No consistent plan for publishing them existed, because an unfortunate controversy arose among the Society's members. Smith forwarded his report to Penn who delivered it to Maskelyne, and it was published in the Royal Society's *Philosophical Transactions*. Biddle's report found its way to Franklin in England who arranged for its publication in the same volume. Ewing withheld his report until later, then forwarded it to Franklin, but it arrived too late for inclusion with the others. Some of the reports were included in a ninety-six-page section of the first volume of the Society's *Transactions* which appeared in 1771. Smith later produced a calculation of the parallax based on observations made at Norriton with comparisons with other observations reported which appeared in the appendix to the *Transactions*.[13]

The response to this first scientific enterprise encouraged the Society to continue astronomical activities and to prepare for the 9 November transit of Mercury. In September Smith and Ewing were delegated "to agree with Mr. Duffield to make a timepiece for

13. John Ewing, "An Account of the Observations of the Transit of Venus...," *Transactions*, 1st ser., 1 (1771): 42–89; *Early Proceedings*, 43; *The Pennsylvania Chronicle and Universal Advertiser, ... 1769*, 3: 162, 171; *The New-York Gazette and Weekly Mercury*, 19 June 1769, p. 3, col. 1: Address of the Society to the Representatives, 22 September 1770, *Pennsylvania Archives*, 8th ser., 7: 6537–38; Hindle, *Pursuit*, 151–52.

the use of the observatory in a plain and cheap manner." Duffield completed the clock on schedule for £15.17.6 and Ewing and Williamson observed the transit from the State House Yard while Rittenhouse, Smith, Lukens, and Biddle were stationed at Norriton.[14]

The Mercury Committee reported on 17 November "that they had opportunity of making accurate observations, the Result of which should be laid before the society at next meeting." They were reported by Ewing and published in the *Transactions*. "Having still the same instruments in our observatory, which we used on the former occasion," he wrote, "together with a new time-piece made by Mr. Duffield of this city, with an ingenious contrivance of his, in the construction of the pendulum to remedy the irregularities arising from heat and cold; we paid the utmost attention to the going of the clock both before and after the transit." They discovered that the clock was too slow for mean time and compensated for it in their calculations. Ewing ended his written report with the statement that if his account contributed "any thing to the advancement of astronomical knowledge, it must reflect an honor on our new observatory, and give pleasure to all lovers of science."[15]

The Society had little time for astronomical activity at Philadelphia during the next two years, and little if any use was made of the observatory. It may have suffered from neglect, because on 17 February 1771, Smith and Ewing were directed to attend to repairs of the observatory, the nature of which is not known.[16]

The success of their first astronomical endeavors led the Society's members to wish for continuation and expansion of these activities on a permanent basis. The existing facility in the State House Yard was obviously inadequate to enable them to compete with their

14. *Early Proceedings*, 43–45, 50.
15. *Gazette*, 16 November 1769, p. 2, col. 1; William Smith, "Account of the Transit of Mercury. . . ," *Transactions*, 1st ser., 1 (1771): 158–62.
16. *Early Proceedings*, 62.

English and European counterparts. Rittenhouse's observatory at Norriton had been abandoned in 1770 with his move to Philadelphia. The costs of constructing, furnishing, and maintaining a properly designed observatory building were beyond the Society's capabilities. The solution was a government supported installation, and accordingly, the Society presented a proposal to the Assembly for the erection of a "Public Observatory." Nothing happened; hence in May 1775 the Society again petitioned the Assembly "to establish a Colonial observatory and appoint Rittenhouse the Director, as he could direct the observations and make the necessary instruments." Political events and other problems facing the province reduced the proposal to a low priority, however, and no action was taken. Meanwhile, the observatory in the State House Yard remained the Society's responsibility and astronomical instruments remained there in storage.[17]

The State House Yard was a popular center for community activities and public gatherings though in undeveloped and unattractive condition. Despite long-standing plans to landscape the large enclosure within the confines of the shingle-sheathed seven foot brick walls, it remained barren with neither plantings nor a planned system of walks for another decade. With Pennsylvania's increasing military activities and the presence of the State House as the capitol, the yard became a convenient storage facility for over fifty cannon and other military equipment. The observatory provided a stage from which public announcements often were read, and it was used with greater frequency as the clouds of war gathered and thickened. People had assembled in the yard to protest the Stamp Act, and protest meetings held in the State House repeatedly overflowed outside. Crowds gathered there in 1773 to register disapproval of the presence of the tea ship *Polly* in the Delaware River, in April 1775 on the day after the encounter on the

17. *Early Proceedings*, 96–97.

Figure 4. The Reading of the Declaration of Independence from the Observatory, 8 July 1776. Magazine illustration, artist not known. From the Picture Post Library (London). Courtesy of Independence National Historical Park.

bridge at Concord, and most recently in May to consider the resolution of Congress recommending that the colonies adopt such form of government as would best suit the people.[18]

On 4 July, the delegates to the second Continental Congress meeting in the State House approved a Declaration of Independence drafted by Thomas Jefferson and revised by the Committee of Five. It was announced that the document would be publicly proclaimed at midday on the following Monday from the observatory platform.

Monday, 8 July, was a warm sunny day and as noon approached, residents from all parts of the city began to collect in the State House Yard. They were joined by others who had traveled into the city from the surrounding countryside. Troops were drawn up in formation for the occasion, and the Committee of Inspection joined with the Committee of Safety made their way as a body to the yard. Forty-nine members of the Congress emerged from the back door of the State House, and Mayor Samuel Powel and other city officials assembled just below the platform.

The sizable crowd waited patiently, and the first signs of restlessness were just becoming evident when shortly after twelve o'clock Philadelphia's sheriff William Dewees arrived and quickly climbed the observatory stairs followed by Colonel John Nixon, who was acting as his deputy. Dewees approached the railing and as he began to speak, a silence fell over the gathering. "Under the authority of the Continental Congress and by order of the Committee of Safety," he began, and went on, "I proclaim a declaration of independence!" Colonel Nixon then stepped forward and read the document. Everyone listened attentively, and when he finished, the troops saluted and the people demonstrated approbation with

---

18. Edward M. Riley, "The Independence Hall Group," *Historic Philadelphia From the Founding Until the Early Nineteenth Century* (Philadelphia: The American Philosophical Society, 1953), 8.

Figure 5. "The Reading of the Declaration of Independence, July 8, 1776." Painting believed to be by Peter Frederick Rothermel, 1861. Present location unknown. From the Stauffer Collection, vol. 7, fol. 532. Courtesy of The Historical Society of Pennsylvania.

Figure 6. "John Nixon Reading the Declaration of Independence, July 8, 1776."
Engraving by Edwin Austin Abbey, late nineteenth or early twentieth century.
Courtesy of Independence National Historical Park.

three great huzzahs.[19] Deborah [Norris] Logan who lived in the
Norris mansion across Fifth Street ventured that the audience "was
neither very numerous or composed of the *most respectable* class of
citizens."[20] Charles Biddle, who was also present, agreed with the
comment "there were very few respectable people present,"
presumably referring to the noticeable absence of individuals of
wealth, family, or position. On the other hand, Christopher Marshall who was also on the scene, described it as "a great concourse
of people."[21]

With the conclusion of the ceremony, the crowd dispersed and
bells began tolling throughout the city. For most the declaration
was not new; it had been published in the Philadelphia newspapers
two days earlier and again that same morning. The delegates to the
Congress filed back into the State House to resume their work.
Some of the crowd followed the speakers to the Court House, where
the declaration was again read, and then observed the King's arms
being removed first from the Court House and then from the State
House. Others made their way to Armitage's Tavern to while away
the afternoon hours.

Not until that evening did the city properly celebrate the event.

19. Silvio A. Bedini, *Declaration of Independence Desk: Relic of Revolution* (Washington, D.C.: Smithsonian Institution Press, 1981), 15–19; Benson J. Lossing, *Pictorial Field Book of the Revolution* (New York: Harper Brothers, 1850–52), 2: 79;
Charles Warren, "Fourth of July Myths," *The William and Mary Quarterly*, 3rd ser.,
2 (July 1945): 237–272; Edmund C. Burnett, *Letters of Members of the Continental
Congress* (Washington, D.C.: Carnegie Institution, 1921–36), 2: 7; "The Truth
About the Signing of the Declaration of Independence," *The Sunday Star* (Washington, D.C.), 4 July 1915, pt. 4, p. 2.

20. Edward Armstrong, ed., *Memoirs* (Philadelphia: McCarty & Davis, 1826),
"Logan and Penn Correspondence," pp. xlv–xlvi HSP; Joseph Jackson, "Opposite Independence Square," pt. 3, *Public Ledger*, 26 October 1913, p. 3.

21. *Autobiography of Charles Biddle, Vice-president of the Executive Council of Pennsylvania, 1745–1821* (Philadelphia: E. Claxton and Company, 1883), 86; *Extracts From
the Diary of Christopher Marshall, Kept in Philadelphia and Lancaster, During the American
Revolution, 1774–1781*, William Duane, ed. (Albany, N.Y.: Joel Munsell, 1877), 83;
R. W. Davis, "Glimpse of Philadelphia in July 1776," *Lippincott's Magazine*, (July
1876): 36.

It was a pleasant night, the sky filled with bright stars, and great bonfires burned brightly all over the city. Many assembled in the Commons where the arms of King George III were brought out, placed on casks and set afire as the watchers cheered. All through the night church bells tolled, applauding the momentous decision made that day.[22]

One of the most vivid accounts of the event appears in a letter writen the following day by John Adams, who had been present as a member of the Continental Congress, "The Declaration was yesterday published and proclaimed from that awfull Stage, in the State house Yard, by whom do you think? By the Committee of Safety,! the Committee of Inspection, and a great Crowd of People. Three cheers rended the Welkin. The Battalions paraded on the common, and gave us the Feu de Joy, notwithstanding the scarcity of Powder. The Bells rung all Day, and almost all night. Even the Chimers [of Christ Church], chimed away. . . ."[23]

With the advent of war the Society ceased its scientific activities, and because the observatory structure was not being used, it was converted into a military guard-house. On 24 July, a little more than two weeks after the Declaration was promulgated, the Council of Safety resolved that the recently elected secretary of the Board of War, Captain Richard Peters, "be authorized to have the Stage in the State House yard fitted up for the accomodation of the Guard, And that he provide a Sufficient number of Camp Kettles for their use."[24] Early in August the State Treasury paid Jonathan

---

22. Duane, *Marshall Diary*, 82–83.

23. *Papers of John Adams*, Robert J. Taylor et al., eds., (6 vols. to date (Cambridge, Mass./London, England: The Belknap Press, 1977–), 4: 372–73.

24. Minutes of the Council of Safety, Council of Safety Chamber, 24 July 1776, Pennsylvania State Archives, Harrisburg; Timothy Pickering's report of a meeting of the Board of War 28 August 1778, and "A Detail of the Guards mounted in or near the City of Philadelphia 28 August 1778," Papers of the Continental Congress, Microscopy No. 247, Roll 157, Item No. 147, Reports of the Board of War and Ordnance 1776–1781, 2 (1777–1779): 221–24, National Archives and

Figure 7. "John Nixon Reading the Declaration of Independence, July 8, 1776." Illustration by Clyde O. DeLand, 1917. Courtesy of Independence National Historical Park.

Britton for "Boards to build Guard House . . . [£.] 164," and
rendered payment to house carpenter William Roberts for "Build.ᵍ
Guard Houses . . . [£.] 182." On 7 August Britton received a second
payment, "for 50,000 feet of 3 Inch W.O. plank . . . [£.] 662 . . 10 . .
0." In addition to remodelling the observatory structure, small
sentry posts were built at the corners outside the yard wall. The
observatory served a military function for the next several years,
used first by the city militia and then by the State militia; in the
summer of 1778, thirty-six men were assigned to guard the State
House and yard. When the British invaded Philadelphia in 1778,
the observatory was utilized by the occupation forces for the same
purpose.²⁵

After the British evacuation the observatory was returned to the
Society for its own use. The modifications made for military purposes and the presence of two armies rendered the structure un-
usable and in a state of considerable disrepair. Early in May 1779
Lukens, Rittenhouse, and Biddle were appointed a committee "to
have the *observatory* repaired and the *instruments* lodged in the
same." They were also instructed to meet with the Speaker of the
House of Representatives in compliance with his request and ad-
vise him that the Society did not favor his proposal to lend the
telescope, which had been purchased for the Society by the Assembly. "It is not proper to lend it to Individuals, as it would be
injurious to the Instrument, and interfere with those views of the

Records Service; "Peters, Richard," *Dictionary of American Biography*, Allen Johnson and Dumas Malone, eds., 20 vols. (New York: Charles Scribner's Sons,
1928–1936), 7: 509–10.

25. Division of Public Records, Cash Book Dec. 1775 – Oct. 1779, entry for
August 1776, pp. 68–69, Records of the State Treasurer, Record Group 28,
Pennsylvania State Archives. No listing can be found for Jonathan Britton but in
1785 a Thomas Britton owned a lumber yard at 501 Riverside Street and in 1796
was listed as a shipbuilder at North Front Street. William Roberts is identified as a
carpenter on Chestnut Street between Sixth and Seventh Streets in 1785. [Alfred
Coxe Prime], *Directory of Craftsmen Compiled by Phoebe Phillips Coxe from Philadelphia
Newspapers 1785–1800*, unpaginated, American Philosophical Society.

Society for which they obtained the use of it from a former Assembly." It is uncertain whether repairs were made, but clearly the observatory was no longer suitable for scientific purposes, and in the following year it was proposed to sell it, and build a new observatory. The Assembly approved the Society's recommendation and a year later, on 7 April, voted £250 "toward expense of erecting an observatory agreeable to the Resolution of the Assembly." Rittenhouse erected an octagonal brick observatory building on his property at Seventh and Arch Streets for his astronomical activities and those of the Society as well.[26]

The State House Yard Observatory was abandoned and when on 18 April 1783 the committee charged with selling it reported an £18 offer, its sale was ordered. On 5 December 1783 the Society requested that Rittenhouse remove all the instruments in the observatory to his own premises. Three months later the account books of the Society treasurer, Francis Hopkinson, noted that the Society had received "By Cash of George Bryan Esq.[r] on William Redigers Order in my Favour being for the Society's Observatory in the State House Square sold to the said Rediger . . . [£] 18." Rediger was a merchant selling wood and candles and during the period in question was employed as the Society's janitor.[27]

Rittenhouse's death in June 1796 raised the question of the disposition of the observatory on Seventh and Arch Streets. His widow expressed "a desire to invest in the Society the Observatory, with a certain quantity of land leading thereto," and deeded the installation to the Society. Upon her death three years later she bequeathed

26. *Early Proceedings*, 102–3; Edward Ford, *David Rittenhouse, Astronomer-Patriot 1732–1796* (Philadelphia: University of Pennsylvania Press, 1946), 106; Hindle, *Rittenhouse*, 226–27.

27. *Early Proceedings*, 102, 108, 118; American Philosophical Society, *Minutes*, 18 April 1783; APS *Treasurer's Book of Accounts, 1782–1791*, Account of Francis Hopkinson Treasurer, entry for 2 March 1784; Comptroller-General's Accounts, Journal AA, p. 746, State Records Office; Ford, *David Rittenhouse*, 116, 118–19; Hindle, *Pursuit*, 168–69; Hindle, *Rittenhouse*, 226, 243.28; APS *Minutes*, 18 April, 26 September 1783; *Early Proceedings*, 118; APS *Treasurer's Book of Accounts*.

$200 to the Society "for the purpose of raising an annual fund to be applied for keeping the observatory in order." In time the development of the area made the observatory increasingly difficult to use and in 1810 the property was returned to Rittenhouse's daughters.[28]

Not until 1816 did the Society's minutes note further consideration of an observatory in Philadelphia. A committee investigated the potential conversion of the pumping station on Centre Square into an astronomical observatory. The building had been abandoned with the construction of the new waterworks in Fairmont Park. The centre house's position was "extremely well calculated for the desired purpose, and moreover the only one in the city that was so; and that all the [required] alterations . . . might be made at a moderate expense." The City Commissioners passed an ordinance that let the building to the Society "for and during the period of seven years from the execution of the said lease, for the yearly sum of one dollar." The Society promptly requested assistance from the City Council "in the erection of an *Astronomical Observatory* under the charge of this Society," and in May $2,500 was appropriated towards the purchase of instruments for the observatory, provided that the Society contributed an equal sum. Years passed and the Society lacked funds and the project was abandoned. A decade later the city considered erecting an observatory in Rittenhouse Square in cooperation with the Society but failed to act.[29] By this time others in Philadelphia were making observations with private or academic facilities, as well as an astronomical observatory of sorts maintained on Front Street by the maker of navigational instruments William H. C. Riggs (later Riggs & Bro.) for rating chronometers of ships in the harbor.[30]

28. *Early Proceedings*, 241–42, 287, 304, 421–22.
29. *Early Proceedings*, 464–65, 476, 669, 672, 688, 699. For a full account of the Centre Square proposal see Whitfield J. Bell, Jr., "Astronomical Observatories of the American Philosophical Society," *Proceedings of the American Philosophical Society*, 108 (February 1964): 9–10.
30. Bell, "Astronomical Observatories," 10–11; *Association of Centenary Firms and Corporations of the United States* (Philadelphia, n.p., 1924), 3rd issue, 181–82.

Meanwhile the observatory in the State House Yard remained forgotten for almost a century and no trace of it survived. It came to public attention once more in the summer of 1875 during preparations for the centennial celebration of the signing of the Declaration of Independence. In anticipation of the event, a committee was appointed by the mayor to oversee the restoration of Independence Hall and its premises. During excavations to improve the sewage system of the former yard now known as Independence Square, a foundation was discovered by Frank M. Etting, a member of the committee. "It appears to have been of circular shape," he wrote, "and was erected about forty feet due west from the rear door of the present Philosophical Hall and about same distance south from the wall of the present (eastern) wing. It would form an eminently appropriate site for a monument for the signers of the Declaration of Independence, so long in contemplation." Etting recorded no measurements, and whether the foundation was of sufficient size to have accommodated the entire observatory structure was not indicated, but was unlikely.[31]

Most likely the foundation was a base for a post erected at the center of the structure upon which the astronomical clock was placed to avoid vibration. The observatory's timepiece, which played an important role in making the observations, was described by Ewing as being "fixed to a large post sunk into the ground four or five feet, secured from shaking by a brick wall at the bottom, and in no ways communicating with the sides of the building."[32]

Lacking contemporary drawings or documented descriptions, all that is known of the Observatory's appearance is derived from several nineteenth-century accounts, only one based with certainty

31. Frank M. Etting, *An Historical Account of the Old State House of Pennsylvania Now Known as the Hall of Independence* (Philadelphia: Porter and Coates, 1891), 64–65.

32. John Ewing, "An Account of the Observations. . . ," *Transactions*, 1st ser., I (1771): 43.

on recollections of a survivor who saw it. The Philadelphia historian John F. Watson indicated that Rittenhouse presumably directed or superintended its construction and quoted Etting's description and location of the structure. Elsewhere in his *Annals*, Watson described the observatory as "a rough frame scaffolding or stage standing midway on the line of the eastern walk, between Fifth and Sixth Sts."[33] In his notes to the Penn and Logan correspondence Edward Armstrong reported "the late Mr. Thomas Pratt, who died in 1869 at the great age of 95, perfectly remembered this structure; and in a conversation with the editor, a few weeks before his death, described it to be as a rough, wooden stage, which stood on the line of the eastern walk, about midway between Fifth and Sixth Streets."[34] Thomas Scharf and Thompson Westcott described the observatory as "about twenty feet high, and twelve to fifteen feet square, at fifty to sixty feet south of the house, and fifteen to twenty feet west of the main walk. It seems to have been used occasionally as a stand for public addresses, it being referred to as such by Stansberry [*sic*], in his militia poem." The reference was to a satiric song, "An historical ballad of the proceedings of a town meeting at Philadelphia, May 24, 25, 1776," by the young Loyalist merchant Joseph Stansbury who maintained a china store opposite Christ Church. Although he sympathized with the colonists, he opposed independence in songs about the kinship and glory of race. Imprisoned for his leanings, released, and again arrested in 1780, Stansbury was permitted to retire behind the British lines. Later he carried Benedict Arnold's first proposals to British headquarters, and continued as Arnold's liaison. In the ballad he mentioned the observatory briefly in the lines

33. John F. Watson, *Annals of Philadelphia and Pennsylvania, In the Olden Time;* . . . . (Philadelphia: Privately printed, 1850), 1: 321; 3 (1909): 225.

34. Armstrong, "Penn and Logan Correspondence," *HSP Memoirs*, 9: xlv–xlvi; Thompson Westcott, *History of Philadelphia* (Philadelphia: Sunday Dispatch, 1867–78, 1886), 2 (1754–1781): Chapter 235, "Events in Philadelphia in 1776 after July 4," unpaginated.

And now the State-house yard was full,
And orators, so grave and dull,
Appeared upon the stage:
But all was riot, noise, disgrace,
And Freedom's sons, o'er all the place,
In bloody frays engage.

Charles Keyser described the structure as "a platform built from an old observatory toward the east side of the yard; it was surrounded by a railing and reached by a stairway from the ground; all around it was open to sunlight."[35]

Westcott's *Historic Mansions of Philadelphia*, cited also by Faris, indicated that "the platform is supposed to have been west of the Biddle Walk and on a line with the present Sansom Street."[36] A work published in London in 1873, referring to "the old State Hall" listed some of its historic artifacts of the Revolutionary period including "also part of the stone step on which John Nixon stood when the Declaration of Independence was read to the assembled crowd outside."[37]

The popularity enjoyed by Independence Hall and Independence Square during the Philadelphia Centennial continued and increased in the following years, and directed the attention of the city fathers to their condition. In 1895 a city ordinance was passed to restore the Hall "to its appearance during the Revolution," and in the following year a resolution of the Councils granted permis-

35. Thomas Scharf and Thompson Westcott, *History of Philadelphia 1609–1884* (Philadelphia: L. H. Everts & Co., 1884), 1: 402; Watson, *Annals* 2 (1857): 303–5; *Dictionary of American Biography* 17: 516–17; Charles Shearer Keyser, *The Liberty Bell* (Philadelphia: Dunlap Printing Co., 1901), 26; John H. Hazelton, *The Declaration of Independence.* . . . (New York: Da Capo Press, 1906), 556, n. 20.

36. Thompson Westcott, *Historic Mansions of Philadelphia* (Philadelphia: Porter & Coates, 1877), 120; John T. Faris, *Old Gardens In and About Philadelphia and Those Who Made Them* (Indianapolis: The Bobbs-Merrill Company, 1932), 29.

37. Julius George Medley, *An Autumn Tour in the United States and Canada* (London: H. S. King & Co., 1873), 109.

sion to the local chapter of the Daughters of the American Revolution to undertake restoration of the council chamber on the second floor of the Hall. Public interest focused on that group of buildings during the next several years and Etting's discovery of the presumed observatory foundation was noted in an article in the *Public Ledger* on 3 August 1897, but there was no further concern for or mention of the old observatory for another decade.[38]

In 1900 the Philadelphia chapter of the American Institute of Architects undertook a detailed study of Congress Hall from documentary sources leading to its restoration. Funds were appropriated in 1910 and two years later work began. Again the Observatory came to notice and became the subject of an exchange of letters between the architect Horace Wells Sellers and Judge Norris S. Barratt of the Court of Common Pleas in which Sellers summarized all information known to that date.[39] Wilfred Jordan, curator of Independence Hall, became interested in the search for the observatory site at this time, and communicated with Henry Darrach on the subject.[40]

No further activity relating to the observatory site is recorded until 1914, when it became a timely issue in relation to the forthcoming Fourth of July celebration. Jordan suggested to the Chief of the Bureau of City Property, William H. Ball, the desirability of a visible marker for the observatory site, and gave the press a story as a means of generating public interest. He requested that Ball encourage the Forestry Division to work with the staff of Independence Hall to

38. Riley, "Independence Group," 39; *Public Ledger* (Philadelphia), 3 August 1897.

39. Riley, "Independence Group," 40; "Extracts from letters to Hon. Norris S. Barrett [sic] – 6 Dec. 1911," Horace Wells Sellers Collection, American Philosophical Society; *Public Ledger*, 21 September 1913; *The Philadelphia Record*, 26 October 1913.

40. Wilfred Jordan to Henry Darrach, 9 December 1911, Sellers Family Papers, INHP.

make the excavation in Independence Square at a point that we know this observatory stood in order to determine the exact location of the foundations of the structure, which, I believe, are still there. After determining their location, I would then propose with your approval of temporarily marking the site for over July 4th, that is, get Mr. Bowen's division to erect a light wooden upright on which we can place a temporary sign neatly lettered, stating briefly the facts in the case, making it large enough however, to be seen by the audience taking part in the 4th of July celebration.

Involving the Bureau of City Property might encourage some society permanently to mark the spot. A week later a *Los Angeles Daily Times* feature based on Jordan's release reported that excavations were begun in the State House yard on 24 June fifty-five feet south of the east wing of Independence Hall in an effort to locate the site of the observatory.[41]

Two days later Ball wrote to John H. Baizley, chairman of the Fourth of July Celebration Committee, stating

Some days ago we made an excavation in Independence Square and found some of the foundation stones of the small observatory which stood there in 1776 from which the Declaration of Independence was first proclaimed. The Bureau of City Property will place over the site of this building in the course of a day or so a temporary wooden marker setting forth the following facts:

On this site stood the observatory erected by the American Philosophical Society to observe the transit of Venus in 1769. From its balcony at noon, on July 8, 1776, under the authority of the Continental Congress and by order of the Committee of Safety of Pennsylvania, the Declaration of Independence was

41. Wilfred Jordan to Chief, Bureau of City Property, 19 June 1914, INHP; "Iconoclast Would Give a New Date to Nation's Birthday," *Los Angeles Daily Times*, 25 June 1914; Benjamin M. Nead, "New Found Tower of State House Preserves History," *Harrisburg (Pa.) Patriot* (?), c. 1914.

read and proclaimed by William Dewees, Sheriff of Philadelphia, Colonel John Nixon acting as his deputy, in the presence of a number of members of the Continental Congress, Samuel Powel, Mayor, and other officials of the City of Philadelphia, and a large concourse of citizens.[42]

Probably in response to Jordan's efforts to popularize the subject, the *Public Ledger* published an article about the old observatory and its historic importance. It went on to mention that no accurate description of the structure existed and from the fact that a half century earlier a stone foundation had been found "in the shape of a circle, the conclusion was reached erroneously that the building was cylindrical, as are some modern astronomical observatories. The foundatio[n], however, was merely for the proper support of the instruments, and the observatory proper was probably frame and rested upon foundations of its own. There is reason to believe, however, that the transit room was somewhat elevated from the yard level, and that it was reached by an exterior stairway, rising to a balcony, or stage."[43]

The Bureau of Public Property, meanwhile, promptly provided a temporary marker for the site in time for the forthcoming celebration, which would be attended by the President of the United States. The day came and Philadelphia's "Greater Fourth of July Celebration" of 1914 proved a notable event, highlighted by the unveiling of the wooden plaque by President Woodrow Wilson.

Jordan took advantage of the interest generated by the presidential visit and suggested to Richard H. Cadwallader, president of the Pennsylvania chapter of the Sons of the American Revolution, that his organization might petition the City Councils for the honor of providing a permanent marker for the observatory site. If

42. William H. Ball to John H. Baizley, 27 June 1914, INHP.
43. *Public Ledger*, 28 June 1914, Jane Campbell Collection, Scrapbooks, vol. 64, INHP.

so, he advised that steps be taken at once because "two other societies are in the field to capture the site."[44]

For an appropriate text for a permanent tablet, Jordan consulted the Librarian of the American Philosophical Society, Isaac Minis Hays. The latter suggested an inscription virtually identical to that which had been proposed by Ball to Baizley for the temporary marker, suggesting that Hays may have written the other as well.[45]

Excavations in search of the site continued and in 1915 W. L. McGee urged that Ball visit them as soon as possible. "As the work has gone so far and as there seems to be a question as to their going further," he advised, "I think you should be satisfied before they stop. I don't think they have gone far enough."[46]

Jordan persisted in his efforts to memorialize the site; time was essential for City Councils processed petitions on a first come first served basis. He recommended that Ball submit a preliminary drawing or sketch of the proposed marker to the Art Jury for approval. He went on

> For the marker itself, we would suggest an old Sun Dial mounted on a circular brick base made up of two elevations in the form of steps, the pedestal from the dial itself to be an octagon. On one of the faces of which, a bronze tablet to be affixed, bearing a description of the spot and the most appropriate form of marker that could be adopted. First for the reason that the building which originally stood on the spot was erected to determine an important astronomical phenomenon, and second a fac-simile Colonial Sun Dial from an artistic point of view would look better in the square at this particular point, than anything else that I can think of.

44. *Report of the Bureau of City Property for 1914*, 15, City of Philadelphia, Department of Records, INHP. Wilfred Jordan to Richard H. Cadwallader, 13 July 1914.
45. I. Minnis Hays to Wilfred Jordan, 26 July 1914, INHP.
46. W. L. McGee to William H. Ball, 14 May 1915, INHP.

Jordan noted that he had provided Judge Barratt with a pre-liminary drawing and copy of the historical facts concerning the site more than a year previously. Jordan believed that the Sons of the American Revolution was the most appropriate organization to undertake that project and estimated that the entire work could be ready for unveiling for less than $2,000.[47]

In due course a design and text for the proposed marker were submitted to the Art Jury which drew a request for additional information to the A.I.A.'s Standing Committee on the Preservation of Historic Sites from the jury's secretary. Sellers responded on behalf of the A.I.A. on 2 July 1915. He summarized the information already known about the old observatory, noting that no records relating to the exact location existed in the files of the American Philosophical Society. However, from the reports made of the transit of Venus Sellers noted

> these observations establish the latitude of the "city observatory" from the surveys of Messrs. Mason and Dixon as 39° 56' 54" north, or 23 seconds north of the southern point of the city which they found to be 39° 56' 29.2". (Upon reference to the original manuscript records of the Mason & Dixon survey the latitude of the south line of the City (South Street?) is stated to be north 39° 56' 29.1" differing only in the decimal of the seconds from the Society's report).

Sellers also noted that in response to an inquiry Professor Doolittle of the Department of Astronomy of the University of Pennsylvania stated "the nature of the observations for which the structure was erected in 1769 did not require them to be of a permanent character for the astronomical instruments employed, the chief consideration being the accurate determination of the interval of time occupied in the transit of Venus across the sun."

Confirmation of this statement appears in the 1771 *Transactions,*

47. Wilfred Jordan to William H. Ball, 14 May 1915 (date not clear), INHP.

in which the only reference to the construction of the observatory related to installation of a post as a support for the clock. Sellers indicated that this precaution "seemed to imply that the building itself was not of sufficient stability to permit placing the time piece on it or its walls or in contact with its floor." He summarized the various descriptions provided in the writings of nineteenth-century Philadelphia historians and in paintings and engravings of the same period. He made the point that although in one it was described as "a commodious building," its cost and the short time required for its construction suggests that it was no more than "a substantial platform designed to accommodate the committee of five and two assistants together with the four telescopes and other instruments as mentioned in the report of the observations." He stated that in opposition to the foregoing was Etting's discovery of foundations of a circular nature and provided measurements of its distance from the State House and other buildings.

It was at this point that the marker had been placed and in view of the assertion that the foundations existed as described,

> . . . the Committee determined to begin its investigations at this site and called upon the Bureau of City Property to make the necessary excavations under its direction. When this labor was available trenches were dug across the area described by Etting, and at a depth of $3^{1}/_{2}$ to 4 feet a line of top soil was disclosed clearly indicating a general surface condition upon which earth had been subsequently superimposed. The excavations when extended to a greater depth below this old or original level showed no indication of artificial filling and disclosed no foundations or materials for same, nor were any such materials revealed by probing with a crow bar below the bottom of the trenches.
>
> At the apparent surface of the ground level, there were indications of a mortar bed or mortar refuse which may have been left from some building operation when the ground was at this lower grade, but as stated, no masonry or other founda-

tion as described by Etting were visible nor was there anything in the appearance of the soil to indicate that such foundations had existed and had been removed at the time the excavations were made in 1876. The trenches where extended outside of this area and to a similar depth, disclosed only the continuance of the same old line of top soil, indicating the former surface of the ground at a depth of 3½ to 4 feet below the present grade, and below this level the trenching and the probing as elsewhere gave no evidence of foundation materials at a lower depth.

After leaving the researches at this site near Philosophical Hall, trenches were excavated on the *east* side of the main path leading to the State House and at the distance from the house described by the authorities noted. Here again the presence of top soil was apparent at about the level already stated but without the inclusion of foundation work or any appearance of building materials except some stone spawls and fragments of brick mixed with the superimposed earth which were evidently deposited when the level of the yard was raised to its present surface, and not suggestive of foundation construction.

Although the committee had intended to extend the investigation to the west side of the main walk, it was prevented from doing so because at this time the jurisdiction of the square passed to the Park Commission, and the investigation ended. The committee concluded that no tangible evidence warranted locating the marker adjacent to Philosophical Hall nor at the point east of the main walk where excavations were also made. Possibly it was situated on the western side of the path near the State House. All evidence, however, suggested that the original foundations were not of a character that would have survived the various changes in the surface of the yard, particularly after the extensive landscaping undertaken after 1784 and later improvements. The committee concluded therefore that the only accurate record of the observatory's actual location was its latitude as established in 1769 and reported in the *Transactions*, but this information was insufficient to identify the precise location within the original area of the yard. The Park Commission apparently lacked the Committee's willingness to continue jointly the investigation.[48]

Sellers decided to inform Ball of the Art Jury's request for a report from the A.I.A. of "the precise location of the Observatory referred to in the inscription of the proposed tablet." He advised Ball that he had complied with the request, and enclosed a copy of his report, noting that a survey and a drawing of the site had been prepared for the committee and placed on file. Inasmuch as no further reference to the project can be found in the files of the city offices, presumably it was permanently abandoned at this time.[49]

The lack of contemporary documentation of the old observatory did not prevent several nineteenth-century artists and illustrators from recreating the scene of the promulgation of the Declaration of Independence with considerable exercise of imagination and varying degrees of accuracy. A magazine illustration, believed to be from the English *Picture Post*, erroneously represented the observatory as a round building made of masonry with a domed roof and a balcony with elaborate balusters approximately halfway up the structure, a conception which in no way coincides with the known

48. Report made by Horace Wells Sellers on behalf of the Philadelphia chapter of the A.I.A. and submitted to Andrew Wright Crawford, secretary of the Art Jury on 2 July 1915, INHP. Copy forwarded to William H. Ball on 21 July 1915. No other copies can be found in the files of the Art Jury, of the City of Philadelphia's Hall of Records, or the Philadelphia or New York Chapters of the American Institute of Architects. Correspondence with Ward J. Childs, City of Philadelphia Department of Records, 11 December 1980, Bruce Laverty, Historical Society of Pennsylvania, 11 December 1980, Martin Yoelson, Independence National Historical Park, 18 December 1980 and 23 January 1981, and Mary Beth Sherrier, Philadelphia Chapter of the American Institute of Architects, 27 April 1981. The latitude and longitude established for the State House Observatory of 39° 56′54″ [sic] and 75° 8′45″ were reported by the Governor of Pennsylvania to the Heads of Inquiry in London on 30 January 1775; see "State of the British Islands in the West Indies," Strachey Papers, 269–270, William Clements Library.

49. Sellers to Ball, 21 July 1915, enclosing copy of report to the Art Jury, INHP.

information. An engraving by Abbey of a painting by an unidentified artist also depicted a building of stone or masonry, although angular with a large structure continuing to some height above the balcony, a structure far too ambitious to relate to reality. A painting, a photograph of which is filed in the Stauffer Collection, depicted a structure which bears little resemblance to what is known of the structure, having an elaborate balustered balcony obviously out of character with the crudeness of the original. The artist was probably Peter Frederick Rothermel, who in 1861 exhibited a painting in Philadelphia of the reading of the Declaration. The most accurate presentation, which nonetheless leaves much to be desired, is a painting by Clyde O. DeLand executed in 1917; the observatory is shown with a much simpler balustrade and a gambrel roof.[50]

The observatory was recreated again, as part of the museum complex erected in about 1929 in Dearborn, Michigan, by Henry Ford. Independence Hall was copied for the museum building at a different scale from the original, and other buildings of the Independence Hall group served as prototypes for other units of the complex. The superintendent of Independence Hall suggested that Ford's architects recreate the observatory as one of the small utility buildings and accordingly they designed and constructed it for Pump House Court No. 4. It is octagonal in shape with a gambrel roof and placed upon a stone base. It was built of wood cut to imitate stone jointing, with a wooden railing at mid-section. It

---

50. From the Picture Post Library, 43–44 Shoe Lane, London, INHP; Stauffer Collection, vol. 7, folio 532 HSP; "Rothermel, Peter Frederick," *Dictionary of American Biography*, 8: 187; MSS., Roberts Collection, Box 720, Miscellaneous Folder, *The Press*, 16 May 1861, Haverford College; "John Nixon Reading the Declaration of Independence, 8 July 1776," illustration by Clyde O. DeLand, 1917, INHP. In 1861 Rothermel exhibited his latest painting "The Reading of the Declaration of Independence July 8, 1776" at Earle's Picture Gallery on Chestnut Street and it was described with great detail in the Philadelphia press.

Figure 8. The recreated Observatory building in Pump House Court No. 4 at the Henry Ford Museum. Courtesy of The Edison Institute Henry Ford Museum and Greenfield Village.

greatly resembles the representation of the observatory in the painting by DeLand.[51]

Despite the survival of only a few fragments of confirmed documentation, it is nonetheless tempting to conjecture the nature and appearance of the State House Yard Observatory. From all accounts, its most dominant feature was a square wooden platform measuring either twelve or fifteen feet in diameter. The observatory was said to rise about twenty feet from the ground, a measurement which probably included the roof of the observing room on the platform. Four braced corner posts with stone or brick footings, possibly with intermediate posts as considered necessary, undoubtedly supported it. Most likely the space underneath the platform was open, except for the substantial post at the center rising through the platform to support the observatory clock. Erected upon the platform must have been a small frame building enclosing the observing room, having a door and several shuttered window openings, and possibly a shuttered opening on a gambrel or hipped roof. It would have been made waterproof inasmuch as the astronomical instruments were stored therein. The platform and observing room were reached by means of open wooden stairs attached to one side of the supporting structure. The limited funds available for materials and labor would have required the simplest possible form.

Britton sold the Committee of Safety 50,000 feet of 3-inch plank for a total of £662.10.0 to be used for "building guard houses." This amount of lumber, which was probably measured in linear and not board feet, was undoubtedly utilized by Roberts to enclose the base of the platform structure on four sides, provide flooring for the base and for the platform, and possibly to construct sentry posts added outside the yard wall. Captain Peters had been ordered to provide a sufficient number of camp kettles, suggesting that the observa-

51. The Edison Institute Henry Ford Museum and Greenfield Village, "Plan of Pump House Court No. 4," Robert O. Derrick, Inc., Architects; 1929 file notes by B. R. Brown, communication from Rosemary Bowditch, 23 August 1984.

tory was to house a number of men. One kettle was assigned for each "mess" or group of six men, so that it is likely that the guard house accommodated several messes, with bunks built one above the other on the walls of the periphery of the interior. The records of the Committee of Safety indicate that in 1778 thirty-six men were assigned duty at the State House.[52]

The exact position of the structure has not been identified with certainty but sufficient evidence suggests that it was situated at a distance of fifty feet from the wall of the east wing of the State House and between forty and fifty feet from the present back door of Philosophical Hall. It was further identified by the latitude established from this point for the city of Philadelphia. The data available should have been adequate to define the site sufficiently for a permanent marker; however, the ground of the yard in the Revolutionary period sloped considerably from the State House towards Walnut Street and a substantial amount of fill was added to level it in subsequent years. The excavations described by Sellers were begun at a depth of three and one-half to four feet and then "extended to a greater depth below this old or original level." Whether the original level was reached remains in question. The post installed to support the observatory clock was sunk five or six feet below the original ground level and placed upon a footing or foundation, presumably the one discovered by Etting. From a review of Sellers's report it seems possible that excavations were not sufficiently deep; hence the foundation may in fact still remain.

Just as little evidence survives of the observatory, few artifacts relating to its use are known. Among the most tangible and documented is the tall case clock in the Society's collections made by Edward Duffield in October 1769 for use in the observatory. It is

---

52. National Archives and Records Service, *Report of the Board of War and Ordnance*, 2: 221–24 (see note 24).

Figure 9. Tall case clock made in 1769 by Edward Duffield for the American Philosophical Society's observations of the transit of Mercury from the State House yard Observatory. Courtesy of the American Philosophical Society.

equipped with a dead-beat escapement and an adjustable but not compensated pendulum. At first it was probably housed in a simple pine case, which was subsequently replaced with the present mahogany case with bell-top hood having three beet-shaped wooden finials.[53]

The historic transit instrument made by the distinguished

53. Robert P. Multhauf, *Catalogue of Instruments and Models in the Possession of the American Philosophical Society.* Memoirs of the American Philosophical Society 53 (Philadelphia: The American Philosophical Society, 1961): 49–50, Figure 20.

Figure 10. Movement of the Duffield clock viewed from the side. Courtesy of the American Philosophical Society.

English instrument maker John Bird in 1763 for the Proprietors of Pennsylvania also survives. It was one of the instruments ordered for the use of Charles Mason and Jeremiah Dixon to establish the boundary between Pennsylvania and Maryland. It measures $33^1/16$ inches in length with a striding level and a small telescope fitted at the center to the rigid axis. Mason and Dixon recorded that they preferred this Bird instrument owned by the Proprietors to the astronomical transit belonging to John Bevis which they had

Figure 11. Astronomical transit made by John Bird in 1763 for the Pennsylvania Proprietors and used by Mason and Dixon to survey the Pennsylvania–Maryland boundary. It is believed it was loaned to the American Philosophical Society in 1769 and used at the State House yard Observatory to observe the transits of Venus and Mercury. View from the top with its striding level. Courtesy of Independence National Historical Park.

brought with them from England. They considered the Bird transit a "less complex and more portable Transit Instrument" more generally useful to them.[54]

54. "Observations for Determining the Length of a Degree of Latitude in the Provinces of Maryland and Pennsylvania, in North America, by Messieurs Charles Mason and Jeremiah Dixon," *Philosophical Transactions*, 58 (1768): 270–273; Bedini, *Thinkers*, 139; Article by Eric Doolittle of the Flower Observatory, 18 September 1912, in an unidentified newspaper, INHP.

Figure 12. Bird astronomical transit. Viewed from the top without the striding level. Courtesy of Independence National Historical Park.

Although only the zenith sector and the transit and equal altitude instrument were listed as the instruments borrowed from the Proprietors in the accounts of the observations, an astronomical transit was also used in the State House Yard Observatory during the observations as noted in Ewing's published account:

Some of us gave particular attention to the regulation of the time-piece, and therefore took the passage of the Sun's limbs over the cross hairs of the transit instrument, both forenoon and afternoon for many days before and after the transit, and particularly on that day. As it had three horizontal hairs fixed in the focus, it afforded us six sets of corresponding altitudes, which generally agreed in giving the time for apparent noon within 2 seconds of each other. . . .[55]

Although Ewing did not positively identify the transit instrument as the one made by Bird and used by Mason and Dixon, no transit instrument is known in the Middle Colonies in this period other than the one made by Rittenhouse and used at Norriton.

The eventual disposition of the various instruments borrowed for the 1769 observations was not recorded and presumably each was returned to its owner. Just prior to their departure for England, Mason and Dixon had consigned the Proprietors's zenith sector, transit and equal altitude instrument, and other scientific equipment they had used to Joseph Shippen's care. Except for "a new Reflecting Telescope" delivered to Dr. Ewing, the remainder of their instruments and equipment was stored in the State House according to Shippen's 22 March 1768 inventory. The astronomical transit may have been among them, possibly the item misidentified by Shippen as "A Sector in all its parts contain'd in 3 Boxes in ye Council Chamber" together with tents and marquees. The records state that the surveyors were equipped with only two zenith sectors, and this would have been a third one.[56]

In 1774 Rittenhouse and Samuel Holland borrowed the Bird zenith sector for the survey of the northern boundary of Pennsylvania, and it was later stored in the State Capitol at Harrisburg. All trace of the Bird transit and equal instrument is lost, while the Bird astronomical transit apparently remained in the State House. For

55. John Ewing, "An Account of the Observations on the Transit of Venus. . . ," *Transactions*, 1st ser., 1, repr., 1789, 42.

56. Thomas Penn Letters, 9 January 1761–20 December 1770. Microfilm Roll No. 9 XR176, HSP: "A Catalogue of Instruments Belonging to the Propriet.ʳˢ of Pennsylvania and Maryland, left by Messrs. Mason & Dixon at the State House the 15th February 1768, some of which were this day removed, and put under Joseph Shippen's care at his own House . . . 22nd March 1768."

many years it was attached to the stone sill of the south window of the tower, and used at noon to take the meridian passage of the sun for establishing official time for the city. Evidence exists that the tower was equipped and used as an "observatory room" containing an astronomical clock by Isaiah Lukens to establish the true time for setting the tower clock.[57]

The transit instrument was lost and forgotten until it came to light again in 1912 during restoration of the tower of the State House. Jordan discovered it hidden beneath the flooring of a platform beside the old supports on which formerly had hung the Liberty Bell. At first the transit was exhibited on a temporary wooden mounting, later replaced by A-frames duplicated from those supporting the Rittenhouse transit instrument in the collection of the American Philosophical Society. Over time the three cross hairs disintegrated. This historic Bird transit instrument, which played an important role in Pennsylvania's history at least once and possibly twice, is the only surviving instrument, although not positively identified, used at the observatory in Philadelphia.[58]

57. Report of Riggs & Bro., March 1966, INHP, relating to the clocks in Independence Hall. Riggs & Bro. maintained the tower clock in Independence Hall during this period.

58. Doolittle, op. cit.; Bedini, *Thinkers*, 139–40; Hubertis M. Cummings, *The Mason and Dixon Line. Story for a Bicentenary 1763–1963* (Harrisburg, Pa.: Commonwealth of Pennsylvania Department of Internal Affairs, 1962), 8–14, 78; Thomas D. Cope, "Zenith Sectors and Discoveries Made with Them, Linked with More Recent Events in Pennsylvania," *Proceedings of the Pennsylvania Academy of Science*, 18 (1944): 72–75; Thomas Penn Letters, 9 January 1761–20 December 1770, Shippen, "A Catalogue . . . ; HSP" Norris S. Barratt to Wilfred Jordan, 14 April 1915, INHP; Wilfred Jordan to Chief, Bureau of City Property, 21 October 1915; Hubertis M. Cummings to David H. Wallace, Curator, Independence Hall National Park, 6 July 1960, Charles E. Smart to David H. Wallace, 27 July 1960, correspondence with David H. Wallace, 1 August and 4 August 1967; Catalogue record of Item No. 11891, Transit. The writer gratefully acknowledges valuable assistance and advice provided by Brooke Hindle and Donald W. Holst of the Smithsonian Institution, Brian P. Doherty and Jane B. Kolter of the Museum Division of Independence Hall National Park, and Elizabeth A. Carroll and

The appearance of the State House Yard Observatory may never be known with certainty, nor its exact location, and time and bureaucracy have deprived it of even the modest marker unveiled by President Wilson. Nevertheless it occupies an important permanent place in the annals of science as the first astronomical observatory erected in the American colonies, one of the makeshift observatories from which the American Philosophical Society made its first scientific observations which earned it a place beside its foreign peers. In American history it is the rostrum from which the Declaration of Independence, the new nation's birth certificate, was first proclaimed. One day future solons may recognize its significance, and by marking the site, permanently memorialize "that awfull Stage."

Murphy D. Smith of the American Philosophical Society, and Rosemary Bowditch, Historical Architect, The Edison Institute Henry Ford Museum and Greenfield Village.

Benjamin Henry Latrobe, "Learned Engineer," The American Philosophical Society, and the Promotion of Useful Knowledge and Works, 1798–1809

EDWARD C. CARTER II

B ENJAMIN HENRY LATROBE (1764–1820), America's first great professional architect and engineer, was born in England and educated in Moravian institutions there and in Germany. He returned to London where he received his professional training first as an engineer (1786–89) under John Smeaton, England's foremost practitioner of that science and art, and his noted assistant William Jessop, and then worked for three or four years in the office of the prominent architect Samuel Pepys Cockerell.[1] From 1792 to 1795, Latrobe strove to establish his own

---

1. Since 1970 Latrobe's multifaceted career has been studied and documented by *The Papers of Benjamin Henry Latrobe*, Edward C. Carter II, editor in chief. Publication has been in two forms: Carter and Thomas E. Jeffrey, eds., *The Microfiche Edition of The Papers of Benjamin Henry Latrobe* (Clifton, N.J.: James T. White & Co., 1976) and the printed edition published by Yale University Press, New Haven and London. In the latter the following books have appeared or are in press: Carter, ed., *The Virginia Journals of Benjamin Henry Latrobe, 1795–1798*, 2 vols. (1977); Carter, John C. Van Horne, and Lee W. Formwalt, eds., *The Journals of Benjamin Henry Latrobe, 1799–1820: From Philadelphia to New Orleans* (1980); Darwin H. Stapleton, ed., *The Engineering Drawings of Benjamin Henry Latrobe* (1980); John C. Van Horne et al., eds., *The Correspondence and Miscellaneous Papers of Benjamin*

private architectural and engineering practice, but the economic depression brought about by the renewal of the war with France and the death of his wife caused him to emigrate in November 1795 to the United States. He hoped to start life anew and capitalize on lands in Pennsylvania that he had inherited from his American mother, Anna Margaretta Antes Latrobe.

Latrobe went first to Virginia, where he was quickly accepted as an educated gentleman by such new friends as Bushrod Washington (APS 1805), later a United States Supreme Court Justice, but failed to establish a profitable professional livelihood despite being consulted on a number of engineering projects. His most important commission was that as architect of the Virginia State Penitentiary at Richmond (1797–99), which must have brought him to the attention of Thomas Jefferson (APS 1780), then the second vice president of the United States and the third president of the American Philosophical Society. Sensing there was little future for him in Virginia, Latrobe first visited Philadelphia in March–April 1798, ostensibly to study the Walnut Street Jail but actually to scout the professional landscape of the Athens of America. Having secured a commission to build the new Bank of Pennsylvania (he later remembered it proudly as his masterpiece), Latrobe returned in early December 1798 to Philadelphia. In rapid succession he submitted his first paper to the American Philosophical Society, on 19 December,[2] and then ten days later presented his proposal for the Philadelphia Waterworks to the city government: *View of the Practi-*

Henry Latrobe, 3 vols. (1984 and forthcoming); Carter, Van Horne, and Charles E. Brownell, eds., *Latrobe's View of America, 1795–1820: Selections from the Watercolors and Sketches* (1985). Hereafter these will be cited as: *The Microfiche Edition, Journals, Engineering Drawings, Correspondence,* and *Latrobe's View of America.*

2. Latrobe published the following papers in the *Transactions* of the American Philosophical Society, all of which but one have been annotated and republished in the Yale edition: "Memoir on the Sand-hills of Cape Henry in Virginia," 4 (1799): 439–443, plate; "A Drawing and Description of the Clupea Tyrannus and Oniscus Praegustator," 5 (1802): 77–81, plate; "On two species of Sphex, inhabiting Virginia and Pennsylvania, and probably extending throughout the

cability and Means of Supplying the City of Philadelphia with Wholesome Water. Both efforts bore fruit almost immediately and had a profound influence on Latrobe's life and career. In March 1799 the city councils accepted his plans for the waterworks and ground was broken for what became Latrobe's greatest engineering achievement. Five months later, on 19 July, Benjamin Henry Latrobe was elected to the American Philosophical Society and thus began a relationship that for the next decade proved intellectually and socially gratifying and most likely benefited him professionally.[3]

Latrobe's interest in science originated in the excellent Moravian education he received in Germany. At the pedagogium at Niesky, he learned mathematics to calculus and was also taught drawing—a skill that distinguished both his scientific and professional work. Next at the seminary at Barby, where science was regarded as complementary to theological studies, there was instruction in physics and botany and students could make use of the cabinet of natural history, an astronomical observatory, and pneumatic apparatus for experiments.[4] Evidently Latrobe continued to indulge his scientific interests upon his return to England, for his Virginia journals and sketchbooks and references to his professional library mark him as a highly competent European amateur—one who could hold his own with almost any American philosopher in the areas of botany, entomology, and geology.[5]

United States," 6 (1809): 73–78, with plate; "First Report of Benjamin Henry Latrobe, to the American Philosophical Society, held at Philadelphia; in answer to the enquiry of the Society of Rotterdam, 'Whether any, and what improvements have been made in the construction of Steam-Engines in America?'" 6 (1809): 89–98, plates, all in *Correspondence* 1; "An account of the Freestone quarries on the Potomac and Rappahannoc rivers," 6 (1809): 283–93, in *Correspondence* 2; "Observations on the foregoing communications [on buildings in India]," 6 (1809): 384–91.

3. Latrobe's Society activities have date citations in the text proper and if not otherwise noted are based on the Minutes of the American Philosophical Society.

4. For a review of BHL's Moravian education see, *Correspondence* 1: 3–10.

5. For an analysis and evaluation of BHL's science and his use of science in engineering, see Darwin H. Stapleton and Edward C. Carter II, "'I have the itch

While science was Benjamin Henry Latrobe's favorite pastime, he was well aware that in the late eighteenth-century Anglo-American world learned scientific societies were loci of power and influence that transcended their formal boundaries of interest. He knew that the initials FRS after one's name brought not only respect but on occasion preferment. Through his father, the Reverend Benjamin Latrobe, and through his own professional and social activities he came to know a number of Fellows of the Royal Society. He signed two of his five American Philosophical Society *Transactions* papers with FAPS after his name (something no other contributing member did at the time), as if to indicate that membership in the newer organization was no less distinguished than that in the older. Although not a member of the Royal Society, Latrobe participated in the other organization that also served as a model for the American Philosophical Society—the Society for the Encouragement of Arts, Manufactures, and Commerce (the Royal Society of the Arts). Its *Transactions* reveal Latrobe's name on the list of contributing members from 1784 to 1795.[6] The American Philosophical Society's own *Transactions* published during the years of Latrobe's active membership contain a mixture of the types of articles (pure science and useful technology) found in the publications of the Royal Society and the Society for the Arts, respectively. In short, Latrobe was exceedingly well prepared to take an active role in the life of the great Philadelphia organization. His American associations and his unrivaled professional qualifications probably dictated that he would have been elected in due course, but it seems that he eagerly desired membership and that from the spring of 1798 onward he advanced his chances of election whenever possible.

---

of Botany, of Chemistry, of Mathematics . . . strong upon me': the Science of Benjamin Henry Latrobe," *Proceedings* of the American Philosophical Society 128 (1984): 173–192.

6. *Transactions* 7 (1789) – 13 (1795). BHL is listed as living at Staples Inn through 1790 and thereafter Francis Street, Tottenham Court Road.

Latrobe strove to establish himself as the first truly professional American engineer. To do so he clearly modeled himself after the greatest civil engineer in England, the man who more than any other person was responsible for elevating his calling from a trade to a profession, John Smeaton, FRS (1724–92). What brought renown to Smeaton and in turn to his profession was the breadth of his ability and intellectual interest. Not only did he execute a large number of engineering projects brilliantly, but also he was a fine applied-scientist who received in 1759 the Royal Society's Copley medal. Of great importance was the fact that his published engineering reports and scientific papers were highly praised as models of clarity and reason. Trained as a scientific instrument maker, Smeaton was elected to the Royal Society in 1753 on the basis of promise; from 1750 to 1788 he published eighteen papers in the *Philosophical Transactions*. His engineering career began in 1753 and continued until two years before his death. The genius of his accomplishment was unmatched; he directed hundreds of projects in all branches of civil engineering: mills and millwork, river improvement and canals, fen drainage, bridges, steam engines, harbors, and his most famous effort, the rebuilding of the Eddystone Lighthouse, the report of which he published in a magnificent volume (1791). In 1812 and 1814 his engineering reports and all of his Royal Society papers were published in four volumes.

Smeaton was a "learned engineer," the first of his kind. It has been said that he was "a philosopher among engineers; but he was also an engineer among philosophers."[7] He commanded respect in boardrooms, Parliament, and the scientific world. What Smeaton had done in England, Benjamin Henry Latrobe attempted to do in the United States. He, too, would attempt to become a "learned

7. A. W. Skempton, ed., *John Smeaton, FRS* (London: Thomas Telford Limited, 1981), 4. This volume contains articles on all aspects of Smeaton's career. There is no adequate biography of Smeaton. See Samuel Smiles, "Life of John Smeaton," *Lives of the Engineers* (London, 1861), 3: 3–90; H. Dorn, "John Smeaton," *Dictionary of Scientific Biography* (New York: Charles Scribner's Sons, 1975), 461–463.

engineer," establish a profession in a nation where professionalism was highly suspect, gain respectability and high status; he hoped, also to become well-to-do if not wealthy. To become the American Smeaton, Latrobe needed the acceptance of American philosophers and a set of initials behind his name too—FAPS.

Latrobe only partially realized his vision. Although he made friends in high places and was generally respected and admired, he died penniless in New Orleans in 1820. If he did not become the American Smeaton (in that he did not establish the profession of civil engineering on a firm basis), he certainly was recognized as America's first professional engineer.[8] Of course, it was in architecture that Benjamin Henry Latrobe made his greatest contribution. The quality, beauty, and sophistication of his buildings and designs; the professional standards that he set for himself, his associates, students, technical staff; and his ability to envision large-scale, integrated, planning projects, such as the United States Capitol, all combined to make him the American John Soane. Latrobe was the father of the American profession of architecture; his two major students, Robert Mills and William Strickland (APS 1820), were the leaders of the first generation of native American professionals. Latrobe wrote excellent engineering and architectural reports for his public and private clients throughout his career that equaled the best efforts of leading English practitioners. Unlike Smeaton, whose works were republished a few years after his death by admiring colleagues, Latrobe's professional plans, drawings, reports, and scientific papers are only now being assembled, studied, annotated, and published in the nine volumes of the Yale edition of *The Papers of Benjamin Henry Latrobe*. In his lifetime, however, he had the satisfaction of seeing six of his scientific papers published in the Society's *Transactions* and knowing that several were reprinted in the journals of European academies.

8. Latrobe's engineering projects and drawings, theory, and practice are set forth in detail in *Engineering Drawings*.

If Latrobe's Virginia sojourn, 1796–98, was professionally frustrating and unrewarding, it did allow him to pursue his scientific interests, especially as a naturalist. He filled his sketchbooks with beautiful drawings of animals, fishes, insects, and plants, and drew sketches and wrote descriptions of the geology and river systems of the tidewater and the piedmont. When he reached Philadelphia, he mined his sketchbooks and journals to prepare four *Transactions* papers. But Latrobe could pursue his natural history interest only so far in Virginia; he complained of a lack of access to scientific works (his own library having been lost in transit from England) and the seeming absence of any amateur, much less professional, scientists. It should be noted that he could not have completed his most enduring contribution to the literature of natural history, his observations on a parasite found in the mouths of fishes called alewives, without recourse to the scientific libraries of Philadelphia.[9]

Late in 1798 and in 1799 he made the acquaintance of three men of geological bent whose friendship accelerated his decision to leave Virginia for Philadelphia: Giambattista Scandella (APS 1798), Constantin François Chassebœuf, comte de Volney (APS 1797), and William Maclure (APS 1799; elected with Latrobe). Latrobe met Scandella and Maclure when they visited Richmond, and he formed close friendships with both men; he encountered

9. "A Drawing and Description of the Clupea Tyrannus and Oniscus Praegustator," supra 2. BHL made careful observations and drawings of both the fish and the parasite, and later consulted the works of Linnaeus, Gmelin, Fabricius and others in Philadelphia. He discovered descriptions of similar parasites but was sufficiently accurate in his observations to doubt that his was identical. Learning that the fish had not been described, he wrote a well organized paper that is considered excellent even by modern standards. The alewife today is known as *Brevoortia tyrannus* (Latrobe), and the crustacean parasite found in its mouth is *Olencira praegustator* (Latrobe). It is reproduced in *Latrobe's View of America*, plate 33. Other Latrobe natural history drawings can be found there also: "Dolphins," plate 5; "Ground Squirrel," plate 15; "Masons, or Dirt daubers," plate 35; "Cyprepedium acaule," plate 46; "Lady Slipper," plate 47; "Dobson fly," plate 50; "Sphex sumac," plate 110; "The Jack," plate 153; and "Ellisia," plate 161.

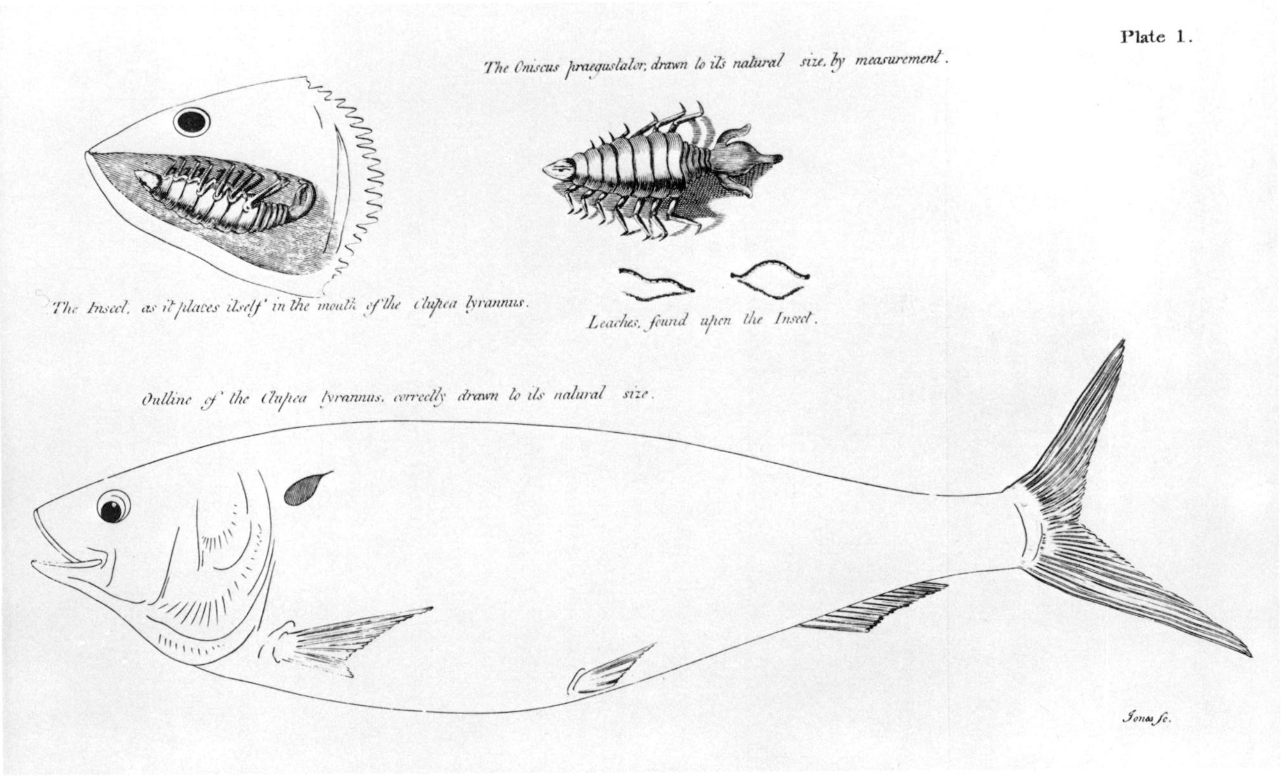
Plate 1.
The Oniscus praegustator, drawn to its natural size, by measurement.
The Insect, as it places itself in the mouth of the Clupea tyrannus.
Leaches, found upon the Insect.
Outline of the Clupea tyrannus, correctly drawn to its natural size.
Jones sc.

Figure 1. A Drawing and Description of the Bay Alewife and the Fish Louse.

Volney in the spring of 1798 during his first brief visit to Phila-delphia. Volney and Maclure helped lay the foundation of Ameri-can geology, and after becoming acquainted with both men Latrobe turned his attention to that infant discipline. Latrobe's last activity in Virginia before departing for Philadelphia on 1 December 1798 was a geological excursion with Maclure to the coal field west of Richmond. They traveled northward together stopping to study "the soil and stratification of the Stone at Fredericsburg . . . and from Georgetown to the little falls, on the Potowmac." Upon his arrival in Philadelphia, Latrobe lived with Maclure's family for the next four months, which must have facilitated his entrance into the city's scientific community.[10]

Latrobe's major concern, of course, was his first great commis-sion, the Bank of Pennsylvania, but another problem he faced was how to gain access quickly to a group of his intellectual, scientific and social peers. Benjamin Henry Latrobe was a proud and formal man who took offense easily if others failed to recognize him as both a gentleman and a highly trained professional. He had already encountered problems dealing with public officials and business-men serving as managing directors of public works projects who showed little deference to his social background or architectural and engineering accomplishments. Numerous observers and his-torians have described the important role that science and medi-cine played in the life of late eighteenth-century Philadelphia, and it is logical that Benjamin Henry Latrobe recognized scientific achievement as a means of social acceptance. When one considers Latrobe's activities immediately after his arrival in Philadelphia in early December 1798, there are reasons to speculate that he sought election to the American Philosophical Society not only for intellec-tual and personal satisfaction, but also as a means of professional advancement. As a "learned engineer," Latrobe would be in a

10. *Journals* 2: 7. Quoted in Stapleton and Carter, "Science of Latrobe," 176.

position to capture other important Philadelphia commissions by impressing the city's leadership groups (in which Society members were present) with not only his professional qualifications but also his wide-ranging intellectual capabilities. It is important that he signed his first *Transactions* paper "B. Henry Latrobe, Engineer," and that he was elected to membership in the Society on 19 July 1799 as "Ben. Henry Latrobe, Engineer." Only doctors and the occasional minister were designated by their professions in the Society's minutes and publications until Latrobe introduced the term "Engineer." It was as if he intended to emphasize the technological and scientific side of his practice—something the nation had particular need of—and not the artistic dimension, even though the Bank of Pennsylvania commission gave him the opportunity to design a revolutionary new type of building, one which traditionally has been thought of as the first great monument of the Greek Revival.

Latrobe's first and perhaps his most important contact with the American Philosophical Society community was through its president, Thomas Jefferson. He wrote to the great Virginian during the spring 1798 Philadelphia trip, met him later in Fredericksburg, and then wrote again in late September from Richmond when Jefferson was at Monticello putting the finishing touches on the Kentucky Resolution. In both communications—the first formal, the second more personal—Latrobe obviously was seeking Jefferson's assistance in obtaining commissions; in each Latrobe pushed his professional qualifications as an engineer and architect, first hinting and then clearly stating that federal work was being denied him because his political associations were with Jeffersonian Republicans. Latrobe also peppered his September letter with news of Virginia politics, flattered Jefferson about his mouldboard improvement in the wooden plough ("I have taken the liberty of having several . . . made for my friends. . . . I have been astonished at their performance"), and sent Jefferson drawings of the "Philadelphia nail cutting machine which I had long made for you."[11] Jefferson had read his paper on the mouldboard at the Society meeting of 4 May and the detailed description appeared in the next volume of the *Transactions* (1799). It is not known whether Latrobe used a copy of Jefferson's paper to make the mouldboards or turned to an earlier version that was fairly well known. In any case, it is obvious that Benjamin Henry Latrobe and his professional qualifications were not unknown to Jefferson when on 21 December 1798 the engineer's first Society paper, "Memoir on the Sand-hills of Cape Henry in Virginia," was read. Latrobe took pains to inform the membership that he intended to send further communications to the Society. At the next meeting on 27 December, it was reported and agreed upon as being worthy of publication and those present were informed by Dr. John Bartram (APS 1769), with whom the engineer had already made contact, that "Mr. Latrobe was willing to etch the engraving."[12]

We may never know who urged Latrobe to submit his paper or whether he acted on his own; in any case, "The Sand-hills of Cape Henry" was Latrobe's entry point to Philadelphia's scientific community and the American Philosophical Society. What is so astonishing is that Latrobe arrived in Philadelphia on about 3 December, took the Bank of Pennsylvania project in hand, and in a little over two weeks' time submitted his paper to the Society (dated 19 December). It seems improbable that Latrobe could have found time to fish the paper out of his sketchbooks and journals in such a short time, but there is no surviving evidence that it was written prior to his arrival.[13] Perhaps his geological survey excursions with Mac-

11. BHL to Thomas Jefferson, 28 March and 22 September 1798, *Correspondence* 1: 81–82, 95–98.

12. For a comparison between BHL's watercolor and the engraving in *Transactions* 4 (1799): 439 see *Latrobe's View of America*, plates 52 and 53. For further illustrative materials upon which this paper was based see ibid., 150–155.

13. BHL had written a long "Sketch of a Letter to Dr. Scandella" in Richmond 4 May 1798 which surveyed his observations of the geology and river systems of

lure stimulated Latrobe to prepare the paper on his arrival. Maclure himself had yet to present a paper to the Society and did not do so until 1809. A very cursory review of the Society's records for the period does not show a similar occurrence in which a newcomer to Philadelphia presented a paper so shortly after his arrival. Perhaps Latrobe was urged to do so by his great patron, Samuel M. Fox, the president of the Bank of Pennsylvania, whose brother George Fox (APS 1784) would later render the engineer important support as a member of the Chesapeake and Delaware Canal Company's board of directors. Fox, the banker, who bent every effort to help Latrobe design and construct what is generally judged one of the most beautiful and influential of all the city's great buildings, may have thought that by promoting Latrobe with the Society he could tie him more closely to Philadelphia.

Latrobe could have submitted a paper on some other branch of natural history, as he did twice later, but he chose to write on geology, a science he utilized brilliantly in his second great project, the Philadelphia Waterworks, and in the process clearly identified himself as an engineer. In November 1797 the city councils had appointed a joint committee to investigate the possibility of bringing pure water into the city. When Latrobe arrived in Philadelphia, very little progress had been made in resolving how best to proceed. Samuel M. Fox, a member of the committee, undoubtedly discussed the entire problem with Latrobe. What transpired next was truly amazing. On the same day that his "Sand-hills" paper was being accepted by the Society, Latrobe accompanied John

---

"the lower part of Virginia, Maryland, and Pennsylvania." The received letter has not been found. Clearly BHL was much involved with geology which was an important adjunct to civil engineering, especially canal building. For a thorough discussion of BHL's geology and the "Sketch of a Letter" see *Journals* 2: 384–414 and Stephen F. Lintner and Darwin H. Stapleton, "Geological Theory and Practice in the Career of Benjamin Henry Latrobe," in Cecil J. Schneer, ed., *Two Hundred Years of Geology in America* (Hanover, N.H.: The University Press of New England, 1979), 107–119.

Miller, Jr., chairman of the committee on watering the city, and other gentlemen interested in the engineer's views on the problem to Spring Mills on the north bank of the Schuylkill to see how supplies of water might be obtained from that source and from the river itself. Miller asked Latrobe for his opinion on the matter, and Latrobe produced a detailed plan for a waterworks, dated two days later, that became the most advanced technological project yet seen in America. "The Philadelphia Waterworks was Latrobe's greatest and most successful engineering work. It was the first urban utility in the United States (in the modern sense), and it was the first convincing display of steam power in the Western Hemisphere."[14] The report was a model of analysis and planning and if really completed in but two days marks Latrobe as a true genius. It demonstrated a thorough comprehension of the underlying geological structure of the area, the topography of the city and its surroundings, the internal measurements of the street system, and the natural history of the city.[15] Published early in 1799 by the order of the corporation of Philadelphia, Latrobe's *View of the Practicability and Means of Supplying the City of Philadelphia with Wholesome Water* revolutionized thinking about the possibilities for a water supply and placed his name and reputation squarely before the city's leadership. Within four months after his arrival in Philadelphia Latrobe, like Smeaton, was on his way to becoming a "learned engineer"; he had written both a scientific paper and a ground-breaking technological report, and he was directing two of the most significant projects of his entire career.

Four months later Latrobe became a member of the American Philosophical Society in a distinguished "class" of seven that included William Maclure, John Redman Coxe, and Thomas Peters

14. *Correspondence* 1: 109.
15. For the report itself see, ibid., 109–125. The entire project is completely discussed and documented and all relevant drawings are published in *Engineering Drawings*, 28–36, 144–198 and plates 5–8.

Figure 2. "West Elevation of the Wall of the Basin; Section of the Basin and of the Coffre Dam thrown up previous to its erection," [1800]. In order to firmly anchor the west wall of the Philadelphia Waterworks' settling basin, Latrobe had the foundation excavated to bedrock. In this pair of sections he was careful to note the surface geology of the bed of the Schuylkill River: blue mud, hard gravel, and loose rock, and granite. (Historical Society of Pennsylvania.)

Smith—"all of Philadelphia." Prophetically, perhaps because of the transient nature of his stated profession, only Latrobe was assigned no city of residence in the Society's minutes. He was listed last after Samuel Elan of Newport, Rhode Island, as "Ben. Henry Latrobe, Engineer."

Once alerted it seems that the active membership of the Society had regard for Latrobe's scientific ability and judgment. As we have noted elsewhere:

> of the seven papers which he presented to the Society after his election five were chosen for publication in the *Transactions*. He also received nineteen committee assignments; most frequently the committees were established to judge the publishability of a paper. For example, he was twice a committee of one to report on papers of the eminent botanist Benjamin Smith Barton. But he was also chosen to serve on committees considering Society operations, including the preparation of volumes IV and V of the *Transactions*.[16]

Latrobe must have found the meetings and duties of the Society both congenial and rewarding as he was fairly faithful in his attendance. The records list him present at eighteen meetings; the press of business and the removal of his residence from Philadelphia in 1803 account for the fact that he attended only four between 1804 and 1808 and none thereafter. When living in Philadelphia and actually available, Latrobe's attendance record qualified him to be thought of as an active member of the Society. But even when absent from the city, he was given numerous committee assignments, mostly to report on the publishability of communications. After moving away, first to Delaware and then to Washington, Latrobe's letterbooks record his attempts to keep in touch with the Society's affairs, mainly by correspondence with his friends Charles

16. Stapleton and Carter, "Science of Latrobe," 179.

Willson Peale (APS 1786) and John Vaughan (APS 1784). These contacts ended after 1808.

Latrobe was twice chosen a Councillor; he helped draft a memorial to the Pennsylvania legislature in January 1800 requesting aid in acquiring "scientific works, and an extensive philosophical apparatus," and served on the committee that recommended the acquisition of the Gilbert Stuart portrait of George Washington that hangs today in the Hall meeting room. But it was in the area of scholarly publication that he made his greatest contribution to the American Philosophical Society. Six of his own papers appeared in the 1799, 1802, and 1809 volumes of the *Transactions* and he passed on at least five others that were published. In addition, as noted above he helped prepare the first of these two volumes for the press. Thus it is fair to say that Benjamin Henry Latrobe played a not inconsequential role in determining the quality and nature of American scientific publication in the first decade of the nineteenth century.[17]

Latrobe's scientific activities were not limited to the Society, for he was also a member of the Chemical Society of Philadelphia founded in 1792 by James Woodhouse (APS 1796), a professor at the University of Pennsylvania. Through his membership in these societies Latrobe became acquainted with a number of the most prominent men of science in America, and his letterbooks and journals reflect his scientific discourse with them. He counted Thomas Jefferson, John Vaughan, Charles Willson Peale, Robert Hare (APS 1803), and Thomas Peters Smith among his friends; he sat on committees and conversed with Robert Patterson (APS 1783), Adam Seybert (APS 1797), and Robert Leslie (APS 1795).[18]

17. From BHL's first paper onward, ninety-two memoirs, experiments, letters, and brief remarks appeared in volumes 4, 5, and 6. BHL either wrote or passed upon about eleven percent of those papers.

18. See BHL's letters to Jefferson, Vaughan, Peale, Hare, Patterson, and Seybert in *The Microfiche Edition*. BHL refers to Smith in his *Journals*, 15 March 1800,

When Latrobe left Philadelphia he retained his scientific interests, but once again as during the Virginia days he faced intellectual isolation. In Washington he made contact with what passed for the civilian scientific community and also the nascent scientific agencies of the federal government such as the Patent Office and the Coast Survey. But generally Latrobe found none of these resources particularly gratifying; science again became the serious avocation it had been for him in Philadelphia. Only in acoustics did Latrobe continue to join science to engineering in an impressive manner. He wrote two papers on the transmission and reflection of sound waves and the problems of echoes and dead spots in auditoriums; the second (an addendum to the acoustics article in the 1812 American edition of *The Edinburgh Encyclopaedia*) showed "that his experience with the halls of the Capitol in Washington gave him more confidence in handling such issues."[19]

If Latrobe was good for the American Philosophical Society, in what ways professionally—and even financially—might the "learned engineer" have profited from membership in the organization? Clearly the most obvious benefit that accrued to Latrobe was Thomas Jefferson's patronage after he became President of the United States in 1801. To Jefferson he would owe the major appointment of his career and also a minor one: as Surveyor of the Public Buildings of the United States, and later as Architect of the Capitol (6 March 1803), and as Engineer of the Navy Department (c. January 1804). Jefferson learned more of Latrobe the man and

---

and in a letter to Gales & Seaton, of 18 January 1817, *Correspondence* 3. Leslie and BHL had been at several American Philosophical Society meetings together and had served on the same committees, yet Leslie was severely critical of BHL's designs of 1802 for Jefferson's proposed naval dry dock at the Washington Navy Yard. Robert Leslie to Thomas Jefferson, 10 January 1803, Jefferson Papers, Library of Congress, Washington, D.C. For documentation on the naval dry dock see *Correspondence* 1: 219–224; 226–227; 233–234; 236–249; and *Engineering Drawings*, 9–11; 116–124.

19. Stapleton and Carter, "Science of Latrobe," 187.

the professional through their contacts in the Society. Both were present at the meetings of 7 January and 7 February 1800 and Jefferson undoubtedly read Latrobe's papers and was well informed of the scope and progress of the Waterworks. That the latter could plan and successfully execute the engineering feat of the period, a sophisticated technological project employing two steam engines, boasting a masonry settling basin anchored to the river bottom, and requiring heavy capital investment, made Latrobe a prime candidate for the most important of federal public works—the U.S. Capitol. On 24 October 1802 Latrobe wrote the president a long letter on the purposes and potential of the proposed Chesapeake and Delaware Canal, reporting that the public response was tepid and that without federal assistance the project's future was unpromising. It was a brilliant exercise in topographical description, technological planning, regional economic analysis, and political reporting. In short, it was exactly the type of state paper that Jefferson relished and admired. Latrobe quite properly mentioned the Society's 1769 expedition that surveyed and leveled the ground between the Christiana River and Elkton.[20] Almost upon receipt of this report, which was also a plea for federal support for this vital public work, Jefferson invited Latrobe to Washington to design the Navy Dry Dock. Although the dry dock was the president's plan, the commission would allow him to verify once more his judgment of the engineer's qualifications, talents, and personal qualities. Latrobe journeyed to the capital without his wife, Mary Elizabeth Latrobe. Thomas Jefferson immediately extended his hospitality to Benjamin Henry Latrobe, who dined twice at the president's house before returning to Philadelphia by Christmas.[21] By then a bond of friendship had formed between these two great men that

20. BHL to Jefferson, 24 March, 1 June, 24 October 1802 enclosed in BHL to Jefferson, 24 October 1802, *Correspondence* 1: 207–219.
21. For delightful description of those dinners, see BHL to MEL, 24 and 30 November 1802, *Correspondence* 1: 232–233; 234–236.

would sustain their artistic and political partnership. Although occasionally strained by the incompatibility of their architectural ideas it resulted in the U.S. Capitol drawing closer to completion, based upon the very highest level of professional competence and taste.

The American Philosophical Society and a number of its interested Philadelphia members were associated with the Chesapeake and Delaware Canal from the project's first stirrings in the 1760s until its completion in 1830. Latrobe was appointed engineer of the company on 25 January 1804, having begun survey work six months previously. The Chesapeake and Delaware Canal was carried forward along the same lines as the Waterworks. Latrobe used much of the same personnel and also employed permanent stone construction. The line of the canal was to cross northern Maryland and Delaware, making possible water transportation between the upper Bay and the Delaware River. Its economic purpose was to direct Pennsylvania agricultural products (via the Susquehanna and the canal) to Philadelphia for consumption or processing and reshipment to European markets. By 1806, when the company exhausted its funds, Latrobe had established the route of the main line and had nearly pushed construction of the feeder to a point where it could provide necessary water for the canal's planned central locks.[22]

This was a project that Latrobe particularly hankered after, and it seems likely that his connection with the Society helped him obtain it. The Committee of Survey that appointed him to undertake the initial work included two Pennsylvania members, both Philadelphians who were shortly to become Society members: Joshua Gilpin (APS 1804), a "uniform friend" of the engineer, and William Tilghman (APS 1805). Gilpin became the leader of five

22. For the complete story of BHL and the C&D Canal, see *Engineering Drawings*, 11–18, 125–129 and *Correspondence* 1 and 2.

Philadelphia directors who were the project's driving force. Another Society member in this group that vigorously supported Latrobe's professional planning and leadership was George Fox, the brother of Latrobe's sympathetic and strong patron, Samuel M. Fox. Latrobe even consulted with the sole remaining member of the 1769 Society expedition, Levi Hollingsworth (APS 1768), the Philadelphia merchant, as to the accuracy of the original survey. Hollingsworth was one of the most active supporters of the project and Latrobe talked with him about it on several occasions.[23] As the Society's minutes of 16 December 1803 show, Latrobe made use of the 1769 expedition's survey maps in carrying forward his own work. After the project was closed down in 1806, the canal company was not revived again until 1822, two years after Latrobe's death. "Significantly, William Strickland [APS 1820], one of Latrobe's pupils who had assisted him on the feeder, was employed by the company for the initial survey of the main line."[24] Once again it was an old friend of Latrobe who rallied Philadelphia and Pennsylvania behind the enterprise, Mathew Carey (APS 1821), the great publisher and ardent American nationalist.

Latrobe used the *Transactions* to promote his own style of steam engineering and that of Nicholas J. Roosevelt, his partner in a number of schemes, over the approach and proposed applications of Oliver Evans, the proponent of the high-pressure engine.[25] Despite misjudging the value of the lightness and greater power of the Evans engine, Latrobe became one of the leading steam engineers of his time, using engines in a variety of commissions, and helped make Philadelphia the center of steam engine building and use in the United States.

23. BHL to Thomas Jefferson, 24 October 1802, *Correspondence* 1: 214 and 218, footnote 8.

24. *Engineering Drawings*, 18.

25. "First Report of Benjamin Henry Latrobe . . . ," dated 20 May 1803, *Transactions* 6 (1809): 89–98. Reprinted and fully discussed, *Correspondence* 1: 301–315.

Perhaps in the end it was more in the matter of reputation that Latrobe gained from his American Philosophical Society ties than from actual commissions. Certainly, Latrobe became known to a wider and more influential world than if he had never been elected to membership. Whitfield J. Bell, Jr., has explored the role and importance of the institution at the time of Latrobe's membership in his excellent and penetrating article, "The American Philosophical Society as a National Academy of Sciences, 1780–1846."[26] The Society that Dr. Bell describes so eloquently and thoughtfully was at the height of its national activity and eminence during the years that Benjamin Henry Latrobe was most involved in its affairs.

Some of Latrobe's detractors attacked him as a hypocrite talking of "*patriotism, disinterestedness*, and *virtue*" while seeking personal economic gain.[27] But as Sam Bass Warner has reminded us, in Latrobe's day there was no clear-cut division between public and private interest as there is today. The controlling ethic was that of privatism wherein the individual sought to do well for himself and the community benefited if sufficient numbers of its citizens succeeded in reaching their economic goals.[28] The American Philosophical Society was not a purely scientific organization. Indeed, it played a very active role in attempting to improve American society by a wide variety of means, many of which encouraged economic growth through the application of advanced technology. Given the nature of the times, it would have been unusual if Benjamin Henry Latrobe had not used his Society association to promote his projects, himself, and the commonweal.

Latrobe was treading a subtle line throughout the first third of his American career as he avoided entrepreneurial involvement in

26. *Proceedings* of the Tenth International Congress of the History of Science (Paris, 1964), 165–175.
27. Thomas Pym Cope, a Philadelphia merchant and member of the Watering Committee, diary, 15 March 1801, quoted in *Correspondence* 1: 176, footnote 8.
28. Sam Bass Warner, *The Private City: Philadelphia in Three Periods of Its Growth* (Philadelphia: University of Pennsylvania Press, 1968).

his public works projects (a role that American capitalists expected technologists to play) and instead preached the doctrine of professional rights and duties (the independent judgment of the engineer or architect guaranteed by predetermined fee payment).[29] At the time there was no definitive recognition and acceptance of the limits of architects' and engineers' behavior even in England. Latrobe was an open and responsive man, always willing to give advice on professional and technical matters to those requesting it. He willingly and successfully advanced the careers and ideas of men he never actually met in an attempt to establish his twin professions in the United States.

Might not a "learned engineer" accelerate this process by associating his technical skills in the public's mind with the American Philosophical Society? At the very least being perceived as a "learned engineer" may have provided Latrobe with the psychological comfort of being personally thought of as both an American gentleman and a professional instead of a hired foreign technologist who drifted from job to job. Accused in 1811 of being merely a "travelling engineer" who, having no commitment to the United States, might return to England the moment his fortune was made, Latrobe responded:

> Your observation applies to me only as far as it is not known that my family for more than 90 years has been settled in Pennsylvania and that altho' I was educated in Europe on the Continent, and for some years practiced my profession in England, I am already an American of the fourth generation. I am therefore a *Travelling Engineer* only by having *travelled home*.[30]

29. According to Daniel Calhoun "Americans seem to have regarded proprietorship as a guarantee of trustworthiness and one's readiness to conform to the interests of an enterprise." *The American Civil Engineer: Origins and Conflicts* (Cambridge, Mass.: M.I.T. Press, 1960), 16.

30. BHL to Thomas Moore, 20 January 1811, *Correspondence* 3. Latrobe's mother, Anna Margaretta Antes (1728–1794), was born in Pennsylvania and moved to England where she studied in Moravian schools and married the Reverend Benjamin Latrobe.

Certainly in the Hall of the American Philosophical Society, Benjamin Henry Latrobe was accepted by his fellow "gentlemen of science" as one of their own and also as a professional. For a decade Latrobe's activities were intertwined with those of the Society; this happy conjunction surely resulted in the promotion of useful knowledge and works and the enrichment of the great American engineer's own life.

Consensus and Conflict About Imprisonment: The Philadelphia Society for Alleviating the Miseries of Public Prisons, 1787–1829*

MARVIN E. WOLFGANG

Introduction

ON 25 October 1829 Charles Williams, an eighteen-year-old youth from Delaware County, Pennsylvania, became the first prisoner admitted to the new Eastern State Penitentiary at Twenty-First Street and Fairmount Avenue in Philadelphia, Pennsylvania. According to the records, he was a "farmer, light black; black eyes; curly black hair; 5'7½"; foot 11"; flat nose, scar on bridge of nose; broad mouth, scar from dirk on thigh; can read." He had been sentenced to imprisonment for two years for having broken into the house of Nathan Lukens in Upper Darby where he stole "1 silver

* The author acknowledges his indebtedness to the archives of the Pennsylvania Prison Society, which celebrates its bicentennial in 1987, being only a name change from the Philadelphia Society for Alleviating the Miseries of Public Prisons. As a former president of the Society from 1960 to 1968 and as a current board member, I am keenly aware of the importance which the Society has had in the history of penal reform. I owe a special debt to my presidential predecessor, the late Professor Negley Teeters, for his prodigious scholarship that aided my research for primary sources for this essay.

I wish to thank Mary Glazier, a doctoral candidate in criminology at the University of Pennsylvania, for her diligent assistance in tracking down primary sources at the Pennsylvania Historical Society. I also thank Selma Pastor, the librarian of the Center for Studies in Criminology and Criminal Law, for her initial research for this essay, her careful editing and typing of the manuscript. Without Mary Glazier and Selma Pastor, this essay would have been deficient.

watch, value $20; 1 gold seal, value $3; 1 gold key, value $2."[1] Thus began the corporeal implementation of the most significant philosophy of prison reform of the late eighteenth and early nineteenth centuries in the United States, England, and the entire European continent.

The history of this turbulent period of philosophical debates, organizational battles, and statutory changes has been well recorded in some of the primary and secondary sources referred to in this essay. We shall not recount that history here, although some of it is necessary as background for our focus, which is on a few of the main controversies faced by the newly created Philadelphia Society for Alleviating the Miseries of Public Prisons between 1787 and 1829. We shall explore two main issues: (1) whether prisoners should be held in separate or solitary confinement, and (2) whether such prisoners should be allowed to engage in labor. To the contemporary mind, these may seem to be banal, archaic concerns. But to many prominent citizens of Philadelphia, New York, Boston, and London, during these early years, these were burning issues that occupied much of their thinking, writing, and personal lives.

It is not our intent to present an equal distribution of the arguments on both sides of an issue. The core of a controversy will be presented, but concentration will be on the positions taken by the Philadelphia Society that led to the construction of the Eastern State Penitentiary.

Let us recall that in the first half of the eighteenth century there was heated debate between advocates of what came to be known as the Pennsylvania, or separate, system and the Auburn, or congregate or silent, system. In the former, prisoners were to be maintained in solitary cells, brought into the prison under black hoods

1. Cited by Negley K. Teeters and John D. Shearer, *The Prison at Philadelphia: Cherry Hill* (New York: Columbia University Press, 1957), 83–84.

so as not to be seen by other prisoners, and never to have social intercourse with them. In the Auburn, New York, prison system, for which the best known model was the Sing Sing prison, there was congregate labor with silence during the day and separate celling at night. The Auburn system became dominant in the United States, except in Pennsylvania; and the Pennsylvania system, with its special architecture, spread throughout Europe and Asia in the nineteenth century.[2]

The driving force behind the philosophy of the separate system was the Philadelphia Society for Alleviating the Miseries of Public Prisons (known later as the Philadelphia Prison Society and, since 1887, the Pennsylvania Prison Society). And the greatest influence on that Society was the Society of Friends. Quakers were about one-third of the membership in the early years; and although Bishop William White, an Episcopalian, was the first president and remained in that office for forty-nine years, he and the Episcopal Church came to accept and adopt the penological perspectives of the Quakers throughout this history of prison reform. It was concern for the conditions in the old Walnut Street Jail, at Sixth and Walnut Streets, Philadelphia, that prompted sensitive citizens of that city to organize. That is where our story begins.

The New Society and the First Penitentiary

The reform movements of the eighteenth century were responsible for concern for "the abject misery" of inmates in the provincial

2. For detailed descriptions of the Auburn-Pennsylvania controversy there are many accounts. Among general twentieth-century histories, see: Harry Elmer Barnes, *The Evolution of Penology in Pennsylvania: A Study in American Social History* (Indianapolis, Ind.: Bobbs-Merrill Co., 1927; Montclair, N.J.: Patterson Smith, 1968), and *The Repression of Crime* (New York: George H. Doran Co., 1926; Montclair, N.J.: Patterson Smith, 1969); Negley K. Teeters, *They Were in Prison* (Philadelphia: John C. Winston Co., 1937), and *The Cradle of the Penitentiary: The Walnut Street Jail at Philadelphia, 1773–1835* (Philadelphia: Pennsylvania Prison Society, 1955); and Teeters and Shearer, *The Prison at Philadelphia.*

jails in Philadelphia. Richard Wistar, a member of the Society of Friends, is commonly associated with the founding of an aborted organization that antedated the famous Philadelphia Society for Alleviating the Miseries of Public Prisons. Just prior to the Revolutionary War, a small group, on 7 February 1776, formed the Philadelphia Society for Assisting Distressed Prisoners. But the British occupation of Philadelphia nineteen months later, in September 1777, caused this parent group to dissolve, leaving no official records.[3]

After the peace of 1783, attention was again given to penal reform by such citizens as Benjamin Franklin and Benjamin Rush, who had long deplored the Code of the Duke of York that had been forced on the colonists in 1718. That Code contained thirteen capital crimes and was considered severely repressive. Efforts by prominent citizens for reform resulted in a new law of 15 September 1786 which, although reducing the number of capital crimes, provided for public punishments, or what Caleb Lownes at the time called "continued hard labor, publicly and disgracefully imposed."[4] Convicts were taken from the Walnut Street Jail to dig ditches, excavate cellars, to grade and fill in ponds. They became known as "wheelbarrow men" and were on public display before the good citizens of the city, who apparently ridiculed them. Commenting on this condition later, Roberts Vaux, one of the most active members of the Philadelphia Society, wrote:

> The sport of the idle and the vicious, they often become incensed & naturally took violent revenge upon the aggressors. To prevent them from retorting injuries still allowed to be inflicted, they were incumbered with iron collars & chains, to which bomb-shells were attached, to be dragged along while

3. Barnes, *The Evolution of Penology*, 80–81.
4. Caleb Lownes, *An Account of the Alteration and Present State of the Penal Laws of Pennsylvania*, 5–6. *Statutes at Large*, 12: 280–81; cited in Barnes, *The Evolution of Penology*, 81 n. 15.

they performed degrading service, under the eyes of keepers armed with swords, blunderbuses, & other weapons of destruction.[5]

His strong opposition to public punishments and to the death penalty caused Dr. Benjamin Rush to read his first pamphlet on this issue at a meeting of the members of yet another group, the Society for Promoting Political Inquiries, held on 9 March 1787 at the house of Benjamin Franklin. "All public punishments," he argued, "tend to make bad men worse, and to increase crime, by their influence upon society."[6] Two months later, under the leadership of Benjamin Rush, the first of the major modern prison reform societies, the Philadelphia Society for Alleviating the Miseries of Public Prisons, was founded on Tuesday, 8 May 1787, in the German School House on Cherry Street.[7] The thirty-seven charter members[8] elected Bishop William White their president.

It was not only opposition to public punishments that resulted in creation of the Society; it was also deep concern about the plight of prisoners and the corruption of penal administration in the Walnut Street Jail, at Sixth and Walnut Streets, directly across from the State House (now Independence) Square. Built in 1773 and used during the war to house prisoners on both sides of the conflict, the jail was in its worst condition in 1787. Perhaps the most graphic description of the jail is found in the comments by George W. Smith, a member of the Philadelphia Prison Society, who was

---

5. Roberts Vaux, *Notices of the Original, and Successive Efforts to Improve the Discipline of the Prison at Philadelphia and to Reform the Criminal Code of Pennsylvania, with a Few Observations on the Penitentiary System* (Philadelphia: Kimber and Sharpless, 1826), 22.

6. Benjamin Rush, *An Inquiry into the Effects of Public Punishments upon Criminals and upon Society* (1787); also cited in Teeters and Shearer, *The Prison at Philadelphia*, 7 n. 7.

7. Vaux, *Notices*, 10.

8. For a list of the charter members, see Barnes, *The Evolution of Penology*, 82 n. 18.

commissioned by it to write a defense of the system of solitary confinement that was later published in 1829:

In this den of abomination, were mingled in one revolting mass of festering corruption, all the collected elements of contagion; all ages, colours, and sexes, were forced into one horrid, loathsome communion of depravity. Children committed with their mothers, here first learned to lisp in the strange accents of blasphemy and execration: young, pure and modest females, committed for debt, here learned from the hateful society of abandoned prostitutes, (whose resting places on the floor they were compelled to share) the insidious lessons of seduction. The young apprentice in custody for some venial fault, the tyro in guilt, the unfortunate debtor—the untried and sometimes guiltless prisoners, the innocent witnesses, detained for their evidence in court against those charged with crimes—were associated with the incorrigible felon, the loathsome victim of disease and vice, and the disgusting drunkard, (whose means of intoxication were furnished unblushingly by the jailer!) Idleness, profligacy and widely diffused contamination, were the inevitable results. The frantic yells of bacchanalian revelry; the horrid execrations and disgusting obscenities from the lips of profligacy; the frequent infliction of the lash; the clanking of fetters; the wild exclamation of the wretch driven frantic by desperation; the ferocious cries of combatants; the groans of those wounded in the frequent frays, (a common pastime in the prison,) mingled with the unpitied moans of the sick, (lying unattended and sometimes destitute of clothes or covering;) the faint but imploring accents for sustenance by the miserable debtor, cut off from all means of self-support, and abandoned to his own resources, or to lingering starvation; and the continual, although unheeded complaints of the miserable and destitute, formed the discordant sounds heard in the *only* public abode of misery in Philadelphia, where the voice of hope, of mercy, of religion, never entered.[9]

9. George W. Smith, *A Defence of the System of Solitary Confinement of Prisoners Adopted by the State of Pennsylvania with Remarks on the Origin, Progress, and Extension of*

One of the major ways the Society sought to, and indeed did, influence public opinion and the state legislature was through memorials, which were resolutions drawn by the Acting Committee and approved by the Society. After a January 1788 meeting of approval, the first memorial was sent to the legislature that both protested public punishments and, more important, suggested a substitute: "punishment by more *private* or even *solitary* Labour, would more successfully tend to reclaim the unhappy Objects. . . ."[10] In the language of these memorials, prisoners are commonly referred to as "unhappy objects," "our erring brethren," "unhappy creatures." In the second memorial, dated 12 January 1789, the Society clearly stated its position that remained unswerving for many decades:

> the committee think it their duty to declare that from a long and steady attention to the real practical state, as well as theory of prisons, they are unanimously of the opinion that solitary confinement to hard labour and a total abstinence from spiritous liquors will prove the most effectual means of reforming these unhappy creatures, and that many evils might be prevented by keeping the debtors from the necessity of associating with those who are committed for trial as well as by a constant separation of the sexes.[11]

The petition had its effect, for on 27 March 1789 the legislature passed an act that made the Walnut Street Jail the reception center for the most serious offenders from all over the Commonwealth. The act caused hardships on distant counties and was difficult to

---

This Species of Prison Discipline (Philadelphia: Philadelphia Society for Alleviating the Miseries of Public Prisons, 1829; reprinted 1833), 11 (emphasis in original).

10. Philadelphia Society for Alleviating the Miseries of Public Prisons, Memorial No. 1, *A Protest against Public Punishments* (Philadelphia, 29 January 1788); reprinted in Teeters, *They Were in Prison*, 447–48 (emphasis in original).

11. Philadelphia Society for Alleviating the Miseries of Public Prisons, Memorial No. 2, *Terrible Conditions Existing in the Walnut Street Jail* (Philadelphia, 12 January 1789); reprinted in Teeters, *They Were in Prison*, 451.

implement. It was the act of 5 April 1790, however, that is historically more important, for it specifically designated solitary confinement for "hardened and atrocious offenders" and, as Teeters and Shearer remark, "marks the legal origin of the Pennsylvania System of prison discipline." "In fact," they claim, "it marks the beginning of the penitentiary system in this country."[12] There were some antecedents abroad for the notion of solitary confinement, as Frederick Wines[13] and Thorsten Sellin[14] have documented. Nonetheless, as Wines has remarked:

> Notwithstanding these various premonitions of the coming revolution in prison construction and management, the real foundation of the separation system can hardly be said to have been laid until, in April, 1790, the Legislature of Pennsylvania directed the County Commissioners of the County of Philadelphia to erect, in the yard of the Walnut Street Jail a suitable number of cells six feet in width, eight feet in length, and nine feet in height, which without unnecessary exclusion of air and light, will prevent all external communication, for the purposes of confining there the more hardened and atrocious offenders, who have been sentenced to hard labor for a term of years, or who shall be sentenced thereto by virtue of this act.[15]

The three-storied cell house that was constructed at the Walnut Street Jail became known as the "penitentiary," where in solitude prisoners were expected to do penance. It is still not clear, however, whether Teeters and Shearer exaggerate by claiming that "it can truly be referred to as the first penitentiary in the world for the housing of convicted felons."[16]

12. Teeters and Shearer, *The Prison at Philadelphia*, 10.
13. Frederick Howard Wines, *Punishment and Reformation: A Study of the Penitentiary System*, rev. ed. (New York: Thomas Y. Crowell Co., 1919).
14. Thorsten Sellin, "The Origin of the 'Pennsylvania System of Prison Discipline,'" *Prison Journal* 50 (1970): 13–21.
15. Wines, *Punishment and Reformation*, 151–52.
16. Teeters and Shearer, *The Prison at Philadelphia*, 10.
The author has found some evidence that Le Stinche, a set of prisons opened in

Consensus for a New Prison

Despite the legislation of 1790, implementation of the separate system was faulty. After examining dockets of the jail for 1795, Sellin found that only 4 prisoners out of 117 admissions were part of the time in solitary confinement; for 1796, only 7 out of 139 were separated.[17] Most prisoners were still in physical and social contact, even at night. As early as 14 December 1801, therefore, the Society delivered its Memorial No. 4 to the Senate and House of Representatives of the Commonwealth of Pennsylvania that in effect requested a new prison to try the experiment of a separate system. The phrasing, however, was more subtle, as the conclusion reveals:

> We are therefore induced to request that you will devise such means as may appear to you most adequate to separate the Convicts from all other designations of prisoners, in order that a full opportunity of trying the effects of Solitude and labor may be afforded.—Signed by Direction of the Society by William White, President.[18]

Conditions worsened over the next decade and the minutes of the Acting Committee reveal the growing disturbance, leading to a specific request for a new prison as the only solution:

> The Com^ee on investigating from time to time the State of the Prison, have to report that Some of the true principles of the Penitentiary System, viz. Classification, Solitary Confinement,

---

1300 in Florence, Italy, may be the major precursor of the penitentiary. See Marvin E. Wolfgang, "A Florentine Prison: Le Carceri della Stinche," *Studies in the Renaissance* 7 (1960): 148–66.

17. Thorsten Sellin, "Philadelphia Prisons of the Eighteenth Century," *Transactions of the American Philosophical Society* 43, part 1 (1953): 329.

18. Philadelphia Society for Alleviating the Miseries of Public Prisons, Memorial No. 4, *Request for Means to Try Separation of Prisoners* (Philadelphia, 14 December 1801); reprinted in Teeters, *They Were in Prison*, 452–53.

and constant employ, are wholly unattainable in the present crowded State of the house.[19]

The Com[ee] see with regret the crowded State of the Prisons & the great want of employment, furnishing opportunities for the prisoners to collect together in companies about the yards & workshops having a tendency to demoralize each other, and afford opportunities to plan for their escape.[20]

The Committee cannot help but think, there is much reason to anticipate similar disturbances while the prison remains in its present crowded State & the prisoners without employment. The only remedy they conceive, is to build a new prison more commodious & better calculated for a Penitentiary.[21]

The design of the penal law of Pennsylvania was the reformation of the criminal and the law expressly Stated that Solitary confinement as far as practicable with hard labour Shall be resorted to.[22]

A Pennsylvania Legislative Committee on the Penitentiary System reported on 27 January 1821 that construction of a new prison was seen as necessary even in 1803 and again in 1818. Among other items, the committee reported:

On the 2d of April, 1803, an act was passed, entitled "*An act to direct the sale of certain unimproved city lots, the property of the commonwealth, in the city of Philadelphia, and to appropriate the proceeds thereof towards the erection of a building for the purpose of more completely carrying into effect the penal laws of this State.*"
    The importance and necessity of having one or more prisons constructed upon the plan of *solitary confinement*, becoming every year more evident to the legislature, an act was passed

19. Philadelphia Society for Alleviating the Miseries of Public Prisons, *Minutes of the Acting Committee*, 10 January 1820, 93.
20. Ibid., 10 April 1820, 95.
21. Ibid., 96.
22. Ibid., 8 January 1821, 102.

the 3^{d} day of March, 1818, authorising the inspectors of the prison . . . to make sale of the Walnut Street prison and the lot upon which it is erected, and to apply the proceeds thereof to the erection of a new prison for the use of the city and county of Philadelphia, to be constructed on the plan of solitary confinement.[23]

Although there were many arguments presented in favor of solitary confinement, by 1821 no true experiment had yet been tried. The legislative report admits this lack but provides one of the most succinct statements of support for the system:

The effects of a pure solitary system have never yet been exhibited in Pennsylvania. Enough has been seen, however, to justify the belief that its effects will be to reform entirely or to deter from the commission of a second offence within the jurisdiction of that state where such a system exists. As man is a social being, whose rights and comforts are protected and cherished by the laws of society, it would seem but reasonable when he violates those laws that he should suffer in that point in which he will feel the most keenly, the loss of social enjoyment. To be shut up in a cell for days, weeks, months and years, alone, to be deprived of converse with a fellow being, to have no friendly voice to minister consolation, no friendly bosom on which to lean or into which to pour out sorrows and complaints, but on the contrary, to count the tedious hours as they pass, a prey to the corrodings of conscience and the pangs of guilt, is almost to become the victim of despair. To a guilty mind, reflection and self examination are painful, but frequently prove salutary. The young offender cannot become more expert in crime while shut up in his cell, because he will want an instructor, and if he is not reformed, he will not be worse than when he entered. The old offender will be out of his element, so unable to plot escapes or future mischief, that he

23. Pennsylvania Legislative Committee on the Penitentiary System, *Report on the Penitentiary System* (Harrisburg, 1821), 4–5.

will gladly fly from a region where conviction produces so much misery.[24]

Thereafter, the conflict between the Pennsylvania–separate and the Auburn–congregate systems grew. The Society's Memorial No. 8 of 22 January 1821, strongly urges the erection of a new prison "in which the benefits of solitude and hard Labour may be fairly and effectively provided." Memorial No. 9, entitled "A Further Appeal for the New Prison" and dated December 1827, reported: "One of the evils of which the Prison Society became painfully sensible, was the pernicious consequences arising from the indiscriminate congregation of all kinds of offenders." Influenced by Memorial No. 8, the state legislature in an act of 20 March 1821, appropriated $100,000 for the erection of the prison in Philadelphia, later to be known as the Eastern State Penitentiary. John Haviland, the architect, had already designed the penitentiary at Pittsburgh under the rubric of *solitary confinement with no labor*, and in July 1826 it was ready for its first prisoners. Not until October 1829 would Haviland's true separate system prison be ready.

The Conflict over Labor

The members of the Prison Society apparently had consensus about solitary confinement and opposition to the Auburn system. At least the minutes of the Society show no dissent on this issue. Some disagreement existed over whether the separate system should permit labor. Before the legislature passed its final major act that designated the philosophy for the new prison, there was vigorous, sometimes acrid, debate, which the legislators had before them in the form of formal testimony, reports of commissions, memorials of the Society, and correspondence from European scholars.

24. Ibid., report of the Board of Inspectors of the Prison of the City and County of Philadelphia, Peter Miercken, President; Thomas Bradford, Secretary, 9.

There was a three-pronged attack. A Board of Commissioners was set up by the legislature in March 1826 to revise the penal code.[25] Three jurists, Charles Shaler, Edward King, and T. J. Wharton, formed the commission. Their long and thoughtful report, filed on 20 December 1827, recommended that both the Philadelphia and Pittsburgh penitentiaries be altered in construction to conform to the Auburn system,[26] much to the dismay of the Prison Society. A second commission, the Building Commissioners, appointed to supervise construction of the Philadelphia prison, was requested by the legislature on 14 April 1827 to report its view on the penal code and on the form of prison discipline. The Building Commissioners filed their report on 4 January 1828, saying: "It is proper here to remark, that the solitary confinement we recommend, is *absolute, without any employment*, except the study of the holy scriptures, connected with affectionate religious instruction; we say, *without any employment*, because less time will be requisite to produce a beneficial result on the mind of the prisoner unemployed than when employed."[27] The Prison Society, as Teeters and Shearer report in more detail,[28] was divided on the issue of labor versus no labor, although they agreed on solitary confinement. By January 1828, when the Building Commissioners filed, nearly three cell blocks of the penitentiary had already been completed, all under the design for solitary confinement. The major issue now seemed to be that of labor.

25. Pennsylvania Legislature, Resolution of the Assembly of 1825–26, *Report of the Commission on the Penal Code* (Harrisburg, 1828), 413; also cited by Teeters and Shearer, *The Prison at Philadelphia*, 20 n. 22.

26. Charles Shaler, Edward King, and T. J. Wharton, *Report on the Criminal Code* (Philadelphia, 20 December 1827), in *The Register of Pennsylvania*, edited by Samuel Hazard (Philadelphia, 8 March 1828), vol. 1, no. 10, 1–149.

27. *Report of the Commissioners Appointed to Superintend the Erection of the Eastern Penitentiary, near Philadelphia on the Penal Code* (Philadelphia, 4 January 1828), in *The Register of Pennsylvania*, edited by Samuel Hazard (Philadelphia, April 1828), vol. 1, no. 16, 262 (emphasis in original).

28. Teeters and Shearer, *The Prison at Philadelphia*, 22–23.

Many of the voices in the argument had already appeared in letters, pamphlets, and essays. The opinions were logically, carefully, rationally (usually) presented by prominent persons here and abroad. For example, Edward Livingston, a defender of the separate system, jurist of New York State and later of Louisiana, raised the importantly debated questions: "Solitude and Labour, then are the two great remedies. How are they to be employed? Is the confinement to be a rigid, unbroken solitude, or only a seclusion from the corruption of evil counsel and example? Is it to be permanent for the whole term of the sentence, or to be mitigated by proofs of industry or amendment? Is the Labour to be forced or voluntary, and is its principal object pecuniary profit to the State or the means of honest support to the convict? These are the great questions to be decided. . . ."[29]

When describing a prison classification scheme, Livingston, who wrote the Louisiana penal code which was not adopted but was praised by Jefferson,[30] denounced the congregate system because "the professors of guilt are suffered to make disciples of those who may be comparatively ignorant,"[31] and after many subdivisions of classes, finally says: "We come, then, to the conclusion that each convict is to be separated from his fellows."[32] Livingston did favor labor, however: "First he must know and feel the unmitigated punishment. . . . He must live on the coarse diet allowed to the unemployed prisoner; he must suffer the tedium coming from want of society and of occupation; and when he begins to feel that labour would be an indulgence, it is offered to him as such . . . in order to produce reformation."[33]

29. Edward Livingston, *Introductory Report to Code of Prison Discipline for Louisiana* (Philadelphia, 1827), 9–10.
30. Teeters and Shearer, *The Prison at Philadelphia*, 25.
31. Livingston, *Introductory Report*, 10.
32. Ibid., 11.
33. Ibid., 53. In a letter to Roberts Vaux, 25 October 1828, Edward Livingston elaborates further on his proposal for labor in solitude:

William Roscoe was a noted English scholar and penal reformer from Liverpool who corresponded with Roberts Vaux in Philadelphia. Roscoe opposed the slave trade and solitary confinement. He condemned the system in general because he assumed there was no labor, based on the Building Commissioners' report. He compared solitary confinement with the Bastille of France and spoke of the system as "the most inhumane, and unnatural, that a tyrant ever invented."[34]

Lafayette was reported in the *Third Annual Report* of the Boston Prison Discipline Society, in 1828, as having expressed earlier his firm opposition to complete solitary confinement:

The opinion of Lafayette is thus expressed in a letter, dated August, 1825, to a gentleman in England:

"As to Philadelphia," says the general, "I had already, on my visit of the last year, expressed my regret, that the great expenses of the new Penitentiary building had been chiefly calculated on a plan of solitary confinement. This matter has lately become an object of discussion. A copy of your letter, and my own observations have been requested; and as both opinions are actuated by equally honest and good feelings, as solitary confinement has never been considered but with a view to reformation, I believe our ideas will have their weight with men, who have been discouraged by late failures of success in the reformation plan. It seems to me two of the inconveniences most complained of might be obviated, in *making use*

As opposed to this system [Auburn] I have ventured to propose one based upon labour in seclusion; as a relief from seclusion without labour; succeeded gradually by instruction, and labour in classified society; labour not coerced, but granted as a favour; and instruction given as the reward of industry and good conduct, not enforced as a task. Whether your opinion or mine agree as to those details I know not, but I am sure we do in the utility of seclusion, accompanied by moral, religious, and scientific instruction, and usual manual labor.

34. William Roscoe, *A Brief Statement of the Causes Which Have Led to the Abandonment of the Celebrated System of Penitentiary Discipline in Some of the United States of America* (London, 1827); cited in Teeters and Shearer, *The Prison at Philadelphia*, 27.

of the solitary cells to separate the prisoners at night, and *multiplying the rooms of common labor,* so as *to reduce the number in each room to what it was, when the population was less dense;* an arrangement which would enable the managers to keep distinctions among the men to be reclaimed, according to the state of their morals and behavior."

And again, as expressed to an American gentleman, in a letter from him dated Sept. 1826:

"The people of Pennsylvania think, said he, that the system of solitary confinement is a new idea, a new discovery;—not so—it is only the revival of the system of the Bastile. The State of Pennsylvania, which has given to the world an example of humanity, and whose code of philanthropy has been quoted and canvassed by all Europe, is now about to proclaim to the world the inefficacy of the system, and revive and restore the cruel code of the most barbarous and unenlightened age.—I hope my friends of Pennsylvania will consider the effect this system had on the poor prisoners of the Bastile. I repaired to the scene, said he, on the second day of the demolition, and found that all the prisoners had been deranged by their solitary confinement, except one;—he had been a prisoner twenty-five years and was led forth during the height of the tumultuous riot of the people, whilst engaged in tearing down the building. He looked around with amazement, for he had seen nobody for that space of time, and before night he was so much affected, that he became a confirmed maniac, from which situation he never recovered."[35]

James Mease was a prominent Philadelphia physician who favored solitary confinement without labor. In 1811, in *Pictures of Philadelphia,* he suggested subjection of "criminals to solitary confinement, both by day and night, without permitting them to labour."[36] In 1828 he wrote in *Observations on the Penitentiary System and*

---

35. Boston Prison Discipline Society, *Third Annual Report* (Boston, 1828), 40–41 (emphasis in original).

36. James Mease, *Pictures of Philadelphia* (Philadelphia, 1811), 27; cited by Teeters and Shearer, *The Prison at Philadelphia,* 28.

Penal Code of Pennsylvania with Suggestions for Their Improvement:

> I refer to the principles upon which avoidance of crime is founded, or repentance ever has, or ever can be brought about in a human being; these are, 1. *Ennui* or a tiresome state of mind from idle seclusion; and 2d. Self condemnation, arising from deep, long, continued, & poignant reflections upon a guilty life. All our endeavors therefore ought to be directed to the production of that state of mind, which will cause a convict to concentrate his thoughts upon his forlorn condition, to obstruct himself from the world, and to think of nothing except the suffering and privations he endures, the result of his crimes. Such a state of mind is totally incompatible with the least mechanical operation, but is only to be brought about, (if ever), *by complete mental and bodily insulation.*[37]

Roberts Vaux, along with Benjamin Rush, was the most prominent member of the Prison Society from its inception. His mercantile family provided him with independent means and, as a member of the Society of Friends, he was a major Philadelphia philanthropist and defender of the separate system of prison discipline. He was a prolific writer on the topic of penal reform, a leading spokesman for the Prison Society, and was appointed by the governor as a commissioner to erect the Eastern State Penitentiary. He was clearly in favor of the separation system, as is expressed in a letter written 21 September 1827, to William Roscoe in England: "Intercourse in prison defeats the claims of justice and the wholesome ends of punishment—it degrades by exposing culprits to the observation of each other, and proclaims the common infamy of their fallen condition—it makes, in reality, little, if any discrimination between offences—it banishes hope—it hardens the heart, and is calculated to quench the last spark of desire for amendment

---

37. James Mease, *Observation on the Penitentiary System and Penal Code of Pennsylvania with Suggestions for Their Improvement* (Philadelphia: Clark and Roger, 1828), 73 (emphasis added).

of life. For these reasons it ought to be regarded as the most cruel and certain exercise of power, to increase and perpetuate every form of wickedness & misery."[38] Vaux was originally in favor of solitude without labor, but vacillated and finally allowed that labor might be reformative.

Gustave de Beaumont and Alexis de Tocqueville reported in *The Penitentiary System in the United States*, in 1833, that the separate system had been tried at the Auburn prison in New York. The prison was erected in 1816, had one cell block of eighty separate cells in 1819, and was occupied in 1821. They spoke of the failure of the experiment at Auburn:

> This trial, from which so happy a result had been anticipated, was fatal to the greater part of the convicts: in order to reform them, they had been submitted to complete isolation; but this absolute solitude, if nothing interrupt it, is beyond the strength of man; it destroys the criminal without intermission and without pity; it does not reform, it kills.
>
> The unfortunates, on whom this experiment was made, fell into a state of depression, so manifest, that their keepers were struck with it; their lives seemed in danger, if they remained longer in this situation; five of them, had already succumbed during a single year; their moral state was not less alarming; one of them had become insane; another, in a fit of despair, had embraced the opportunity when the keeper brought him something, to precipitate himself from his cell, running the almost certain chance of a mortal fall.[39]

Such was the trouble with no labor allowed. And such was the reason that Dr. Franklin Bache, physician to the penitentiary in Philadelphia, advocated *work within solitary confinement*:

38. Roberts Vaux, *Reply to Two Letters of William Roscoe, Esquire, of Liverpool, on the Penitentiary System of Pennsylvania* (Philadelphia, 1827), 11–12.

39. Gustave de Beaumont and Alexis de Tocqueville, *The Penitentiary System in the United States* (Philadelphia, 1833), 5; quoted by Teeters, *They Were in Prison*, 187.

In regard to labour in connexion with separate confinement, it may be interesting to inquire, 1st, whether it would be proper; and if proper, 2ndly, whether it would be practicable. This appears to me to be the most difficult point of the inquiry and one respecting which my mind has been most wavering & unsettled. I incline, however, to the opinion, that as a general rule, the prisoner should be engaged in some useful employment in his cell, to be withdrawn at the discretion of the principal of the prison, when necessary for the enforcement of discipline. I would advocate this course, not with a view to the productiveness of the labour, but because it seems important that a habit of industry should be formed, if possible, during the prisoner's confinement as this may have a favourable influence on his future conduct when discharged.[40]

As the *Second Annual Report* of the Boston Prison Discipline Society pointed out in 1827, however, "in regard to labor, it is not yet decided whether it shall be introduced or not. If it is introduced, what kind it shall be; who shall teach how the work is to be done. . . ."[41] Francis Gray, writing in 1847, upon reviewing the legislative history, pointed out that the Eastern State Penitentiary "was originally intended . . . for solitary confinement without labor" but that when convicts were received in 1829 there was a provision "for the introduction of solitary labor."[42]

Before the state legislature and the governor made a decision about solitary confinement versus the congregate system, an important distinction was made between the separate system and solitary confinement. The Auburn system kept prisoners silent and in congregate labor during the day, but the warden exercised corporal punishment, flogging, the lash, which the Prison Society, especially through the ninety-four-page essay of George W. Smith,[43]

40. Franklin Bache, *Letter to Roberts Vaux* (Philadelphia, 13 March 1829), 10.
41. Boston Prison Discipline Society, *Second Annual Report* (Boston, 1827), 76.
42. Francis C. Gray, *Prison Discipline in America* (Boston: Charles C. Little, 1847; Montclair, N.J.: Patterson Smith, 1973), 32.
43. Smith, *A Defence of the System.*

quickly dismissed as unethical and unequal treatment. Was there, however, a distinction between Lafayette's solitary confinement in the Bastille and the Pennsylvania separate system? The best descriptive difference came sixteen years after the Eastern State Penitentiary was opened, from an English writer, Joseph Adshead:

What Is Separate Confinement?

It is totally different in its nature from solitary confinement. It differs from it in the following particulars: In providing the prisoner with a large well-lighted and well ventilated apartment, instead of immuring him in a confined, ill-ventilated and dark cell;—in providing the prisoner with every thing that is necessary for his cleanliness, health and comfort during the day, instead of confining him to bread and water;—in alleviating his mental discomfort by giving him employment;—by the regular visits of the Officers of the prison, of the Governor, Surgeon, Turnkeys or Trades' Instructors, and particularly of the Chaplain, instead of consigning him to the torpor and other bad consequences of idleness, and the misery of unmitigated remorse, resentment, or revenge;—in separating him from none of the inmates of the prison except his fellow prisoners, instead of cutting him off, as far as may be from the sight and solace of human society;—in allowing him the privilege of attending both chapel and school, for the purpose of public worship and education in class (securing on those occasions his complete separation from the sight and hearing of his fellows) instead of excluding him from the Divine Service and instruction;—in providing him with the means of taking exercise in the open air, whenever it is proper and necessary, instead of confining him to the unbroken seclusion of his cell. The object of Separate Confinement is the permanent moral benefit of the prisoner—an object which he can plainly see that the system has in view.[44]

There was another issue raised by those who advocated separate

44. Joseph Adshead, *Prisons and Prisoners* (London, 1845), 8; quoted by Teeters, *They Were in Prison*, 190.

confinement without labor. They contended that confinement without employment was more severe and could produce a similar reformation in a shorter time than separate confinement with labor. In the literature of the time, interesting ratios were suggested that have not previously been brought together.

For example, the Board of Inspectors of the prison in Philadelphia in 1821, calling for solitary imprisonment without labor in the county jails until the state penitentiary could be built, suggested a ratio of one year in solitary confinement *without* labor as equivalent to three years in solitary confinement *with* labor. Their report, in part, reads as follows: "Three years is now the usual sentence for grand larceny. Under the solitary system one year will produce more and better effects. It costs ninety dollars per annum, for the support of a prisoner, without work, on the lowest calculation. Three years, at 90 dollars a year is $270.00; one year at 90 dollars, 90.00; gain on each prisoner, $180.00."[45] The reasoning was that "employment diminishes in a very large degree the tediousness of confinement and thus mitigates the punishment, [thus] it may be a question whether labour ought not to be abandoned altogether, except as an *indulgence* to penitent convicts and as a relaxation from the much more painful task of being compelled to be idle."[46] This report was signed by Peter Miercken, President, and Thomas Bradford, Secretary.[47]

45. Pennsylvania Legislative Committee, *Report on the Penitentiary System*, 9–10.
46. Ibid.; cited also by Teeters and Shearer, *The Prison at Philadelphia*, 20.
47. Pennsylvania Legislative Committee, *Report on the Penitentiary System*, 20.
It was not until the Eastern State Penitentiary was functioning for over a year that there appear reports of the kinds of labor that were performed by prisoners in their solitary cells. Barnes, *The Evolution of Penology*, 169 n. 323, has the most detailed description:
The industrial distribution of the Eastern Penitentiary down to 1835 was as follows:
"In 1831 there were 87 convicts in custody in the institution; 43 were engaged in weaving and dyeing; 18 in shoemaking; 4 in blacksmithing; 3 in carpentering; 2 in carving; 2 in lockmaking; 2 in wool-packing; and 1 each in

In his 1827 *Letter on the Penitentiary System of Pennsylvania*, addressed to William Roscoe, the English critic of the system, Roberts Vaux wrote: "*By separate confinement*, other advantages of an economical nature will result; among these may be mentioned a great reduction of the terms of imprisonment; for instead of from three to twenty *years*, and sometimes longer, as many *months*, except for very atrocious crimes, will answer all the ends of retributive justice and penitential experience. . . ."[48]

In Franklin Bache's letter to Roberts Vaux, 13 March 1829, he suggests "that prisoners on this [the Pennsylvania] system are supposed to require a shorter sentence for any given crime, than on the prevailing system a circumstance which will cause a very considerable saving of expense."[49] Earlier, the Board of Commissioners, reporting on 8 January 1829, which recommended solitary confinement without labor, wrote: "We say *without* any employment, because *less time* will be requisite to produce a beneficial result on the mind of the prisoner unemployed than when employed."[50] In his 1828 essay, James Mease suggested a one-to-ten ratio: "I will venture to say that one year passed in this way [soli-

---

carriage-making, tailoring, cooking and washing." *Senate Journal*, 1831–32, Vol. II, p. 447.

"In 1832, with a total of 97 confined, 43 were employed in weaving and dyeing; 32 in shoemaking; 5 as blacksmiths; and 4 as carpenters. The remainder were distributed in sundry minor occupations and in the domestic service." *Senate Journal*, 1832–33, Vol. II, p. 515.

"In 1833 there was a total of 154 convicts confined. Out of this number 59 were occupied in weaving, warping, dyeing and spooling; 52 as shoemakers; and five as carpenters. The rest were distributed as above." *Senate Journal*, 1833–34, Vol. II, p. 418.

"In 1834, with a total of 218 incarcerated, 83 were employed in making shoes; 70 in spinning, dyeing, weaving and dressing; and 6 each as carpenters, blacksmiths and 'sewers.'" *Senate Journal*, 1834–35, Vol. II, p. 473.

48. Cited by Sellin, "The Origins of the 'Pennsylvania System,'" 15 (emphasis in original).

49. Bache, *Letter to Roberts Vaux*, 11.

50. *Journal of Prison Discipline and Philanthropy* 1 (January 1845), 10 (emphasis in original).

tary confinement] would have more effect upon criminals, than ten years passed in the continual society of numerous fellow convicts, where reflection is prevented by the bustle of the work in the day, and drowned at night by idle or wicked conversation."[51]

Although almost all parties concerned with the labor/no labor issue believed, without historical test, that the separate system without labor would cause earlier repentance and reformation than work in a singular cell, most writers, including the majority of the Prison Society, favored labor in solitude. The controversy was brought to a conclusion by the Pennsylvania act of 23 April 1829 that firmly stipulated "separate or solitary confinement at labour," just six months before Charles Williams entered the Eastern State Penitentiary as its first prisoner.

Epilogue

There were other conflicts concerning the architectural plans for the new prison at Twenty-First Street and Fairmount Avenue, known in 1829 as Cherry Hill. The English architect, John Haviland, who became a citizen of the new country, won the competition over William Strickland, a well-established Philadelphia architect. That interesting story requires a special study.[52]

Concern with the conflicts that were to be resolved through the experiment in the new state prison was still reflected in the *First and Second Annual Reports* of the Eastern State Penitentiary. Charles S. Coxe, president of the Board of Inspectors, cautiously wrote in the *First Annual Report*: "Indeed, although nothing seriously militating against the system has yet occurred within their observation, and much that is favourable has struck them, they would rather suffer a longer time to elapse, and await the operation of the institution upon a larger body of prisoners, before they should feel them-

51. Mease, *Observation on the Penitentiary System*, 17.
52. Teeters and Shearer, *The Prison at Philadelphia*, 33–53.

selves authorized to express to the legislature a decided judgment; for, or against, the system of solitary confinement with labour."[53]

History beyond the *First and Second Annual Reports* of the new prison is a record of slow deterioration of the Pennsylvania separate system in Philadelphia. At the same time, however, the system of prison discipline and John Haviland's architecture became a model for cultural diffusion throughout Europe and Asia.[54] During the rest of the nineteenth century, the Eastern State Penitentiary surely became the most famous prison in the world.

In the 1980s prisons in the United States are vastly overcrowded and many lawsuits have been presented to the courts to reduce this condition. The voices of Benjamin Rush, Roberts Vaux, George W. Smith, and others from the eighteenth and early nineteenth centuries are echoed today. The separate "apartment" (a lovely euphemism of that time) may be the better solution and is often the request of prisoners and prison reformers. Under our present prison system, there are rapes, sodomies, robberies, and more homicides in prison for some prisoners than occur outside of prison.

History suggests that human intentions often produce inhumane consequences. Charles Dickens's *Notes* from his visit to the Eastern State Penitentiary on 8 March 1842 seem to say so.[55] But was he

53. Pennsylvania Legislature, *First and Second Annual Reports of the Inspectors of the Eastern State Penitentiary*, made to the Legislature at the Sessions of 1829–30 and 1830–31 (Philadelphia, 1831), 5.

54. Norman Johnston, *The Human Cage: A Brief History of Prison Architecture* (Philadelphia: American Foundation, Inc., Institute of Corrections; New York: Walker & Co., 1973).

55. Charles Dickens, in *American Notes for General Circulation* (London: Chapman and Hill, 1842), 119–20, wrote about the Pennsylvania separate system after his visit:

In its intention I am well convinced that it is kind, humane and meant for reformation; but I am persuaded that those who devised the system and those benevolent gentlemen who carry it into execution, do not know what it is they are doing. . . . I hold this slow and daily tampering with the mysteries of the brain to be immeasurably worse than any torture of the body; and because its ghastly signs and tokens are not so palpable to the eye and sense of touch as

right? The Prison Society challenged him. The Pennsylvania system of solitude and separateness allows for privacy but also for visits from positive, constructive outside members of society although there was no social intercourse with other inmates.

Put in the best philosophic posture of the eighteenth century, should inmates of today be offered private apartments, with visits and contacts from friends, relatives, counselors, potential employers, but no interaction with other prisoners, who may assault, rape, or kill them? Were George Smith, James Mease, Roberts Vaux unreasonable? The congregate prisons of today are probably more criminogenic than they are reformative.[56] Congregation still produces contamination. A review of the Pennsylvania system probably deserves some of our current attention for developing future penal policy.

---

scars upon the flesh, because its wounds are not on the surface, and it extorts few cries that human ears can hear; therefore I denounce it as a secret punishment.

56. Still, there is concern about solitude in prison. It is more than a curious note that a federal district judge, at the time of this writing, expressed his concern about confinement in idleness:

A federal district judge has ruled that Tennessee's death row is a "prison within a prison" and treatment of inmates there amounts to cruel and unusual punishment [the Eighth Amendment].

The judge, John T. Nixon, said Friday that he would appoint a special master to work with the state on improvements at Unit VI of the Tennessee State Prison.

"The principal concern of the court is that the *inmates remain idle and confined in their small cells for so much time*," said the judge, who made a surprise inspection on April 19 of the unit.

The judge ordered the state to submit a plan within 90 days of the appointment of the special master, on how the problems would be corrected (*New York Times*, 26 May 1985) (emphasis added).

Cotton Textiles and Industrialism

Thomas C. Cochran

IN the last five centuries energy from water, wood, coal, or oil has
far more than ever before been applied to making what people
want and operating what they use. Probably the most fundamental
advances were in the use of waterpower in early modern Europe
and the gradual evolution of steam as a source of power from the
late seventeenth to the nineteenth century. Accompanying the
early advances in using natural sources of power went advances in
designing mechanical devices ranging from improved ships to extraordinarily complex clocks. Better houses, better utensils for
homes and fields, cheaper cloth and better transportation all developed over the centuries from mechanical improvements.

English historians of the last century selected the improved spinning of cotton fibers for cloth-making and improved steam engines
in the late eighteenth century as revolutionary elements in this
gradual development. Because England was relatively lacking in
waterpower and wood, much of the new spinning machinery,
specifically the Arkwright frame, came to be operated by coal
burning steam engines. Some of the steam engines used were of an
improved type, associated with the name of James Watt of England, but engines of the older Newcomen type remained the chief
source of steam power.[1] This combination of steam-powered spin-

---

1. John Fitch and Henry Voight had a Watt-type engine made in Philadelphia
in 1786 although as Fitch wrote in his 1790 patent application: "not a person could
be found who was familiar with the minutiae of Boulton and Watt's new engine."
James Flexner, *Steamboats Come True: American Inventors in Action* (New York: Viking

ning mills was called in 1884, by English historian Arnold Toynbee, "The Industrial Revolution": The increase in the use of cotton cloth both domestically in England and for export to the world was in itself truly revolutionary from about 1790 to 1830, but this particular "revolution" was in only one product, not in the broad range of the utilities of life. Why cotton thread that to about 1815 had to be handwoven into cloth should be more important than waterpowered woodworking machines, better stoves, or factory made nails was not discussed. In 1984 David Landes argued that the clock makers were the leading mechanicians and pace setters of advancing industry in the period before the mid-eighteenth century. If the unquestioned acceleration in the use of water and steam power from 1750 to 1850 is to be regarded as a "revolution," perhaps the most revolutionary leaders would be the machine tool makers, chiefly in Great Britain and the United States, and their most "revolutionary" period of advance would be about from 1800 to 1830.

To explain these exaggerations in historiography, it should be realized that there was a dramatic quality to the English steam textile factory altogether lacking in the waterpowered textile, flour, saw, or paper mill. The early cotton spinning mills still needed large quantities of unskilled labor, including that of children; hence a crowded factory. The English scene was made dramatic by grime and soot from steam engines burning soft coal in crowded cities, and horrendous noise from machinery run by gear wheels. Few could be moved to either awe or consternation by a lonely flour, paper, or saw mill using leather belting instead of gear wheels and turning out large quantities of products in a quiet country setting. But the "dark satanic" mills of Leeds or Manchester were something new and alarming to the quiet world of waterpower and

---

Press, 1944), 103. See also: Frank D. Prager, ed., *The Autobiography of John Fitch* (Philadelphia: American Philosophical Society, 1976). In 1786 Fitch and Voight designed a 3″ double acting cylinder, p. 169.

handwork; the obnoxious mills themselves more than their product were revolutionary. Similarly the waterwheel, even if very large, was a part of the world's peaceful, natural heritage; the smokey, chugging steam engine and the grinding gears seemed something risen from the infernal regions.

A balanced historical view at the present day indicates that with increasing world trade and migration in the eighteenth century, the northwestern countries of Europe and their colonial areas in North America all increased their degrees of mechanization. Where there were feudal efforts to preserve the status quo, those of urban guilds to preserve hand manufacturing monopolies, or a powerful aristocratic society that looked down on trade or manufacture, change was slow; where a business, entrepreneurial middle class was stronger as in England progress was more rapid. In America, where nearly a half century of British restriction on manufacturing was suddenly ended by revolution, it was, after 1781, most rapid of all. As Adam Smith wrote in his *Wealth of Nations* "Though North America is not so rich as England, it is much more thriving and advancing with much greater rapidity to the further acquisition of riches."[2]

If the misleading phrase "Industrial Revolution" is still to be used, it might better be applied to both Britain and the United States from 1780 to 1815. The seacoast areas of the mid-northern British colonies in America had reached a stage of economic maturity by 1750 where they would soon manufacture a large part of their own goods. In fact, left to itself, the economic situation would probably have evolved toward a mutual interchange on the basis of comparative advantage.[3] Although the colonies would have remained exporters of raw material, Connecticut and Pennsylvania rifles might, for example, have gone to England in exchange for broadcloth and worsteds. As James Henretta says, "by 1775, the

2. (London: Cannon Edition), 91.
3. See: Edwin J. Perkins, *The Economy of Colonial America* (New York: Columbia University Press, 1980).

commercial expertise, artisan skills, and capital resources for domestic industrial development had been generated in the northern cities and their surrounding areas."[4] He also notes that a one-third increase in export wheat prices between 1750 and 1770 stimulated manufacturing in the middle states. Alice H. Jones concludes that "in financial assets the sturdy Middle Colonies were outstanding."[5]

These considerations alter the historical picture of a rapid transfer of technological knowledge from England to the United States in the decades immediately following the Revolution. The basic transfer had been gradual and in the form of the embodied skills and in the interests of immigrants over the whole colonial period. Not only artisans from the British Isles, but German, Dutch, and French craftsmen brought their skills to America. Challenged by a rich, but demanding environment, they developed new ways of doing things. In this sense there never was an "Industrial Revolution" in America, but rather a continuous growth. While by 1775 there was still an American need for artisan-mechanics, the essential transfer of basic European skills to America had taken place. Immigrant mechanics from Great Britain continued to reinforce American progress, but they were not essential to the advance. David Jeremy finds that "in the late eighteenth century American carpenters, cabinet makers, wheel wrights, ironmasters, blacksmiths, wire drawers and clock makers . . . could equal most of their British counterparts."[6]

Before discussing the character of expanding industrialization, it is necessary to understand some of the American geographical and social background. At the end of the Revolution, the States were a loose confederation and after 1789 a federation where most of the dealings with business, except for uniform tariffs, were within the

4. *The Evolution of American Society, 1700–1815* (Lexington, Mass., 1973), 70–79.
5. "Wealth and Growth of the Thirteen Colonies," *Journal of Economic History* 44 (1984): 251.
6. "British Textile Technology . . . ," *Business History Review* 47 (1973): 37.

jurisdiction of the state, with local law courts friendly to the active entrepreneur. In the states destined for precocious industrial development this meant easy incorporation for business purposes and state subscriptions to corporate stock, particularly in banking and transportation.

Corporations as devices for gathering capital were much more important to the rather rapidly expanding American states than to the settled nations of Europe. Capitalist Americans tended to be land poor or tied to mercantile ventures. The country lacked the individual accumulations of money such as those of the wealthy English brewers, who often went into private banking. Hence the ability to raise capital from many small subscriptions was a great aid in America, particularly in providing the infrastructure of improved transportation needed for industrialization. The great utility of the legal device to the young United States is shown by the fact that by 1800 this country had more than ten times as many corporations as both England and France.[7]

The regions of the United States that first industrialized were relatively compact, running along the Atlantic Coast from Baltimore, Maryland, to Portsmouth, New Hampshire, well connected by water transportation, and about the same total area as England and Wales. The Northeast seaboard had some advantages that Britain lacked, such as abundant wood and waterpower. Consequently, the absence of soft coal in the regions east of the Susquehanna River and the peculiarities of using anthracite coal were only minor disadvantages. Charcoal smelted and hammered iron was little more expensive at New York or Philadelphia than British imports and was judged by American users such as Mathias Baldwin of locomotive fame to be better for machinery than foreign coke smelted and rolled iron. It was certainly preferred for farm and other metal implements by local blacksmiths.

7. See: Joseph S. Davis, *Essays in the Earlier History of American Corporations*, 2 vols. (Cambridge: Harvard University Press, 1917).

While resources and adequate transportation were essential to the rapid growth of an American industrial economy, they were less nationally distinctive than the social and cultural systems that allowed a large degree of freedom of action to the individual and placed a high value on physical activity and financial success, and a population expanding at an unprecedented rate. A three percent a year increase in population in the early nineteenth century was in itself spectacular, but combined with the fact that immigrants were chiefly adults anxious to buy or build houses and equip farms added to a market demand for construction and crude durable goods probably unknown in world history. The estimate by present-day statistical economists such as Paul David that United States per capita gross national product grew only about half of one percent a year during the early nineteenth century can be misleading compared either to other nations or to modern times. That it grew at all after absorbing a three percent annual increase in people needing transportation and basic equipment for home and farm-making is a mark of high achievement in meeting this growing demand, and also a tremendous stimulant to the improved production of inexpensive durable goods.

Furthermore, in contrast even with Britain, the whole range of social strata of the Northeastern population was drawn into efforts to increase production. Even the great landlords of the Hudson Valley engaged in trade and industry. Robert Livingston of New York, one of the greatest landed "aristocrats," financed the first commercially successful steamboat.[8] A large part of all American artisans apparently devoted their thought to how to do or make things more cheaply. So many men were "inventing" the needed tools and machines that at times the contest for priority, as for example in steamboats, became severe and uncertain. The products of large numbers of little machine shops of 1800 (called blacksmith

8. He was a lineal descendant of the earls of Linlithgow.

shops in the census) were, however, continuously being improved by small changes, made by mechanics supervising their use, in general without secrecy or patent protection.[9]

Factors, largely absent in Britain or Europe, were the continuous immigration of workers of all types to the United States and high domestic fertility together with migration within the country.[10] These resulted in forces strongly favorable to mechanization either invented on the spot or transferred from current Western World knowledge and skills. Most stimulating of all was a steadily increasing demand for cheap durable goods, particularly for new homes and their equipment. Probably increases in demand were more of an incentive to mechanization in important fields such as woodworking than was relatively scarce labor. Britain's lead in transportation by canal in the late eighteenth and early nineteenth century was largely compensated for in America by small canals around rapids in rivers, as on the Susquehanna and the Mohawk, and the use of first the sailing shallop and then the river steamboat.

If no single type of machine production leads inevitably to continuing industrialization, is there any causal element that seems indispensable? Nathan Rosenberg seems to have correctly defined this as "machines for making machines," or in practice a machine tool industry.[11] The developers of boring, milling, planing, and cutting devices were the largely unsung heroes of continuing industrialism. Without John Wilkinson's improved boring device James Watt's steam cylinder would not have worked. The contributions of the machine shops such as those of Nathan Sellers and

9. The patent laws up to 1836 were very poorly drawn and only an inventor whose device was truly unique (a rare thing) had much chance of collecting royalties. The 1800 Census listed over 200 "blacksmith" shops in Philadelphia, which included shops from horseshoeing to making steam engines.

10. See for example: Edward C. Carter II, "Naturalization in Philadelphia 1789–1806," *Pennsylvania Magazine of History and Biography* 94 (1970): 333 ff.

11. See: Nathan Rosenberg, *Perspectives on Technology* (Cambridge: Cambridge University Press, 1976).

his descendants in Philadelphia not only had many "firsts" such as annealing brass and iron and a better straightening board for papermaking, but the Sellers family continued to improve all sorts of devices and spawn nearby shops run by relatives or former employees. In the early 1820s Sellers and Pennock were exporting fire engines to a world market, and young George Escol Sellers rated American heavy shop machinery the most advanced in the world.[12] Even British commissions that examined American technology in the 1850s admitted that America had originated the basic concepts and practices of mass production.[13]

An important factor in preserving the view that America was dependent on Great Britain for technological advance was the difference in communication between the two nations. British knowledge was eagerly embraced by American mechanics, and British workmen frequently emigrated to the United States. In contrast many new American advances were either partly inappropriate to Britain, as with woodworking machinery, or menaced entrenched craft groups as with factory-made nails. Also, few Americans emigrated to the British Isles. Great Britain advanced even in the nineteenth century through more and more division of specialized labor, while America progressed more by machines for eliminating workers. Hence American ideas that would upset labor relations were generally disregarded, even when advanced by an American inventor living in England, such as Jacob Perkins. The net result has been that many American technologists admitted a debt to Great Britain but only a few Englishmen before 1850 recognized any inspirations from America.

The socioeconomic differences between Great Britain and the United States influenced many processes. Toolmaking, for exam-

12. *Early Engineering Reminiscences (1815–1840) of George Escol Sellers*, ed. Eugene S. Ferguson (Washington: Smithsonian, 1965), 112, 126.
13. See: Nathan Rosenberg, ed., *The American System of Manufacturers: The Report on Machinery in the United States, 1855* (Edinburgh: Edinburgh University Press, 1969), 1.

ple, diverged rather widely. Finishing iron or steel parts by milling probably developed at John Hall's rifle plant at Harper's Ferry between 1818 and 1826, but was not generally adopted in England until the 1890s, although some British lathes were probably superior to those in the United States.[14] But such variations in needs and methods did not prevent the general rise of industry in the toolmaking Western World. Regardless of volume of production, advances in machine tools were the firm growing roots of industrialism.

In spite of the necessary underpinning of practical machine tools in the early 1800s, America's unique contributions were in the construction industry and agricultural machinery. More and more woodworking machines were innovated, ranging from the many sawmills of the Colonies on through the cheap machine made nail, the circular saw, the screw auger, and ready cut two-by-fours or other sizes of construction materials of the 1830s. In the census of 1810 there were 2,016 sawmills listed for Pennsylvania which were credited with producing 74,538,640 board feet of lumber.[15] "With their high speed and endless modification the wood machines demanded a higher grade of ingenuity and skill in their construction than machines for cutting and shaping metal," and the United States per capita consumption of wood was many times that of Great Britain.[16] A million houses were built in the United States from 1800 to 1820, and by 1815 Philadelphia had steam driven furniture factories.[17]

14. See: Paul Uselding, "Studies of Technology in Economic History," in Robert Gallman, ed., *Research in Economic History Supplement* (Greenwich, Conn.: Jai Press, 1977), 171–173.

15. James Elliot Defebaugh, *History of the Lumber Industry of America*, 2 vols. (Chicago: The American Lumberman, 1907), 477.

16. Nathan Rosenberg, "America's Rise to Woodworking Leadership," in Brooke Hindle, ed., *America's Wooden Age* (Tarrytown, N.Y.: Sleepy Hollow Restoration, 1975), 40–41.

17. Eugene S. Ferguson, *The Americaness of American Technology* (Wilmington, Del.: Hagley, 1975), 7; for house estimate: and V. Clark, *History of Manufactures in*

The construction industry offers an example of gradual industrialization without a factory system. Lumber reached cities such as Philadelphia by being floated down the Delaware, or came to rural areas and towns from local sawmills. Urban lumber distribution was a wholesale trade, as were shingle and brick manufacture, but from there on the industry was in the hands of "master builders" and subcontracting specialized workers. Industrialization came in the form of improved machines and materials for all processes. Among the most important of these were the machine manufacture of nails, bricks, and shingles after 1800, the circular saw after 1815, and the balloon-frame house from 1833 on. The latter did away with corner posts and heavy framing by using a cage-like construction of two-by-fours nailed to each other. It also greatly decreased the cost of two-story buildings. Thus this major improvement, from the standpoint of costs, depended on American factory-made nails in place of Old World morticing and tenoning of the members of the frame. The application of steam to urban sawmills in the 1820s and 1830s also led to a sharp decrease in the cost of city buildings.

But the main emphasis here is on building construction as offering a continuous demand for not only houses, but all types of house fittings and the machines to make them. While farm families might build for themselves with the aid of a local carpenter, they needed factory-made nails, stoves, metal work, and utensils. In the 1820s and 1830s central heating began to make urban buildings still more dependent on pipes and machinery.[18]

With the surplus of land in the early Republic, Americans were interested in mechanical devices for saving manpower in agriculture. The iron plow was increasingly perfected from 1800 to 1820 by which time the blades had become interchangeable, cheap, and

the U.S. (McGraw Hill, 1929), 1: 473, for use of steam: Robert M. Vogel, "Building in the Age of Steam," in Charles Peterson ed., *Building in Early America: Contributions Toward a History of a Great Industry* (Radnor, Pa.: Chilton, 1976), 121–122.

18. See: Charles E. Peterson, *Building in Early America*, passim.

machine (rather than blacksmith) made. Harrows and hayrakes were developed during the same period, but the agricultural innovation that was to attract the greatest world attention was the mechanical reaper created by Cyrus Hall McCormick in western Virginia in 1831. By 1833 a New Yorker, Obed Hussey, had also developed and patented a reaper. Offering difficulties in use on rolling rocky fields, the reaper did not achieve great national success until McCormick established a factory in Chicago in 1847 and won the business of prairie farmers.

Because of long rivers available for inland transportation such as the Hudson in the East or the great central Mississippi River system, America led in the utilization of the steamboat. One of the paradoxes of economic as distinct from technological history is the fact that the first commercial steamboat was operated on the Delaware River in 1790 from Philadelphia to Trenton, an area that because of ample stagecoach service had no immediate need (demand) for moderate speed steam navigation. John Fitch's only full commercial season in 1790 cost £30 for operation and took in only £20 in fares. If Fitch had built his boat on the Hudson or the Raritan, he would probably have been able to make it pay, just as Robert Fulton with a much bigger but slower boat did a generation later. Fitch never realized that successful technological innovation almost inevitably needs to work with, not ahead of economic demand.[19] A factor that went far toward spreading steamboats to rivers all over America was the replacement of the heavy, bulky, low pressure, Watt-type, steam engine by the high pressure type, perfected in this country by Oliver Evans.

19. See: James Flexner, *Steamboats Come True* (New York: Viking, 1944); and *Autobiography of John Fitch*, ed. Frank D. Prager (Philadelphia: American Philosophical Society, 1976). France is said to have lagged in industrial development primarily because of lack of demand. See: Alan S. Milward and S. B. Saul, *The Economic Development of Continental Europe, 1780–1870* (Totowa, N.J.: Rowman and Littlefield, 1973).

Evans was America's greatest inventor in the period following the Revolution. Born near Newport, Delaware, he came in his thirties to live in Philadelphia. In 1772, at the age of seventeen, he designed a Watt-type condensing steam engine, but found no one interested in this expensive source of power when rivers offered their power freely to the builder of a waterwheel.[20] Being a member of a flour milling family, working in America's largest pre–Civil War industry, he turned his genius to eliminating all labor from inside the mill. By the mid-1780s grain had only to be poured in at one end and flour drawn off into barrels at the other. Unlike the early textile mills which contained relatively little machinery useful for other industries, Evans' mill was almost entirely made up of machines generally useful for conveying goods and transmitting power: the hopper boy, conveyors, lifts, leather belting, and other devices were to be parts of modern mass production. In addition application of new devices to very big flour mills taught builders much about the large-scale utilization of waterpower that was useful in many other industries such as textiles.

Meanwhile Evans continued experiments with steam, and in 1787 received patents from Maryland and New Hampshire for a steam carriage. He also sent plans for a high pressure steam engine to England but there was no commercial response because of the Watt patent. Any Anglo-American comparisons of early steam engines must observe that Britain needed them far more for industrial power than did America which had cheaper power from waterwheels. Evans devoted his many energies to flour milling and collecting royalties on his patents. In 1795 he published *The Millwrights Apprentice* that by 1860 had had eleven editions in various languages. He also ran a store and a machine shop in Philadelphia that was the predecessor of his famous Mars Works for steam

20. See: Greville and Dorothy Bathe, *Oliver Evans: A Chronicle of Early American Engineering* (Philadelphia: Historical Society of Pennsylvania, 1935).

engines built in 1808. After Evans's death in 1818 a son and his partner continued the plant. Oliver Evans had perfected the high pressure steam engine to a point where in early 1803 he filled the world's first commercial order. In 1812 his son started an engine plant in Pittsburgh. No matter what type of metal work was going on, "it was shops such as these that transformed the industrial capacities of the United States and developed the know-how of the modern age."[21]

Before 1812, Carroll M. Purcell says, in America a steam engine could only be built in New York or Philadelphia.[22] Furthermore, Dianne Lindstrom points out that in a developing economy, a large city such as Philadelphia surrounded by prosperous farm-land becomes a generating complex for continuous improvement in processes of production.[23] The existence of a large part of the buyers of machines near enough to the producer to come directly to him for repairs provided important stimulation. When railroad equipment was needed, after 1830, there were Philadelphia shops making stationary steam engines that could immediately turn to locomotives and other equipment; similarly there were shipyards ready to make iron hulls and install screw propellers.[24]

Such "heavy industry" developments form the mainstream of American industrialization, as well as that of other advanced nations. The unique contribution in income from cotton textiles was a peculiarly British development that was mirrored only to a smaller degree in the United States. The existence of some big cotton mills

21. Eugene S. Ferguson, *Oliver Evans: Inventive Genius of the American Industrial Revolution* (Greenville, Del.: Hagley, 1980), 46.

22. *Early Stationary Steam Engines in America: A Study in the Migration of Technology* (Washington: Smithsonian, 1969), 56.

23. *Economic Development in the Philadelphia Region, 1810–1850* (New York: Columbia University Press, 1978).

24. See: David B. Tyler, *The American Clyde: A History of Iron and Steel Shipbuilding in the Delaware River from 1840 to World War I* (Newark: University of Delaware Press, 1958).

by 1820 from the Baltimore area to that of Boston should not be confused with the continued growth of other more basic machine processes essential to industrial progress.

Great Britain's late eighteenth-century overall growth in productivity is no longer regarded as unusually rapid; it was perhaps less than that of France or the United States.[25] To reflect even partially the quantitative rate of industrialism in the latter country, national statistics should be broken down into those covering only the areas where a rapid shift to machine-processes was taking place. Such quantitative comparisons, however, are not the focus of my discussion.

In view of all preceding facts, what explains the survival of ideas that cotton textiles were largely responsible for industrialization, and that New England was the early center of American industry? The pervasive answer is the fact that southern New Englanders were more literate and careful in creating and preserving records than the people of the Middle States, particularly those of Quaker-influenced Philadelphia. Thus not only did New England produce more early historians, but offered relatively far more records to use. In a fit of despondency, for example, Oliver Evans in Philadelphia burnt all his papers, and there is no evidence that early records of the famous Sellers family of Philadelphia mechanics and inventors avoided an accidental but thorough destruction in the twentieth century.[26] In 1832 most eastern Pennsylvania manufacturers failed to fill out the questionnaires of the McLane Report on Manufactures, whereas returns seem quite complete for southern New England. In addition both Connecticut and Massachusetts made state

25. See Harley C. Knick, "British Industrialization before 1841: Evidence of Slower Growth during the Industrial Revolution," *Journal of Economic History* 42 (1982): 267–289. Peter H. Lindert, "Remodeling British Economic History: A Review Article," *Journal of Economic History* 43 (1983): 486–492; and Alan S. Milward and S. B. Saul, *The Economic Development of Continental Europe, 1780–1870* (Totowa, N.J.: Rowman and Littlefield, 1973), 29.
26. Conversation with Nicholas Sellers of Radnor, 1984.

surveys of manufacturing in the 1830s and 1840s, while Pennsylvania and New York, by far the largest manufacturing states, financed no statistical collections.

The lack of early records had a cumulative effect because historians necessarily work where the records are, and the manufacturing records were chiefly in Connecticut and Massachusetts. The first specialized, scholarly industrial history written in America was New Englander Samuel Batchelder's *Introduction and Early Progress of the Cotton Manufacture in the United States*, published by Little, Brown of Boston in 1863. Batchelder strives to be comprehensive, noting early developments in Philadelphia and elsewhere outside of New England, but the far greater explicitness of the latter's records inevitably take up the bulk of the book.

Compounding the bias of records, but having the same cultural sources, was the early twentieth-century interest of New England colleges and universities in economic history. Necessarily their faculties and graduate students used the local records, which meant that while statistically the Middle States might be properly placed, the examples that constituted the main body of the text were from New England. Among the able works of the early twentieth-century New England scholars was Caroline F. Ware's *The Early New England Cotton Manufacture: A Study of Industrial Beginnings*, published by Houghton Mifflin in Boston in 1931. She notes that New England mills had initially to draw skilled workers from the Philadelphia area (p. 15) and that in 1816 a Lancaster, Pennsylvania, factory produced slightly better sheeting by power looms than the famous factory at Waltham (p. 71). She also notes that in the first quarter of the nineteenth century, except for Connecticut there was little New England manufacture of hardware or general machinery (p. 159). "The want of other fields, then, perhaps as much as the merits of the business, sent capital into textile manufacture when commerce failed" (p. 15). Quite properly Ware's account centers on New England. No author has written as detailed a history of textiles centering on early eastern Pennsylvania, although in 1810 Alfred Jenks at Holmesburg (Philadelphia County) had the nation's first independent factory specializing in textile machinery.[27] The histories of woolen clothing and carpet manufacture follow much the same pattern of noting some events and statistics from Philadelphia, the carpet center, but drawing narrative and illustrative material chiefly from New England.[28]

The net effect has been to see industrialism in the United States as originating in the giant textile mills at Lowell in the 1820s rather than in the mills and machine shops of Philadelphia and its environs in the last quarter of the eighteenth century. In addition the rapid changes in the Philadelphia area were not in textiles but in steam, machinery, and general construction.

The Philadelphia history shows that industrial growth was not always revolutionary in the sense of new machines suddenly altering older methods of production. In textiles, for example, Philadelphia and many British centers found cheap labor more flexible and economical than investment of fixed capital in machinery such as the waterframe. As the Philadelphia Almshouse illustrates, the factory system and the beginnings of mechanized industrialism are also not the same. There had been the gathering together of workers in large numbers under supervisors throughout modern history. For example, while the Philadelphia Almshouse had some three hundred spinners and weavers, some paupers and some hired, producing cloth by hand methods, only a few blocks away, Oliver Evans's "blacksmith shop" with less than thirty workers

27. Edwin T. Freedley, *Leading Pursuits and Leading Men: A Treatise on the Principal Trades and Manufactures of the United States* (Philadelphia: Edward Young, 1856), 335. Philip Scranton, *Proprietary Capitalism, the Textile Manufacture of Philadelphia 1800–1885* (Cambridge, England: Cambridge University Press, 1983) deals only briefly with the early period.

28. See: Arthur H. Cole, *The American Wool Manufacture*, 2 vols. (Cambridge: Harvard University Press, 1926); and Arthur H. Cole and Harold F. Williams, *The American Carpet Manufacturer* (Cambridge: Harvard University Press, 1941).

was pioneering the modern high pressure steam engine—the prime mover of early Western World industrialism.[29] Similarly in manufacturing suburbs such as Germantown the change from handicraft to machine production was gradual from the eighteenth to the nineteenth century.[30] In his book on southeastern Pennsylvania James Lemon cites several towns where by 1775 craftsmen or manufacturers "far outstripped" agricultural tradesmen as a percentage of taxpaying citizens.[31]

Judging by history some causal elements seem indispensable for this pre-1850 industrial progress. First a cultural tradition of interest in practical work and an elite acceptance of the value of improving productive physical devices is required. This type of value was weak, absent, or even negated in Latin and Eastern Europe, and strong by 1775 only in England and the United States. Governments providing reasonable security for capital investment were present in the latter two nations and this seems a requisite for early progress. It should be emphasized that the period before 1850 was that of workbench technology, one dependent for improvements on ingenious mechanics, not on deductions from scientific theory. In fact, while the United States had a few scientists, neither nation had formal schools of engineering or science, and there seems little connection between men like Franklin or Priestley and the progress of industry. Second, available resources need to be sufficient for supporting the particular technological improvements. The relative abundance of coal in Britain and iron, wood, and waterpower in the northeastern United States were essentials for rapid development. Third, a moderate density of population and

29. See: Cynthia Shelton, "Textile Production and the Urban Laborer: The Proto-Industrialization Experience of Philadelphia, 1787–1820," *Working Papers from the Regional Economic History Research Center* 5, no. 4 (1982).

30. See: Stephanie Grauman Wolf, "Artisans and the Occupational Structure of an Industrial Town: 18th Century Germantown, Pa." *Working Papers from the Regional Economic History Research Center* 1, no. 1 (1977): 46–89n.

31. Wolfe, 34.

means of transportation are needed to create demand and encourage new endeavors. For example, even in the late colonial period there were large enough inland towns such as Coatsville, Lancaster, Columbia, and York in eastern Pennsylvania to lead to trade in iron and other products between them.[32] On a larger scale both northeastern America and Britain had the ocean, bays, and rivers reaching well inland.

The approach advanced here for the rise of *early* industrialism therefore may be called geo-cultural, the proper values or ethos combined with sufficiently benign geography. Obviously, all development requires demand; a high rate of domestic fertility plus generally increasing levels of northern European immigration assured a continuously rapid increase in consumption, a factor peculiar to the northeastern United States.

In the long view of history the rise of industrialism seems rapid from 1750 to 1850; yet, as we have seen, it appears that it was the result of many causes operating differently in different areas. To move from simple beginnings to complex advanced industry for making heavy metal components both for machine tools and finished products was necessary. In the United States this type of work developed first at a large number of small machine shops in the Philadelphia area in the late eighteenth century, and this locality continued until the late nineteenth century to be the center for the machine tool and heavy industry that was the backbone of industrialism. Comparisons between the United States and Great Britain are those between apples and oranges. By 1770 both countries were prepared to develop rapidly a supply of the utilities for which there was a strong demand. In both areas the demands and the resources to meet them differed. America had a greater need for river transportation and building construction; Great Britain, for

32. Paul F. Paskoff, *Industrial Evolution: Organization, Structure and Growth of the Pennsylvania Iron Industry* (Baltimore: Johns Hopkins University Press, 1983), 2.

exports and particularly war material between 1793 and 1815. There is no reason to expect the history of the effects of these demands to be similar, and it was not. Specific "technology transfers" that took place in both directions across the ocean were less important and more dependent on indigenous needs for their adaptation than has often been assumed.

The American Industrial Revolution Through Its Survivals

BROOKE HINDLE

THREE-DIMENSIONAL survivals of the American Industrial Revolution offer a different access to understanding, different from the documentary approach. The importance of machines and mechanisms in this story has always been recognized but, from the elder Arnold Toynbee who is credited with originating the term Industrial Revolution to recent historians, the written record has been relied upon too exclusively.[1]

Today the term Industrial Revolution is under challenge on the grounds that what happened took place over too long a period of time and in too many countries to be regarded as revolutionary. It is notable, however, that those opposing the term have additional axes to grind; Rondo Cameron, for example, emphasizes the continental story and reduces the role generally accorded to England.[2] Paul Paskoff, who substitutes the term "Industrial Evolution" does so in terms of the American iron industry which was conspicuously slow to adopt and forward British innovations in this field.[3]

It is true enough that the technological, economic, and organizational changes which we associate with the classical Industrial

---

1. Arnold Toynbee, *The Industrial Revolution* [1884] (Boston, 1956).

2. Rondo Cameron, "The Industrial Revolution: A Misnomer," in Jürgen Schneider, ed., *Wirtschaftskrafte und Wirtschaftwege: Festschrift für Hermann Kellenbenz* (5 vols., Stuttgart, 1981), 5: 367–76.

3. Paul F. Paskoff, *Industrial Evolution; Organization, Structure, and Growth of the Pennsylvania Iron Industry, 1750–1860* (Baltimore, 1984).

Revolution of the late eighteenth and early nineteenth centuries had long been developing. The theaters of machines published in France, Italy, and Germany—but not in England—from the sixteenth century through the eighteenth pointed toward the coming changes with surprising clarity.[4] Mechanization grew gradually over many centuries; indeed two of the five textile production processes had already been mechanized before the eighteenth century revolution that is often seen as beginning with the flying shuttle of 1733 and the spinning jenny of 1770. Factory production appeared at least as early as the fourteenth century. More directly pertinent, the transition of England to a coal-based economy, which directly underlay the classical Industrial Revolution, was largely an achievement of the sixteenth and seventeenth centuries.[5]

Even after taking note of the long course of technological change, the sharpest turn still appears to be that taken in the late eighteenth and early nineteenth centuries. Our present world is a direct fulfillment of the direction of that movement. Its clearest beginnings were in England, and the American development took off from transfers from England. Americans imported the technological and organizational advances selectively and at an uneven rate— but with growing avidity. We took the ball and ran with it on our own turf, playing a somewhat different game.

The perspective on this history offered by preserved artifacts and drawings has largely been neglected. That is the objective of a projected exhibit and book at the Smithsonian Institution where the best collection of pertinent survivals is held, and this paper is a précis of that effort.[6] The study of artifacts must begin with the

---

4. E.g. *The Various and Ingenious Machines of Agostino Ramelli*, ed. by Eugene S. Ferguson and Martha Gnudi (New York, 1978); Jacob Leupold, *Theatrum Machinarum* (10 vols., Leipzig, 1724–39).

5. John U. Nef, *The Conquest of the Material World* (Chicago, 1964), 220–21.

6. The exhibit, entitled *Engines of Change, the American Industrial Revolution, 1790–1800*, is scheduled for fall 1986. The book will have the same title.

question of what sort of understanding can be derived from them. The answer turns out to be that the best insights available concern those contexts most closely related to the objects. Best illuminated of all are technology, material culture, and those individuals who produced or used the artifacts.

It is useful to distinguish two ways in which people have experienced technology. The first is the experience of the inventor, entrepreneur, and builder in conceiving and producing devices. This embraces the fundamental creative process and includes social and economic as well as intellectual and experimental inputs. To many, this is the most exciting part of the story. Most people experience technology one step removed—as did a mill girl in operating a spinning frame or a passenger in traveling by train, or more externally still as the consumer benefiting from cheaper cotton goods and the farmer experiencing lower freight costs. Clearly, the further the experience is removed from the hardware, the less help the study of objects can offer the historian. The distant social relationships of technology may indeed be the most important of all, but material culture history has the least value there. It is most helpful for the internal story of technology and for its direct relationships.

Material culture history is much less developed than history based primarily upon documents. History from the written record enjoys accepted standards of validation and evaluation; today it is being extended by the newly successful application of social science techniques. Material culture study and even the study of the history of technology are newer; they do not have equally developed standards. Three-dimensional sources must be used with particular care.

There are many problems. Preserved artifacts were saved under wildly differing circumstances, often for reasons that produce a less representative corpus than surviving written sources. The value system under which each object was identified as worth memorializing may long since have been displaced. Besides, each has had a distinctive life cycle, has gone through alterations and usage in a sequence of contexts. It may have played very different roles at different points of time. The meaning of what is preserved is seldom immediately obvious. Moreover, some of the information contained is encoded and can be extracted only by the application of techniques unfamiliar to most historians. And, even when such efforts as chemical analysis, examination of microstructure, or the application of carbon dating are carried through, the yield in usable information is small.

But the greatest contribution the study of material culture can offer does not lie in such painfully extracted details; it offers the otherwise unattainable opportunity to experience a piece of the past. A preserved object that was there, in time past, is also here in our own time.[7] Naive collectors and untutored curators exaggerate the opportunity to see the past through artifacts, but they understand something fundamental. A historical artifact, especially when restored and interpreted as well as we are able, is a solid piece of the past in a way that no quotation can ever be.[8] Words have a marvelous virtuosity and ability to convert to their symbolism all manner of human experiences—but no word ever means precisely the same thing to two individuals. It is a symbol whose meaning depends not only on the character, set of mind, and context of the writer, but also on the knowledge, experience, and direction of thought of the reader. Words are nothing but symbols while objects are both symbols and something more basic.

Objects express deeply the purpose and manner of their manufacture and, usually, their use as well. Seeing, walking around,

---

7. Thomas J. Schlereth, paper on "Material Culture" delivered at Organization of American Historians meeting, 5 April 1984.

8. Brooke Hindle, "How Much is a Piece of the True Cross Worth," in Ian M. G. Quimby, ed., *Material Culture and the Study of American Life* (New York, 1978), 353–76.

touching, and hearing a machine in operation yields a direct acquaintance and understanding unattainable from words alone. It is a kind of knowledge or way of knowing that is essential to the historical researcher and just as valuable to the museum visitor. It rests upon visual thinking.

The meaning of visual or spatial perception has been probed by such scholars as Cyril S. Smith and was brought before the American Philosophical Society at its 1984 annual meeting.[9] This mode of thought has come much more into attention since the work of Roger W. Sperry in the 1960s which established its physiological base.[10] Spatial, multidimensional thinking takes place primarily in the right hemisphere of the brain differentiating it functionally from verbal, linear, logical thinking. This underlines the historian's need to understand and practice the spatial thinking that has always been the creative process behind both making and using three-dimensional devices.

Spatial or visual thinking lies behind the mechanization and three-dimensional aspects of the Industrial Revolution. Obliquely, this was recognized in its day in the celebration accorded the creativity of individual inventors. Technological change was perceived to be the result of invention, and each change required the eponymic identification of the inventor responsible. This understanding underlay the patent system and was supported by it. It is well expressed in Christian Schussele's 1862 painting entitled "Men of Progress" (Figure 1). This gathers together in a single room the recognized inventors of the ante bellum era: among them McCormick, Colt, Goodyear, and at the center in the place of honor, Samuel F. B. Morse. Today, we would say that these inventors were effective in the mental manipulation of images as they thought through various ways of accomplishing objectives. Logic and sci-

9. Symposium on the Visual Arts and the Visual Sciences, American Philosophical Society Annual Meeting, 20 April 1984.
10. See Brooke Hindle, *Emulation and Invention* (New York, 1981), 85–107.

ence were minor factors, although it was generally agreed that science should lead to new technology. That aspiration is expressed here in the straight line that can be drawn from the portrait of Benjamin Franklin, "Grandfather of all Inventors," through Joseph Henry, the only scientist in the room, to Morse. Henry's work did lie behind Morse's and the telegraph has sometimes been called the first science based invention.

Today we tell ourselves that we have a better understanding of technological change than this eponymic view—but the view is not yet dead, and perhaps for good reason. A new company journal, entitled *Financial Enterprise*, celebrates individual entrepreneurs and inventor-entrepreneurs in much the same eponymic way, to-day.[11] Of course, this caricatures a complex and extended process, but the individual creativity it recognizes was cherished in the early nineteenth century and still needs to be in our day.

The processes by which England used mechanization and new organization to speed up production were admired in America even before independence. Few of these changes were successfully introduced until the new nation was established, and then they came in unevenly. From the start, the United States enjoyed advantages in developing its own Industrial Revolution, advantages based upon its great undeveloped resources, the sparsely settled continent, and its freedom from political and economic constrictions. Another advantage was sharing a common language and most aspects of a common culture with the British. This encouraged the immigration of British mechanics and it eased efforts of visiting Americans to learn about the new technology.

Moreover, Americans were not technologically naive—as can be too easily assumed. The fact that nine-tenths of them were engaged in farming or farm related activities meant that they were more, not less, mechanically minded than the bulk of urban dwell-

---

11. *Financial Enterprise*, published by the General Electric Credit Corporation.

Figure 1. Christian Schussele, *Men of Progress*, 1862. (Courtesy of the National Portrait Gallery)

ers. The farmer was forced to rely on his own mechanical ingenuity, repairing his tools, making his furniture, processing his food,
building his house, and doing many jobs that would have been
taken to craftsmen in less isolated communities and by people with
more money. Indeed, many farmers engaged in some productive
craft to compensate for their deficient exchange. This further increased their receptivity to new technologies.

The American patent system followed Britain's but differed from
it. It was based upon a desire to encourage mechanical creativity
for the purpose of improving the national economy. Patents had a
double cutting edge: they protected the inventor with a short term
monopoly, and they then made the invention available so all could
benefit from it. Contrary to the advice of President Washington,
patents were not offered to those who introduced or transferred
ideas from other countries as they were in Britain. The biggest
difference was not stabilized until 1836; then, a practicable examination system was introduced.

Three-dimensional working models or patent models survive for
some early patents, such as Whitney's cotton gin. Colt's revolver,
Stevens's steamboat, and Evans's grist mill.[12] The history of each
must still be based upon the written record, but the machines or
models add a special kind of understanding.

The process of transferring technology had, of course, begun
with the first American settlements, and that story is full of unsuc-
cessful transfers. So is the story of transferring the improved mechanization that marked the Industrial Revolution. One of the most
clear-cut examples of this is the failed effort to introduce the Newcomen engine to America. The Newcomen engine was a piston and
cylinder steam engine in which steam was condensed to produce a
vacuum into which air pressure pushed down the piston. It ap-

12. Models or elements of the original of each of these devices are held by the
National Museum of American History.

Figure 2. Newcomen Engine. (Benjamin Martin, *Philosophia Britannica* [London, 1747], Plate XXIV)

peared in England in 1712 and was brought to America in 1753. It worked here for pumping water out of a deep copper mine in what is now North Arlington, New Jersey. But it did not take; it did not spread because it was voracious of fuel and cranky to keep running.

It was used in England for coal mines where fuel was no problem; in America there were no deep coal mines and no other needs that could justify it (Figure 2).

This, therefore, is a failed transfer and it may be a question of whether or not this specific cylinder, all that remains from the engine, helps anyone in understanding it (Figure 3). It may not be an original part of the engine and may even be a piece of the pipe used to raise water—rather than the actual cylinder. Still, its pedigree is good; it was a part of the engine at some point. That makes it a "piece of the true cross" and we know that such association is important to museum visitors and in some degree to everyone. Its historical value, however, is limited.[13]

Other transfers worked better. Much technology was transferred silently in the minds and hands of mechanics who migrated to this country and were employed because of what they knew. A few such figures have become celebrated for their contributions—most prominent among them, Samuel Slater. Slater's role in beginning successful mechanized cotton spinning in this country has received ample attention and his spinning frame and carding machine, now held by the Smithsonian Institution, are fine three-dimensional monuments to this eponymic view of a complex change (Figures 4, 5). His contribution deserves memorializing, but focusing upon his machines minimizes the role of precursors, of parallel and conflicting developments, and of the non-technological aspects of the changes entailed. Those dimensions can be dealt with adequately only in books. The Slater story is an important episode in the transfer of technology, and its drama highlights the difficulty of providing an adequate context for these memorials.

The *John Bull* is an example of still another form of transfer—the importation of a foreign-built machine and its adaptation to the American scene. The bare fact that it survives solves little but calls

13. Hindle, *Emulation and Invention*, 60–61.

Figure 3. 1753 Newcomen Engine Cylinder, North Arlington, N.J. (Courtesy of the National Museum of American History)

for extracting and presenting the best history related to it. It is the oldest operable steam locomotive in this hemisphere, perhaps in the world. It was not only successful but influential, becoming the prototype for American-built locomotives that followed it.[14]

The first problem is one common to many artifacts. The locomotive has been altered to a form it never had during its service life

14. See John H. White, Jr., *The John Bull: 150 Years a Locomotive* (Washington, D.C., 1981).

Figure 4. Samuel Slater Spinning Frame. (Courtesy of the National Museum of American History)

Figure 5. Samuel Slater Carding Machine. (Courtesy of the National Museum of American History)

(Figure 6). This is the form to which it was "antiqued" for the Philadelphia Centennial of 1876. Its appearance in actual use, with a cab and conical, wood burning stack made it look much more like what evolved as the standard American locomotive (Figure 7). This state of the locomotive, as well as its original imported appearance, can both be easily projected from the locomotive as it now exists.

A more elusive difficulty is the meaning of the addition of the cow catcher. This turns out to be something different from what its first appearance suggests. The two wheels were added as pilot wheels—

Figure 6. The *John Bull*. (Courtesy of the National Museum of American History)

Figure 7. The *John Bull*, After Withdrawal from Service. (Courtesy of the National Museum of American History)

Figure 8. The *John Bull*, The Pilot Truck. (Courtesy of the National Museum of American History)

not to support a cow catcher. They were needed on the sharper, steeper, and less well engineered American tracks, in order to keep the locomotive on the rails (Figure 8). More than that, the two forward large wheels were uncoupled from the two drive wheels and connected rigidly to the two small pilot wheels. A slot was cut behind the bearing of each of the forward large wheels—permitting them, with the pilot wheels, to swivel slightly like a truck[15] (Figure 9). The slit is less than an inch wide, but this gave a play of over an inch to the pilot wheels. These changes were probably made about 1833, just after John Jervis is credited with inventing the bogie truck.[16] Most engines built thereafter followed the *John Bull* in using a forward truck and only two powered wheels. The *Brother Jonathan* built at the West Point Foundry is typical (Figure 10).

Yet, long before this transfer, Americans had pioneered in other technologies. They had, of course, made many early railroad trials that did not succeed; as early as 1811 John Stevens experimented with steam locomotives.[17] They succeeded first in other developments, including agriculture and food processing.

An American success of 1784, patented in 1790, was Oliver Evans's automated grist mill[18] (Figure 11). Grist mills, of course, were ubiquitous in the American scene and fundamental to the life and prosperity of the country. In Evans's case, the only late transfer was the transfer of enthusiasm for mechanical improvement. That, the Americans had received well before the American Revolution.

Evans began with the old, familiar grist mill. Nothing about its basic technology had to be changed. The dam, the waterwheel, the

15. William H. Withuhn, "Testing the John Bull, 1980," in ibid., 107–21.
16. On John Jervis, see Angus Sinclair, *Development of the Locomotive Engine* [1907] (Cambridge, Mass., 1970), 58–59.
17. See [John Stevens], *Documents tending to Prove the Superior Advantages of Rail-Ways and Steam Carriages over Canal Navigation* (New York, 1812).
18. Oliver Evans, *The Young Mill-Wright and Miller's Guide* (13th ed., Philadelphia, 1850).

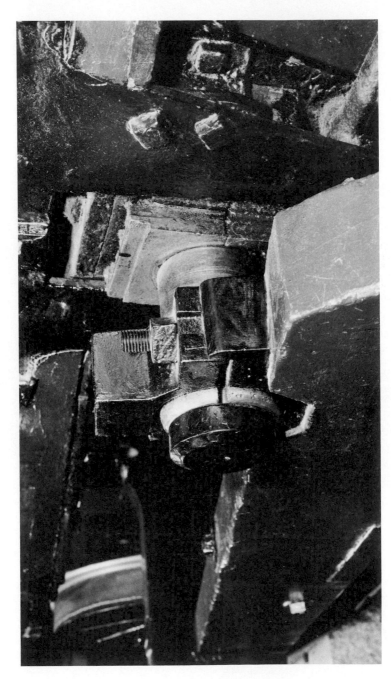

Figure 9. The *John Bull*, Slot Permitting Truck to Swivel. (Courtesy of the National Museum of American History)

Figure 10. *Brother Jonathan* Model, Detail Showing Truck. (Courtesy of the National Museum of American History)

head race and tail race all remained the same; so did the mill stones, the gearing, and the transmission shafts. His improvements involved moving the grain and meal from one point to another by mechanical means without anyone ever having to carry it or handle it. He achieved total mechanization of this process, a genuine continuous flow operation.

To move the grain and milled meal vertically, Evans relied upon an ancient chain of buckets which he called "elevators" (Figure 12). The buckets or cups were made of wood in his day and the belt of leather or canvas. Later constructions and restorations have substituted metal cups and canvas, currently even plastic. His most original design he called the Hopper Boy because it performed the function previously carried out by a boy (Figure 13). This was a rotary rake that spread the meal out to dry and then,

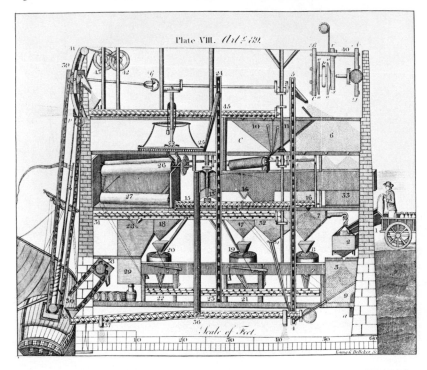

Figure 11. Drawing of Oliver Evans Mill. (Oliver Evans, *The Young Mill-Wright and Miller's Guide* [13th ed., Philadelphia, 1850], Plate VIII)

reversing the process, scraped it together so it could be sent on. To move meal and grain horizontally, he followed Jonathan Ellicott in applying the Archimedean screw.[19]

A large number of mills in the middle states region introduced Evans's improvements—usually not including the Hopper Boy. That was the least used of his ideas. Despite his patents, extended efforts, and much bitterness, he was never able to profit much from the royalties he asked. His effort to export his improvements to

19. Eugene S. Ferguson, *Oliver Evans: Inventive Genius of the American Industrial Revolution* (Greenville, Del., 1980), 25–26.

Figure 12. Evans Mill Elevator Cup. (Pierce Mill, Washington, D.C., photograph by author)

England was even less successful. The English felt no need for a continuous flow mill to replace their own small and irregularly operated mills.

The same attitude of mind may also have made them more contented with their crop producing technologies; at any rate, they imported very little of the newly mechanized agricultural technology that, before 1860, began to change the American scene. It was not that Englishmen failed to seek such improvements. Many new plows, threshing machines, and reapers had been attempted—but none was very successful.

Figure 13. Evans Mill Hopper Boy. (Colvin Run Mill, Great Falls, Va., photograph by author)

What was achieved was largely a matter of American effort and development. Plows had evolved even in the colonial period away from the European imports; most conspicuously, the northern deep plow and the southern shallow or shovel plow. The most important change in terms of productivity came in the nineteenth century with the introduction of the cast iron and then, in 1837, John Deere's "singing" steel plow, which was actually mostly polished iron.[20]

20. John Schlebecker, *Whereby We Thrive; A History of American Farming, 1607–1972* (Ames, Iowa, 1975).

McCormick's Reaper.

Figure 14. McCormick Reaper. (Courtesy of the National Museum of American History)

The most celebrated mechanization was the reaper, with Cyrus McCormick garnering most of the credit—as well as the largest reward. McCormick's followed a long train of reapers and the version he patented in 1833 worked a little better than its predecessors but was not satisfactory (Figure 14). It came into conflict primarily with the Obed Hussey reaper which had certain superior features. The two were finally combined and the reaper became a practical reality by the 1850s.

The thresher had a similarly long history of partial successes until the Pitts patented one in 1834 that worked reasonably well. Powered usually by steam, the thresher came to be used by farmers on an itinerant basis. Thresher owners moved from one farm to another working on the harvested grain in season.

Farming was directly related to industrialization, specifically to the mechanization of transportation and production. The move to factory production required a larger farm output to sustain those who left the farm for the factory. In fact, farm output did increase rapidly, beginning in the antebellum era. It now appears that the early increase did not result from the mechanization of farming and food processing but largely from opening more productive lands and, indirectly, from such improvements as transportation.

There are many ways of getting at rising anticipations once the possibilities of industrialization began to be perceived. A fascinating vignette is provided in the story of two immigrant engineers from Germany who came to the United States in 1831 on the same ship. They were John Augustus Roebling and John Adolphus Etzler. Roebling is known to everyone for the great suspension bridges he designed, culminating in the famous Brooklyn Bridge. Etzler is not as widely known, although students of American literature have been introduced to him in a backhanded manner through a review Henry David Thoreau wrote of his book, *The Paradise Within the Reach of All Men, Without Labor, By powers of Nature and Machinery.* Thoreau is credited with demolishing Etzler's pretensions. But although published some ten years after the book, it beat an already dead horse.[21]

Actually Etzler and Roebling had much in common. They came to the United States to found separate communities of Germans seeking a new life. Each community had some of the characteristics of the utopian communities springing up at that time throughout the country. Etzler's failed quickly, as did a second attempt he made in Venezuela. Roebling's suffered most of the same vicissitudes, but was held together by sheer hard work. It did endure, although Roebling himself became disenchanted, undertook various engineering projects, and finally moved away—never to return.[22]

The material culture approach reveals a significant contrast between the two men. Many material objects remain as testimony to Roebling's work, from his bridges, through patent models, to the buildings left at Saxonburg, Pennsylvania, his model community. Nothing material remains from Etzler, absolutely nothing. At first glance, that might seem mere chance but it is more than that. Etzler never proposed, sketched, or discussed a machine or material project that worked. Every project failed.

The difference in their thinking is clearly spelled out in the reactions of the two men to the sailing ship on which they crossed the Atlantic. Each looked at it through the eyes of an engineer, Etzler seeing an inefficient, labor intensive way of furling and unfurling sails, setting them, and, in general, of harnessing the power of the wind. He set out to replace that system with one based upon a

21. John A. Etzler, *The Paradise Within the Reach of All Men, Without Labor, By Powers of Nature and Machinery* (Pittsburgh, 1833); Henry David Thoreau, "Paradise (To Be) Regained," *The United States Magazine and Democratic Review* 13 (1843): 451–63.

22. Brooke Hindle, "Spatial Thinking in the Bridge Era: John Augustus Roebling versus John Adolphus Etzler," in *Bridge to the Future; Proceedings* of the New York Academy of Sciences (*Annals* of the New York Academy of Sciences, vol. 424. New York, 1984), 131–47.

new look at the needs and possibilities. He designed a wholly different sailing vessel in which all the handling of sails would be done mechanically—not by the exercise of human muscles (Figure 15). He replaced the three masts with a single mast and a single sail, one that opened and shut like a Japanese fan. It could be operated and the mast could be rotated by ropes powered by a small windmill, mounted on the stern of the ship.[23]

Not content with this change, Etzler sketched out a device to harness wave power to propel the ship by mechanized oars or paddle wheels (Figure 16). This was to be accomplished by a platform supported beneath the ship that would move up and down relative to the vessel as the ship's buoyancy lifted and dropped it. The only trial of this plan failed totally when a small version of it sank quickly; the drag of the platform pulled it under.[24]

Etzler had some good ideas; surprisingly, he employed wind, water, tide, and wave power at a time when the steam engine was the great hope. But, he never worked through the details to produce workable devices—as did Roebling. Roebling was equally fascinated by the ship and the sails that harnessed the power of the wind to propel it, but he spent the whole voyage trying to learn all he could about this long used design. He was particularly taken by the system of stays that held the masts in place against all the forces of wind that might be encountered. He talked at length with the mate about the stays, wrote elaborate descriptions, and left behind a sketch (Figure 17). Unquestionably, this study led him to the diagonal stays that became the hallmark of his bridges from Niagara to Brooklyn (Figure 18).[25]

23. John A. Etzler, "Description of The Naval Automaton," (Philadelphia, n.d.) in Joel Nydahl, ed., *The Collected Works of John Adolphus Etzler* (Delmar, N.Y., 1977).

24. Patrick R. Brostowin, "John Adolphus Etzler: Scientific Utopianism during the 1830's and 1840's" (Ph.D. dissertation, New York University, 1969), 153.

25. John A. Roebling, *Diary of My Journey from Muehlhausen in Thuringia via Bremen to United States in North America in the year 1831* (Trenton, 1931), 31–32.

Figure 15. The Naval Automaton Showing Sail and Windmill. (John A. Etzler, Patent #2533, "Navigating and Propelling Vessels by the Action of Wind and Waves" [1842] [Microfilm, Woodbridge, Conn., n.d.])

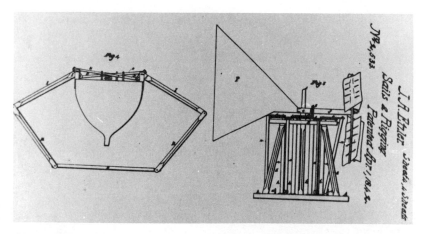

Figure 16. Naval Automaton Platform. (Ibid.)

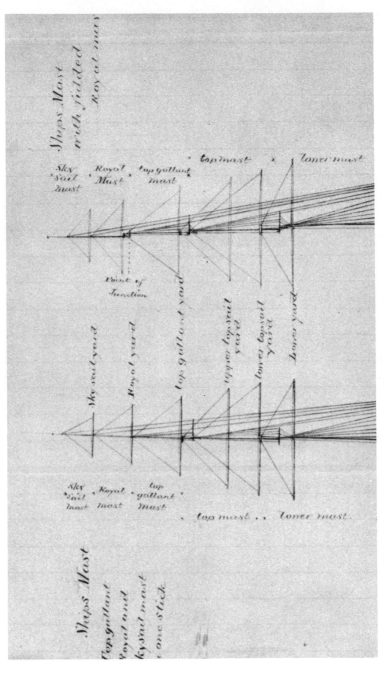

Figure 17. Roebling Drawing of Ship's Stays. (Courtesy of Rensselaer Polytechnic Institute Archives)

Roebling's bridges seemed to be something new under the sun. They reached farther and carried more traffic than others had; they became monuments to man's rising technological capabilities—but they were achieved by a step by step study, analysis, and trial that marked all of Roebling's work. Etzler's science fiction vision looked farther into the future than Roebling, but Roebling was concerned with doable projects; he forced them through all obstacles until they worked.

Similar careful study and limited change based upon components he understood lay behind his patents. For the most part, he patented devices closely related to his work, especially to bridge building. Notable was his wheel for laying the strands of the cable in place on the bridge and a method for anchoring the cables at their ends. His wire rope making patent was particularly successful, providing the foundation for the prosperity of the Roebling Company[26] (Figure 19).

The American System of Manufactures has been labeled one of the most clear-cut contributions made in this country to the process of mechanization. It is a designation given by the British after the 1851 Crystal Palace fair to manufacture by interchangeable parts or by the uniformity principle. This had long been promoted by Eli Whitney, but not until Edwin H. Battison took apart Whitney muskets and examined them in detail was it clear that he never attained the goal himself.[27] The traditional method required each piece to be filed and shaped to fit with others to produce such a component as the gun lock. The Whitney locks carry tell-tale numbers—each corresponding to a particular musket and not fitting a different gun (Figure 20).

The achievement of interchangeability was finally accomplished in the manufacture of guns only because cost became a secondary

26. John A. Roebling Patent, *"Machine for Making Wire Rope,"* 16 July 1842.
27. Edwin H. Battison, "Eli Whitney and the Milling Machine," *Smithsonian Journal of History* 1 (1966): 9–34.

Figure 18. Brooklyn Bridge Stays. (Photograph by author).

Figure 19. Wire Rope Making Machine Patent Model. (John A. Roebling, Patent #2720. Courtesy of the National Museum of American History)

consideration for the government. Commercial gunmakers long had to use cheaper methods, but interchangeability opened a wonderful chance to repair damaged ordnance in the field without having to return the pieces to the ministrations of a skilled gunmaker. The achievement depended upon the development of the true milling machine, the use of many special purpose machines each designed to perform a very specific operation, and the application of meticulous gauging of each part that had to fit precisely.

The Blanchard lathe is a surviving example of special purpose

Figure 20. Whitney Gun Lock Showing Numbered and Fitted Elements. (Edwin A. Battison, "Eli Whitney and the Milling Machine," *Smithsonian Journal of History* 1 [1966], p. 25, figure 16).

machinery, designed to produce gun stocks which were precise copies of a model stock which the machine followed in cutting the blank (Figure 21). Thomas Blanchard put it into use in the Springfield Armory, the major center for the development of interchangeability in guns.[28]

Gauging was developed to a point where it was so well controlled that gun manufacture could be contracted out to commercial firms. William Thornton was a government inspector who went about testing the guns being manufactured, sampling random pieces at each factory. His chest of gauges gives a good sense of the manner

28. See David A. Hounshell, *From the American System to Mass Production, 1800–1932* (Baltimore, 1984), 35, 38–39; Merritt Roe Smith, *Harpers Ferry Armory and the New Technology* (Ithaca, 1977).

S.A. - 1960

Figure 21. Blanchard Lathe Replica. (Courtesy of the National Museum of American History)

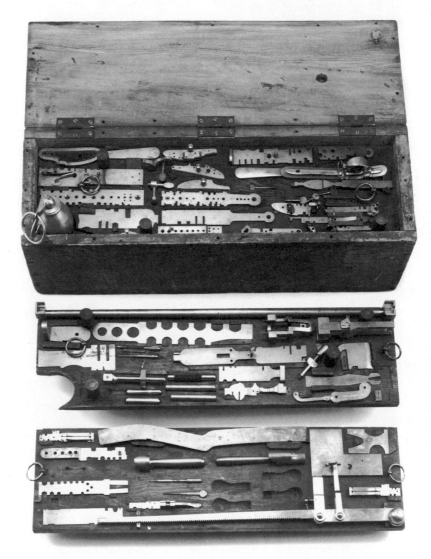

Figure 22. Thornton Gauge Chest. (Courtesy of the National Museum of American History)

Figure 23. Robbins and Lawrence Rifle Detail. (Courtesy of the National Museum of American History)

in which control was exercised over this precision manufacture (Figure 22).

Many examples of early interchangeability survive. The Hall rifle, made at the Harpers Ferry Armory was a technological success, although not regarded favorably by those who used it in the field. The Robbins and Lawrence rifle so impressed the British that they bought gunmaking machinery from that small Vermont firm for the Enfield Armory (Figure 23).

Improvements in mechanization depended upon the ability to improve technology beyond the existing state of the art—but it was not solely a question of technological capability. Many things could be done technologically that were not feasible economically or socially. More than that, there are always several ways of accomplishing any technological objective, and the specific manner used involved interactions with external factors. Government support was necessary to carry the R and D needed for interchangeable gun manufacture. On the other hand, a parallel, and much simpler achievement of interchangeability, was accomplished commercially in clockmaking. Here far looser tolerances were required

Figure 24. Eli Terry Clock. (Courtesy of the National Museum of American History)

and Eli Terry succeeded early in making wooden clocks on the "uniformity principle" (Figure 24). There is no clear evidence of a direct connection between gun- and clockmaking as each moved toward interchangeability.[29]

Perhaps the most celebrated presentation of American mecha-

29. John Joseph Murphy, "The Establishment of the American Clock Industry: A Study in Entrepreneurial History" (Ph.D. dissertation, Yale University, 1961), 29–49.

Figure 25. Model of *The America*. (Courtesy of the National Museum of American History)

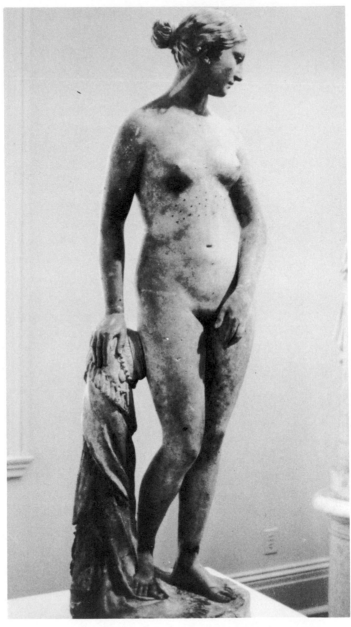

Figure 26. Hiram Powers, *The Greek Slave*, Plaster Model with Measuring Sockets. (Courtesy of the National Museum of American History)

nization and technological advance was in the American exhibits at the first world fair, the 1851 Crystal Palace in London. American exhibits were few compared with those of Britain and France but they evoked surprise and attention, especially from English manufacturers.

Many American exhibits represented traditional, craft technology. Farm products showed little mechanization but even locks, pianos, and scientific instruments had been made in craft fashion. The yacht *America*, that won the cup in an associated race, was a product of traditional technology (Figure 25). The more exciting entries, however, involved the new technology. The Colt revolver and Robbins and Lawrence rifle expressed manufacture by interchangeable parts, or the American System. McCormick's and Hussey's reapers showed off an American fulfillment of an objective long pursued in England.

Particularly interesting was *The Greek Slave* by the American sculptor Hiram Powers. The fair intentionally included works of art, in which the United States was but little represented. The marble *Greek Slave*, however, represented a process of production that was not entirely alien to the uniformity principle and gauging at the center of the new technology (Figure 26). Plaster casts of the statue were first produced—with many small, metal sockets imbedded at skin level. Measuring devices then permitted a carver to produce multiple marble copies. The scale could be changed readily, and many copies at both full and reduced scale were produced.

This examination of a few of the survivals of the American Industrial Revolution has an exciting impact. The objects do provide direct contact with some of the most central aspects of that history. They permit a comprehension of three-dimensional problems, solutions, and cultural impacts that are not accessible from other approaches.

The understanding of each object, however small or limited in scope, requires the best understanding of the written record. Even more obviously, the interpretations of the larger context within which that object flourished must depend primarily upon the written record. But large interpretations or syntheses have become elusive as specialization encourages one subdiscipline to break off after another.[30] The history of technology and material culture history represent such subdisciplines that must be integrated into the larger story of the American Industrial Revolution. Other active fields, such as labor history, business history, economic history, and the new social history, have similarly central contributions to make. None of them alone can offer an adequate general synthesis.

The study of survivals of the American Industrial Revolution is rewarding in itself. Its great importance, however, lies in opening these insights to inclusion in our broadest understanding of history.[31]

---

30. See Brooke Hindle, " 'The Exhilaration of Early American Technology,' A New Look," in David A. Hounshell, ed., *The History of American Technology: Exhilaration or Discontent* (Wilmington, Del., 1984), 7–17.

31. The most helpful recent essay in the diminishing category of synthesis is Thomas C. Cochran, *Frontiers of Change; Early Industrialization in America* (New York, 1981).

One Professor's Chief Joy:
A Catalog of Books Belonging to
Benjamin Smith Barton

Joseph Ewan

WHO in America assembled the largest library of natural
history in Jefferson's time? The Bartrams, father and son,
acquired a number of choice titles, mostly by gift. David Hosack
and Samuel Latham Mitchill[1] had notable libraries. Jefferson's
wide-ranging collection totaled nearly 5,000 volumes, including 38
strictly botanical titles, of which only one survives.[2] The largest
natural history collection in America before 1815, however, was
assembled by Benjamin Smith Barton (1766–1815) of Philadelphia.
Physician, professor, naturalist, student of Indian tribes, biblio-
phile, citizen of the world, Barton spent his happiest hours in

---

1. David Hosack's library of over 4,000 volumes was "one of the best private
libraries in the country," Harriet Martineau, *Retrospect of Western Travel* (London,
1837), 1: 309. The medical books were dispersed by auction; 205 botanical books
are at NYBG. A. M. Vail, *Journal of the New York Botanical Gardens* 1 (1900): 22–26.
No catalog of Mitchill's is available.

2. E. M. Sowerby, *Catalogue of the Library of Thomas Jefferson*, 5 vols. (Washington,
1952); J. Ewan, *Missouri Botanical Gardens Bulletin* 64 (June 1976, Suppl.), 8 pp.
Jefferson owned nine Linnaean titles; Barton, twelve.

My thanks to Mrs. Caroline Morris, librarian-archivist, at the Pennsylvania
Hospital, who patiently aided and abetted the search for Barton's books since
1952. Then, too, the staffs of the American Philosophical Society, the Academy of
Natural Sciences, the Library Company and the Boston Public Library, have
joined in this recovery of a lost chapter in our history. From that first census my
wife Nesta has toiled to make this record complete and accurate.

botany. His library contained at least 372 titles by our count, in the fields of botany, zoology, anthropology, travel, and a few in medicine, often the earliest copies to arrive in the United States. When *Indian Zoology* arrived from Thomas Pennant, Barton remarked "I have reason to think this is the first copy of the work that has ever reached this country." Cognizant of values, he commented on the rarity of his just acquired *True and Exact History of the Island of Barbadoes* (1673) by Ligon.[3] The oldest book still in his library is *Margarita philosophica* of 1512 (274 in this catalog).

Collectors must rely on the cooperation of booksellers who, in friendship and loyalty, snatch a wanted book from the flowing market stream they so closely scan. Because Barton sought special titles, it is not surprising that his booksellers were more often European rather than American. Barton met Samuel Paterson (1728–1802) in Edinburgh while a student, and he continued to buy into the third generation of Patersons—as late as September 1811.[4] London bookseller Samuel Price forwarded a "small packet" of books left with him in 1815 on Barton's last visit to England.[5] But it was Charles Dilly, publisher and bookseller of London who supplied most of his books. Often Barton hoped his want list would be credited against his own titles. In June 1799, for example, he wrote George Robinson of Paternoster Row that he was shipping 320 copies of *Fragments* "to dispose of, for me, to the best advantage, retaining the usual commission for yourselves."[6] He wanted twenty-four titles. In Philadelphia, Jean Louis Fernagus of "the French

3. *Philadelphia Medical and Physical Journal* 1 (1804) pt. 1: 134. On another occasion, "Mr. Lawson's work is not easily met with." *Discourse* (1807): 87.

4. Samuel Paterson, 21 November 1808, "as the ship was detained severall weeks owing to a dispute I gave him another parcel ..." and 17 July 1809, to BSB.

5. Samuel Price, London, 30 September 1815, to BSB.

6. BSB, June 1799, to George Robinson (1737–1801), "king of booksellers," and two sons, George (d. 1811), and John (d. 1813). List of wanted books in BSB Misc. papers, APS.

Bookstore,"[7] John Conrad, and James Parke, son of the hospital's Dr. Thomas Parke, for examples, favored him.[8]

Visitors to distant parts were implored to obtain specialties. "Mrs. —— informs me that you intend to proceed to Germany," and five wanted titles followed.[9] His correspondent A. C. de Freytos of Funchal sent his copy of Loureiro's *Flora cochinchenensis*. The American consul at St. Petersburg, Levett Harris, negotiated Gmelin's *Flora sibirica*, and Barton asked Willdenow when his new edition of the *Species plantarum* would be published.[10] He sent his *Materia Medica* to Professor Gouan in Montpellier, noting four books he very much wished to possess. John Vaughan, in an excellent position to follow the market, was entreated to obtain certain Italian titles.[11] Too, Barton was a subscriber to Jeremy Belknap's *History of New Hampshire*, Robert Thornton's botany of 1803, and Alexander Wilson's *American Ornithology*.

When planning his trip to Europe in 1815, Barton listed twenty-four titles he hoped to procure. José Correa da Serra evidently received the books as they arrived in Philadelphia, we do not know how many, but a receipt for Barton's check of $400.00 survives. "Tomorrow I will myself accompany the books and see them safe in your house."[12]

Following his death 19 December 1815, the Board of Managers of Pennsylvania Hospital appointed Zaccheus Collins and Samuel W. Fischer "to examine [Barton's books] and make such purchases

7. J. L. Fernagus, 16 November 1813, to BSB.

8. Central is W. J. Bell, Jr., "Old Library of the Pennsylvania Hospital," *Bulletin Medical Library Association* 60 (1972): 543–550.

9. One of the five books was Persoon's *Synopsis plantarum* (1805) which Barton borrowed from Jefferson in 1810 after he had failed to obtain a copy.

10. BSB, 12 June 1803 to Carl Ludwig Willdenow (Archives of the Botanical Garden Library, Copenhagen).

11. BSB, 31 May 1802, to John Vaughan (Barton letters, APS).

12. J. Correa da Serra, 20 May [1815] (Boston Public Library, T. P. Barton autograph, coll. no. 11).

as they may deem useful additions to the [Hospital's] library." The purchase price of $2770.00 was made in two installments, the first in April 1816, and the balance on 12 May 1817. An undetermined number of books not "deemed proper" for the medical library were to be sold at auction the ensuing autumn; its date has not been ascertained.[13] On behalf of the American Philosophical Society John Vaughan purchased 134 books at the initial sale. The receipt dated 13 April 1816 for $151.60 was signed by young Thomas Pennant Barton for Mary P. Barton, administrator.[14] A few Barton books came later to the Society through John Vaughan who evidently attended an auction conducted by Titon Greland of Philadelphia, 30 January 1822, or (?) 1823.[15] Society records note that he deposited certain books 16 April 1824, and officially presented these to its library 5 May 1826. Certain Barton books now at the Library Company of Philadelphia were purchased by Joseph Parker Norris for $72.50 as authorized 4 April 1816 by the Managers of the Hospital.[16] Thomas Pennant Barton kept a few of his father's books, and these with his famous Shakespeare collection are now at the Boston Public Library.

That Barton was ungenerous in sharing his books has been intimated in the writings of Francis Harper and Jeanette Graustein.[17] Yet medical students of this busy professor had access to literature for their dissertations on the medical uses of plants. Thomas Jeffer-

13. Pennsylvania Hospital, Minutes of Managers, 12 May 1817, p. 275, and 28 July 1817, p. 278. C. S. Rafinesque promptly reported the sale of Barton's library in *American Monthly Magazine & Critical Review* 2 (1817): 84. T. G. Morton, *History Pennsylvania Hospital 1751–1895* (1897), 348, and F. R. Packard, *Some Account of the Pennsylvania Hospital* (1938), 67, both erroneously cite 1797 as the year of the sale.

14. American Philosophical Society Archives, No. 42.

15. *Catalogue of a valuable private library . . . from the libraries of Priestley, Wistar, and Barton . . .* (Philadelphia, no yr.), 12 pp. 340 items. Copy at Library Company of Philadelphia.

16. Library Company of Philadelphia, Minutes of the Directors, vol. 4.

17. "[Muhlenberg's] attitude was the antithesis of Barton's in being thoroughly altruistic" (F. Harper, *Bartram's Travels*, Yale, 1958, xxxi). "[Barton] lacked generosity of spirit" (J. E. Graustein, *Thomas Nuttall, naturalist*, Harvard, 1967, 40).

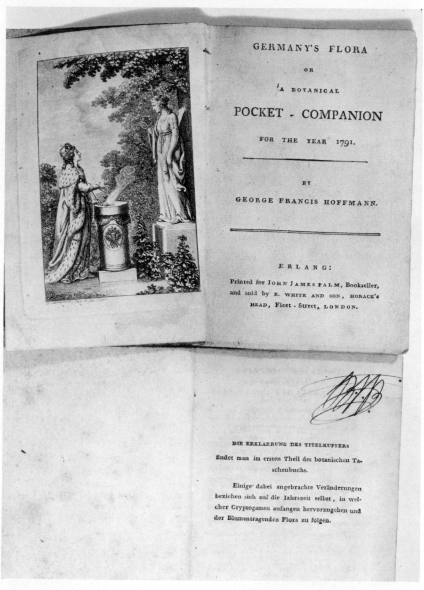

GERMANY'S FLORA

OR

A BOTANICAL

POCKET - COMPANION

FOR THE YEAR 1791.

BY

GEORGE FRANCIS HOFFMANN.

ERLANG:

Printed for JOHN JAMES PALM, Bookseller,
and sold by B. WHITE AND SON, HORACE's
HEAD, Fleet - Street, LONDON.

DIE ERKLAERUNG DES TITELKUPFERS

findet man im ersten Theil des botanischen Ta-
schenbuchs.

Einige dabei angebrachte Veränderungen
beziehen sich auf die Iahrszeit selbst, in wel-
cher Cryptogamen anfangen hervorzugehen und
der Blumentragenden Flora zu folgen.

Barton signed his books. Thomas Nuttall carried this *Flora*—only two copies
traced in U.S.—when he explored for Barton about the Great Lakes.

son and he exchanged source books on Indian life and language. Frederick Pursh and Thomas Nuttall gained their familiarity with American plants with his books at his home, and Nuttall carried Barton's copy of Hoffmann's *Flora* on his journey to Michelmacki-nack and back. William Bartram was requested to keep Pennant's *Indian Zoology* "until you [have] had leisure to pursue it."[18] Manassah Cutler promised to return the books he had borrowed, and in 1794 André Michaux transcribed sections on quadrupeds and birds of *Systema naturae* using Barton's copy. Meriwether Lewis carried Du Pratz's *History of Louisiana* in his saddlebag to the Pacific and back.[19] William Hamilton, C. F. H. Denke, Mathew Carey (the publisher), and Zaccheus Collins, all borrowed his books. There may well have been a personal antipathy not discernable in our time which made Barton cool toward sharing his resources with Rev. Henry Muhlenberg of Lancaster. Muhlenberg complained that although he had had the pleasure of seeing Abbot's *Rarer Lepidopterous Insects of Georgia* and on a later visit, Plukenet's works at Barton's home, he was not permitted to linger over them.

Barton probably intentionally purchased few medical books, depending on the hospital's library which in 1806 held 5,828 books. He received dissertations from grateful students, but whatever reference books were already duplicated in the hospital library were probably picked up by fellow physicians or students after his death.

With the change in emphasis in medical education and practice, the rise and development of the Academy of Natural Sciences, and the publication of American texts, interest in Barton's library soon waned. Who of later generations would expect a fabulous natural history collection lining a hospital reading room? Yet, when the

18. BSB, 30 December 1792, to Wm. Bartram (Bartram Papers 1: 8. Historical Society Pennsylvania).
19. P. R. Cutright, "Lewis and Clark and Du Pratz," *Bulletin Missouri Historical Society* 21 (October 1964): 31–35.

United States Exploring Expedition was being organized in the 1830s, three of Barton's books, perhaps known to Titian Ramsey Peale, were taken along, and returned (numbers 40, 44, and 207 in this catalog).

In short, Benjamin Smith Barton assembled what at the time of his death in 1815 was the largest and most extraordinary natural history library in America. During the last half of his foreshortened life he acquired, mostly by purchase, often carefully planned through requests to his friends, 372 titles. Among his books are many association copies, the gift of his colleagues or by the acquisition of books from William Byrd's library. Barton annotated some of his books internally: Catesby (No. 73), Jacquin (166), Kalm (172), Michaux (227) and Zeisberger (372). From a check of the holdings of national libraries many of his books prove to be of exceptional rarity. Most of his books are in the library of the Pennsylvania Hospital (177) and the American Philosophical Society (126) and the Library Company (23). It is gratifying to substantiate this facet of an American naturalist of diversified talents.

NOTES ON THE CATALOG

All the books listed carry Barton's signature except that we have not seen those marked with an asterisk. These have been entered from manuscript or inferential evidence that the title was once in his library. The order of the entry: item number; author's full name; short title; place of publication; date; significant annotation or the like; reference to bibliographies where fuller information will be found; present location.

BIBLIOGRAPHIES CITED BY ABBREVIATION

Austin:	Austin, Robert B. *Early American Medical Imprints, a guide to works printed in the U.S., 1668–1820.* Washington, D.C., 1961.
Blacker-Wood:	*Dictionary of the Blacker-Wood Library of Zoology and Ornithology.* 9 vols. Boston, 1966.

Countway: *Author-Title Catalogue of Frances A. Countway Library of Medicine.* 10 vols. Boston, 1973.

Dryander: Dryander, Jonas. *Bibliothecae historico-naturalis Josephi Banks.* 5 vols. London, 1798–1800.

Evans: Evans, Charles. *American Bibliography: A Chronological Dictionary.* 13 vols. Chicago and Worcester, 1903–1934, 1959.

Pritzel: Pritzel, Georg August. *Thesaurus literaturae botanicae.* 2d ed. Leipzig, 1872–[1877].

Sabin: Sabin, Joseph. *Bibliotheca americana.* 29 vols. New York, 1868–1936.

Soulsby: [Soulsby, B. H.] *Catalogue of the Works of Linnaeus . . . in the libraries of the British Museum.* 2d ed. London, 1933.

TL–1:   Stafleu, Frans A. *Taxonomic Literature.* Regnum Vegetabile. vol. 52. 1967.

TL–2:   Stafleu, Frans A., and Richard S. Cowan. *Taxonomic Literature.* 2 ed. Regnum Vegetabile. Vols. 94, 98, –. 1976–.

Wolf: Wolf, Edwin, 2nd. "Dispersal of the Library of William Byrd of Westover." *Proceedings of the American Antiquarian Society* 68 (1958): 19–106.

Wood: Wood, Casey. *Introduction to the Literature of Vertebrate Zoology.* London, 1931.

LOCATION ABBREVIATIONS

ANSP: Academy of Natural Sciences, Philadelphia.
APS: American Philosophical Society, Philadelphia.
BPL: Boston Public Library.
LCo: Library Company of Philadelphia.
PH: Pennsylvania Hospital, Philadelphia.

Abbot, John. *See* Smith, James Edward.
 1. Acharius, Erik. *Methodus qua omnes detectis lichenes.* Stockholm, 1803. Pritzel 9. PH.

2. Acosta, Joseph. *Naturall and Morall Historie of the East and West Indies.* London, 1604. BSB: "1789." Pritzel 14; Sabin 131. APS.

3. Adair, James. *History of the American Indians.* London, 1775. Sabin 155. APS.

4. Aiton, William. *Hortus Kewensis.* 3 vols. London, 1789. BSB loaned to M. Cutler (C, *Life* 2: 290). Pritzel 78. ANSP.

5. Albers, Johann Abraham. *Preisfrage, worin besteht eigentlich das Übel, das unter dem sogenannten freywilligen Hinken der Kinder bekannt ist?* Vienna, 1807. APS.

6. Allioni, Carolo. *Flora pedemontana.* Torino, 1785. Pritzel 108. PH.

7. Anderson, Andrew. *On the Eupatorium perfoliatum of Linnaeus.* New York, Van Winkle, 1813. Countway 1: 249. APS.

8. Anderson, James. *On the culture of Silk on the Coast of Coromandel.* Madras, 1792. Inscription: "Prof. Smith Barton, from the author." APS.

9. Anghiera, Pietro Martire, d'. *Decades of the Newe World.* London, 1555. Sabin 1561. LCo.

10. [Anonymous.] *Literary Memoirs of Germany and the North.* London, 1759. APS Lib. Comm. Rept. 1977 (1978), 141. APS.

11. [Anonymous.] *New System of the Natural History of Quadrupeds, Birds, Fishes, and Insects.* 3 vols. Edinburgh, 1791–1792. Blacker-Wood 6: 51; Wood 198. PH.

12. Ariminensi, Ianco Planco. *Fabi Columnae Lyncei.* Florence, 1743. PH.

13. Aublet, Jean Baptiste Christophe Fusée. *Histoire des Plantes de la Guiane Française.* 4 vols. Paris, 1775. Pritzel 277; TL-2 206. PH.

14. Aucourt e Padilha, Pedro Norberto de. *Raridades da natureza, e da arte, divididas pelos quatro elemento.* Lisbon, 1759. PH.

15. Audebert, Jean Baptiste, and L. P. Vieillot. *Oiseux dorés ou à reflets métaliques: Histoire naturelle et générale des Colibris, Oiseaux-mouches, Jacamars et Promerops.* 3 vols. Paris, 1802. "BSB. Dec. 4, 1813." PH.

16. Avellar Brotero, Felix de. *Flora Lusitanica seu plantarum.* 2 vols. Lisbon, 1804. "Received August 11, 1813, B.S.B." Pritzel 1194; TL–2 812. PH.

17. Azara, Felix d'. *Essais sur l'Histoire Naturelle des Quadrupeds de la Provence du Paraguay*. 2 vols. Paris, 1801. no atlas. "B.S.B. March 27, 1809." Sabin 2537. PH.

18. [Bailey, Nathaniel.] *Dictionarium rusticum, urbanicum et botanicum*. 2 vols. 3d ed. London, 1726. LCo.

19. *Bancroft, Edward. *Experimental researches concerning the Philosophy of permanent colours*. London, 1794. BSB loaned to Joseph Priestley (J.P. to J. Vaughan, 30 May 1796. APS archives).

Banks, Joseph. *See* Houstoun, W.

20. Barbeyrac, Charles de, and Hermann Boerhaave. *Dissertations nouvelles sur les Maladies de la Poitrine*. Amsterdam, 1731. PH.

21. Bard, Samuel. *An enquiry into the Nature, Cause and Cure of the Angina suffocativa, or Sore Throat Distemper*. New York, Inslee and Car, 1771. Austin 125; Evans 11977. APS.

22. Barrelier, Jacques. *Plantae per Gallian Hespanicam et Italiam*. Paris, 1714. "B.S.B. 1813." Pritzel 423 PH.

23. Bartholinus, Thomas. *Anatome*. Leiden, 1673. PH.

24. Bartolozzi, Francesco. *Ricerche istorico-critiche circa alle scoperte d'Amerigo Vespucci*. Florence, 1789. Sabin 3800. APS.

25. Bartram, William. *Travels through North and South Carolina*. Philadelphia, 1791. Pritzel 447; Sabin 3870. APS.

26. Beatty, Charles. *Journal of a Two Months Tour with a View of Promoting Religion*. London, 1768. Sabin 4149. LCo.

27. Beck, Theodoric Romeyn. *On Insanity*. New York, 1811. Presented by author. APS.

28. Belknap, Jeremy. *American Biography*. 2 vols. Boston, 1794–1798. APS.

29. Bergmann, Torbern Olof. *Manuel du Mineralogiste*. Paris, 1784. APS.

30. Berkenhout, John. *Synopsis of the Natural History of Great Britain and Ireland*. 2 vols. 3d ed. London, 1795. TL–2 467; Blacker-Wood 1: 416. PH.

Bieberstein, L. B. Friderico Marschall. *See* Marschall a Bieberstein.

31. Bigelow, Jacob. *Florula bostoniensis*. Boston, 1812. Pritzel 772. APS.

32. Bivona-Bernardi, Antonio. *Sicularum plantarum, centuria prima*. Palermo, 1806–[1807]. Pritzel 805. PH.

33. Björnlund, Benedictus. *Materia medica selecta.* Aboae, 1797. APS.

34. Blanckaert, Steven. *Lexicon medicum renovatum.* 3d ed. Halle, 1739. PH.

35. Bloch, Marcus Elieser. *Abhandlung von der Erzeugung der Eingeweidewürmer.* Berlin, 1782. With sig. "1806." Countway 1: 811. ANSP.

36. Blumenbach, Johann Friedrich. Von der Federbuschpolypen in den Göttingischen gewässern. *Göttingisches Magazin* (1780) 11 pp. APS.

37. Blumenbach, Johann Friedrich. *Geschichte . . . der Knochen des menschlichen Körpers.* Göttingen, 1786. Cf. Countway 1: 820. APS.

38. Blumenbach, Johann Friedrich. *Beyträge zur Naturgeschichte.* Göttingen, 1790. APS.

39. Blumenbach, Johann Friedrich. *Abbildungen naturhistorischer Gegenstände.* Göttingen, 1796. 1st Heft only? Countway 1: 820. PH.

40. Blumenbach, Johann Friedrich. *De generis humani varitate nativa.* 3d ed. Göttingen, 1795. Borrowed by the US Exploring Expedition from Pennsylvania Hosp. Countway 1: 820. PH.

41. Blumenbach, Johann Friedrich. Über die Liebe der Thiere. *Göttingisches Magazin* (1781) 15 pp. APS.

42. Boate, Gerard, and Thomas Molineux, and others. *Natural History of Ireland in three parts.* Dublin, 1755 [1725, 1726]. Blacker-Wood 1: 578. PH.

43. Boerhaave, Hermann. *Index alter plantarum quae in horto academico Lugduno-Batavo aluntur.* (reprint of 1720) 2 vols. Leiden, 1727. Pritzel 931; TL−2 593. PH.

44. Boháč, Jan K. [Bohadsch, Johannes Baptista.] *De quibusdam animalibus marinis.* Dresden, 1761. "Loaned to the US Exploring Expedition from the Penn. Hospital." Blacker-Wood 1: 585. PH.

45. Boldo, Baltasar Emmanuel. *In insulam Cubensem nunc legatus Villanova Thomae.* Havana [1798]. PH.

46. Bonannio, P. Philippo. *Rerum naturalium historia.* Pars prima et secunda. Rome 1773 and 1782. PH.

47. Bonnet, Charles. *Oeuvres d'histoire naturelle et de philosophie.* Amsterdam, 1780. Cf. Blacker-Wood 1: 596, and cf. Pritzel 983. PH.

48. *Book of Common Prayer.* New ed. London, 1787. Sabin 6351. APS.

49. Bordiga, Benedetto. *Storia delle piante forestiere . . . medico ed economico.* 2 of 4 vols. Milan, 1791–1792. Pritzel 1000. PH.

50. Born, Ignaz von. *Index rerum naturalium Museum Caesarei Vindobonensis.* Vienna, 1778. Dryander 2: 320. PH.

51. Bossu, Jean Bernard. *Travels through . . . Louisiana.* 2 vols. London, 1771. Sabin 6466. APS.

52. Brackenridge, Hugh Henry. *Incidents of the Insurrection in the western Parts of Pennsylvania in the year 1794.* Philadelphia, 1795. Sabin 7189: "a scarce book." APS.

53. Bradley, Richard. *The History of Succulent Plants.* London, 1716–1727. 5 vols. in 1. Sig. "1801." Pritzel 1075; TL–2 699. PH.

54. Bradley, Richard. *New Improvements of Planting and Gardening.* 4th ed. London, 1724. Pritzel 1076. PH.

55. Brand, Thomas. *Case of a Boy mistaken for a Girl.* London, 1787. APS.

56. Brerewood, Edward. *Enquiries Touching the Diversity of Languages.* 4th ed. London, 1674. APS.

57. Brisson, Mathurin Jacques. *Ornithologie.* 6 vols. Paris, 1760. Dryander 2: 112; Blacker-Wood 1: 682. PH.

Brotero, Felix de Avellar. *See* Avellar Brotero.

58. Broussonet, Pierre Marie Auguste. *Ichthyologia sistens piscium descriptiones et icones.* London [n.d. Paris, 1782]. Inscribed on t.p. "ex dono auctore" the name erased. Blacker-Wood 2: 11; Dryander 2: 175. PH.

59. Bruyn, Cornelius de. *Travels into Muscovy.* London, 1759. Cf. Dryander 1: 136. LCo.

60. Buchoz, Pierre Joseph. *Dictionnaire raisonné universal des plantes.* 4 vols. Paris, 1770. Pritzel 1323; TL–2 874. PH.

Buffon, G. L. L. de. *See* LeLarge de Lignac, J. A.

61. Buffon, George Louis Leclerc de. *Natural History.* 2 vols. London, 1781. Barton's gift to William Bartram. LCo.

62. Bulliard, Jean Baptiste François. *Herbier de la France ou Collection complette des plantes indigenes de ce royaume.* 13 of 16 vols. Paris, 1780–[1793]. Pritzel 1356; TL–2 905. PH.

63. Bulliard, Jean Baptiste François. *Histoire des plantes venéneuses et suspectes de la France.* Paris, 1784. Pritzel 1354. PH.

64. Buonanni, Filippo. *Rerum naturalium historia . . . Museo Kircheriano.* 2 vols. Rome, 1773–1782. "B.S. Barton, August 10, 1813. Paid 14 dollars." Blacker-Wood 2: 49; cf. Dryander 1: 227. APS.

65. Burmann, Nicolai. *Flora indica . . . Prodromus florae Capensis.* Leiden and Amsterdam, 1768. Pritzel 1396; TL–2 935. PH.

66. Bustamante Carlos, Calixto. *El Lazarillo de Ciegos Caminantes desde Buenos-Ayres hasta Lima.* Gijon [probably Lima], 1773. Sabin 9566. APS.

67. Callender, James Thomson. *Sketches of the History of America.* Philadelphia, 1798. Sabin 10070. APS.

68. Carili, Giovauni Rinaldo. *Lettres Americaines dans lesquelles on examine l'origine, l'etat civil . . . des habitans de l'Amerique.* Tomes I, II. Boston, for Buisson, Paris, 1788. LCo.

69. Carlander, Christophoro. *Quid Linnaeo patri debeat medicina, dissertatione academica breviter adumbratum.* Upsala, 1784. Soulsby 990. APS.

70. Carver, Jonathan. *Three Years Travels . . . North America,* Philadelphia, 1796. Evans 30169. APS.

71. Casström, Samuel Nikolaus. *Dissertatio entomologica novas insectorum species.* Upsala, [1781]. Parts 1, 5, 6. APS.

72. [Castiglioni, Luigi.] *Storia delle piante forestiere le piu importante nell' uso medico.* 4 vols. Milan, 1791–1794. Pritzel 1596. PH.

73. Catesby, Mark. *Natural History of Carolina, Florida and the Bahama Islands.* 2 vols. London, 1771. "B.S.B. September 17, 1810." Pritzel 1602; Sabin 11509. PH.

74. Cathrall, Isaac. *A medical Sketch of Synochus maligna.* Philadelphia, 1794. Countway 2: 467. APS.

75. Cavanilles, Antonio José. *Icones et descriptiones plantarum.* 6 vols. Madrid, 1791–1801. Pritzel 1616; TL–2 1061. PH.

76. Cavanilles, Antonio José. *Observaciones sobre la historia natural . . . del Reyno de Valencia.* 2 vols. Madrid, 1797. Pritzel 1617. APS.

77. Chaisneau, Charles. *Atlas d'histoire naturelle ou Collection de Tableaux, relatifs aux trois règnes de la nature.* Paris, an XI [1802]. Inscribed on half title: "Mr. S. Bachelot Souligne, P.M." and on preface "B.S.B." PH.

78. Chalmers, Lionel. *Account of the Weather and Diseases of South Carolina.* London, 1776. Countway 2: 500. APS.

79. Charas, Moise. *New Experiments upon Vipers.* London, 1673. Countway 2: 520. APS.

80. Charleton, Walter. *Onomasticon zoicon . . . animalium . . . fossilium.* London, 1668. "1804." Dryander 2: 5. PH.

81. Charlevoix, Pierre François Xavier de. *A Voyage to North America.* Dublin, 1766. Vol. 2. Sabin 12143. LCo.

82. Chastelleux, François Jean, Marquis du. *Travels in North America.* 2d ed. London, 1787. Sabin 12229. APS.

83. Claesse, Lawrence. *Morning & evening Prayer, Litany . . . & several Chapters of the Old & New Testament translated into the Mahaque Indian Language.* New York, Wm. Bradford, 1715. APS.

84. *Clavigero, Francesco Saverio. *History of Mexico.* 2 vols. London, 1787. Charles Cullen transl. Sabin 13519.

Clusius. *See* L'Escluse.

85. Colonna, Fabio. *Fabi Columnae Lyncei.* Florence, 1744. Cf. Pritzel 1822. APS.

86. *Columbian Magazine.* Philadelphia, 1787–1790. Vols. 1, 2, 4. Sabin 14869: "extremely rare magazine." APS.

87. Conover, Samuel Forman. *An inaugural Dissertation on Sleep and Dreams . . . submitted to . . . Rev. William Smith, S.T.P. Provost . . . for the Degree of Doctor of Medicine.* Philadelphia, 1791. Evans 23290. LCo.

88. Cornut, Jacques Philippe. *Canadensium plantarum.* Paris, 1635. Received by the hospital 12 August 1837. Pritzel 1894; Sabin 16809; TL–2 1233. PH.

89. Coxe, William. *Travels in Switzerland.* 3 vols. London, 1789. APS.

90. Crameri, Johann Andreas. . . . *Elementa artis docimasticae.* 2 vols. Leiden, 1739. Inscribed in both vols. "Abr. Chovet M.D., Benjamin Smith Barton." PH.

91. Crantz, Heinrich Johann Nepomuk von. *Stirpium austriacarum.* 2d ed. Vienna, 1769. Pritzel 1954; TL–2 1265. PH.

92. Cruikshank, William Cumberland. *The Anatomy of the absorbing Vessels of the Human Body.* London, 1786. Countway 3: 76. APS.

93. Currie, William. *A Dissertation on the autumnal remitting Fever.* Philadelphia, 1789. APS.

94. Curtis, William. *Botanical Magazine.* London, 1798. Vol. 25. Z. Collins paid $80.00 for PH; rest missing. Pritzel 2007; TL–2 1290. LCo.

95. Cuvier, Georges. *Extrait d'un ouvrage . . . de Quadrupèdes.* Paris, 1800. "From the author." Publ. [November 17, 1800]. APS.

95a. *Cuvier, Georges. *Recherches anatomiques sur les reptiles.* Paris, 1807. Bell, W. J., Jr., *Bull. Med. Libr. Assoc.* 60 (1972): 548.

95b. *Cuvier, Georges. *Recherches sur les ossemens fossiles de Quadrupèdes.* 4 vols. Paris, 1812. loc. cit.

96. Darwin, Erasmus. *Phytologia.* Dublin, 1800. Barton's gift to William Bartram. Pritzel 2062. LCo.

97. Davies, John. *History of the Caribee Islands.* London, 1666. Sabin 72332. APS.

98. Dickson, James. *Fasciculus plantarum cryptogamicarum britanniae.* London, 1785–1801. Plates drawn by Sowerby. Pritzel 2224. PH.

99. Dillenius, Johann Jakob. *Hortus Elthamensis.* 2 vols. London, 1732. Pritzel 2285; TL–2 1471. PH.

100. *Dillwyn, Lewis Weston. *British Confervae.* London, 1809. Account, PH purchase, 3 May 1817. Pritzel 2287.

101. Doerner, Christian Friederich. *Observationum, ad historiam embryonis facientum, pars prima.* Tübingen, [1797]. APS.

102. Douglas, James. *Dissertation on the Antiquity of the Earth, read at the Royal Society, 12 May 1785.* London, 1785. LCo.

103. Douglass, William. *A Summary . . . History of British Settlements in North America.* 2 vols. Boston, 1749–1751. Sabin 20726. APS.

104. Drake, Daniel. *Notices concerning Cincinnati.* Cincinnati, 1810. Inscription: Dr. Barton . . . compliments of . . . the author. Sabin 20825. APS.

105. Duhamel du Monceau, Henri Louis. *Physiques des Arbres.* 2 vols. Paris, 1758. Pritzel 2468. APS.

106. Duméril, André Marie Constant. *Zoologie analytique.* Paris, 1806. Blacker-Wood 2: 725. PH.

107. Edwards, George. *Essays upon Natural History.* London, 1770. Blacker-Wood 3: 48. APS.

108. Edwards, Jonathan. *Notion of the Freedom of Will.* Wilmington, 1790. 4th ed. Sabin 21930. APS.

109. *Eliot, John. *Holy Bible . . . translated into the Indian Language.* Cambridge, 1663. Cf. *New Views* (1798), xii, where BSB believes he owns the second ed. Sabin 22154 or 22158.

110. *Ellis, John. *Natural History of Corallines.* London, 1755. Account, PH purchase 3 May 1817. Dryander 2: 339.

111. Ellis, John. *Directions for bringing over Seeds and Plants . . . Dionaea muscipula.* London, 1770. Pritzel 2665; Sabin 22319. PH.

112. Ellis, John. *Copies of two Letters 1. to Dr. Linnaeus 2. to Wm. Aiton.* London, 1771. Pritzel 2666. PH.

113. [Engel, Samuel.] *Essai sur cette question: quand et comment l'Amérique a-t-elle été peuplée d'hommes et d'animaux?* 5 vols. Tome Premier [-cinquième] Amsterdam, 1767. Sabin 22568. LCo.

114. Evelyn, John. *Silva, or Discourse.* York, 1776. Pritzel 2766. PH.

115. Fabricius, Johan Christian. *Entomologia systematica.* Copenhagen, 1792–1794. vols. 2, 3, 4. PH.

116. Fabricius, Johan Christian. *Systema rhyngotorum.* Braunschweig, 1803. From the author, acc. letter to BSB. Blacker-Wood 3: 281. PH.

117. Fabricius, Johan Christian. *Systema piezatorum.* Braunschweig, 1804. Inscribed "From the author." PH.

118. *Farquar, John. *Sermon of John Farquar.* Dublin, 1782.

119. Fernel, Jean. *Universa medicina.* 5th ed. Frankfurt, 1592. Inscribed on t.p., "John Tanner." Countway 3: 665. PH.

120. Fischer, Gotthelf von Waldheim. *Notice d'un animal fossile d'Siberie inconnu aux naturalistes.* Moscow, 1808. PH.

121. Flayer, John (Part 1) and Baynard, Edward (Part 2). *History of cold Bathing, both ancient and modern.* 5th ed. London, 1722. PH.

122. Fordyce, George. *Dissertation on simple Fever.* London, 1794. Countway 4: 4. APS.

123. Forster, Johan Reinhold. *Enchiridion historia naturali inserviens.* Halle, 1788. Dryander 1: 187; TL–2 1827. PH.

124. Franklin, Benjamin. *Expériences et observations sur le l'électricité faites à Philadelphie en Amérique.* 2d ed. 2 vols. Paris, 1756. transl. by Thomas-François Dalibard. "I have been able to find only

Volume II of this edition," P. L. Ford, Franklin Bibliog. (1886), 81. PH.

125. Gage, Thomas. *A New Survey of the West Indies*. 4th ed. London, 1699. Sabin 26301. APS.

126. Garcia, Gregorio. *Origen de los Indios del Nuevo Mundo*. 2d ed. Madrid, 1729. Sabin 26567. APS.

127. Garcilaso de la Vega. *Histoire de la conquête de la Floridae*. . . . Tome Premier [-Tome Seconda]. Leiden, 1731. Sabin 78748. LCo.

128. Garriga, Joseph. *Descripcion del esqueleto de un Quadrupedo*. Madrid, 1796. PH.

129. Gass, Patrick. *Journal of Voyages and Travels . . . of Capt. Lewis and Capt. Clarke*. Pittsburgh, 1807. Sabin 26741. APS.

130. Geoffroy, Etienne Louis. *Histoire abrégée des insectes . . . aux environs de Paris*. 2 vols. Paris, 1764. APS.

131. Geralin, Johanne Ericus. *Dissertatio botanica Protea*. Upsala, 1781. Student thesis attributed to Thunberg. TL-1 1299 PH.

132. *Gerard, John. *The Herball* [probably the emaculata, T. Johnson ed. London, 1633]. Missing from PH although Z. Collins records paying $5.00 for it. Account, PH purchase 3 May 1817. Pritzel 3282.

133. Gertanner, Christoph. *Aufangsgründe der antiphlogistichen chemie*. Berlin, 1792. APS.

134. Giseke, Paulus Dietrich. *Caroli Linne, Praelectiones in ordines naturales plantarum*. Hamburg, 1792. Pritzel 5434; Cat. Bib. Nat. Paris 61: 9; TL–2 2029. PH.

135. Gmelin, Johann Georg. *Flora sibirica*. 4 vols. St. Petersburg, 1799. Pritzel 3381; TL–2 2047. PH.

136. Gookin, Daniel. *Historical Collections of the Indians in the North East*. "Apollo Press," 1792. Gift of J. Belknap to BSB, see letter BSB to Belknap 28 August 1793. Sabin 27959. APS.

137. Gouan, Antoine. *Histoire des Poissons*. Strasbourg, 1770. Ex dono auctoris, 1804. Blacker-Wood 3: 603. PH.

138. Grant, Charles. *History of Mauritius or the Isle of France*. London, 1801. APS.

139. Greenwood, James. *An Essay towards a practical English Grammar, describing the Genius and Nature of the English tongue*. 2d ed. London, 1722. "Nov. 8, 1807." LCo.

140. Griffin, Edward Dorr. *Kingdom of Christ; a missionary Sermon, preached in Philadelphia, May 23, 1805.* Philadelphia, 1805. APS.

141. Gronovius, Johann Frederick. *Flora virginica.* Leiden, 1739–1743. "Cadwalader Colden, Ex Dono C. L. Gr. Fr. Gronovié 1744" on flyleaf. Pritzel 3607; TL–2 2189. PH.

142. Gronovius, Johann Frederick. *Flora virginica.* 2d ed. Leiden, 1762. T.p. and et seq. removed for repro in BSB's own edition of *Flora virginica.* Library copy lacks folding map. Pritzel 3607; TL–2 2189. PH.

143. *Gronovius, Johann Frederick. *Index supellectilis lapidae.* Leiden, 1740. BSB told A. Bruce, 8 September 1810, in letter, that he owned a copy. Dryander 4: 26.

144. Groot, Hugo de (Grotius). *Annales et historiae de rebus belgicius.* Amsterdam, 1658. APS.

145. Haij. Isak. *Dissertatio entomologica sistens insecta svecica . . . partem quintam.* Upsala, [1794]. Thesis under Thunberg. APS.

146. *Haller, Albert von. *Bibliotheca botanica.* Zurich, 1771–1772. 2 vols. Account, PH purchase 3 May 1817. Pritzel 3727.

147. Haller, Albert von. *Historia stirpium indigenarum helvetiae inchoata.* Bern, 1768. "Bought of Dr. Andrew Ross Dec. 3, 1798." Pritzel 3725; TL–1 473. PH.

148. Hanin, L. *Cours de botanique et de physiologie vegetal.* Paris, 1811. Pritzel 3762. PH.

149. Harris, Thaddeus Mason. *Natural History of the Bible.* Boston, 1793. PH.

150. Harris, Thaddeus Mason. *Journal of a Tour . . . account of Ohio.* Boston, 1805. Sabin 30515. APS.

151. Haüy, René Just. *Sur l'électricité des minéraux.* Paris, n.d. Cf. *Dictionary of Scientific Biography.* APS.

152. Hermann, Paul. *Paradisus batavus.* Leiden, 1705. Bought at auction of Wm. Byrd books, October 1781, by Du Simitière for 5 shillings. Pritzel 3994. Wolf 117. PH.

153. Hernandez, Francisco. *Opera.* 3 vols. Madrid, 1790. BSB loaned to C. Wistar in 1806 (T. P. Barton in BPL). Pritzel 4001; Sabin 31517. APS.

154. Herrera, Antoine de. *Description des Indes occidentales. Qu'ou . . . Nouveau monde.* Transl. do Espinol. Amsterdam, 1622. Sabin 31543. APS.

155. *Hill, John. *British Herbal.* London, 1756. Account, PH purchase 3 May 1817. Pritzel 4063.

156. Hillary, William. *Observations on the Changes of the Air.* London, 1766. APS.

157. Hoffmann, Georg Franz. *Germany's Flora or A Botanical Pocket Companion for the Year 1791 [and 1795].* 2 vols. London and Erlangen, [1792–1796]. BSB loaned to T. Nuttall for expedition to "Northwest." Cf. Pritzel 4132. See Figure PH.

158. Holmes, Abiel. *Life of Ezra Stiles.* Boston, 1798. Sabin 32582. APS.

159. [Home, Henry, Lord Kames.] *Gentleman Farmer.* 4th ed. Edinburgh, 1798. BSB's gift to John Bartram Jr. LCo.

Hooker, William. *See* Salisbury, R. A.

160. Hornemann, Jens Wilken. *Hortus regius botanicus hafniensis in usum Tyronum et Botanophilorum.* Copenhagen, [1815]. "Celeberrimo Bartonio Auctor J. W. Hornemann." Pritzel 4265. PH.

161. Houstoun, William. *Reliquiae Houstounianae: seu plantarum in America meridionali a Gulielmo Houstoun.* London, 1781. Pritzel 4290; TL–1 553. Probable gift of Joseph Banks, its sponsor. PH.

162. *Hudson, Wm. *Flora anglica.* London, 1762. Account. PH purchase, 3 May 1817. Or 3d ed. 1798.

163. *Hull, John. *Elements of Botany.* 2 vols. Manchester, 1800. Account. PH purchase, 3 May 1817. Pritzel 4323.

164. Humboldt, Alexander von, and A. Bonpland. *Voyage aux régiones équinoxiales du nouveau continent.* Paris, 1807–1814. BSB bought Part 2 "Travels" 13 Feb. 1811 from E. Earle & Co. Cf. Sabin 33752. APS.

165. Ingenhousz, Jan. *Experiments upon Vegetables.* London, 1779. Pritzel 4435. PH.

166. Jacquin, Nikolaus Joseph. *Selectarum stirpium americanarum historia.* Vienna, 1763. Pritzel 4362; Sabin 35521. PH.

167. Jefferson, Thomas. *Notes on the State of Virginia.* London, 1782. Cf. *New Views* (1798), xx. Sabin 35894. APS.

168. Jones, Samuel. *On Hydrocele.* Philadelphia, 1797. Austin 1088. APS.

169. Juncker, Joannes. *Conspectus medicinae theoretico-practicae, tabulis CXVI omnes primarios morbos methodo.* Halle, 1718. Cf. Countway 5: 558. PH.

170. Jussieu, Antoine Laurent de. *Genera plantarum.* Torino, 1791. "Amos Gregg, Jr. Jan. 17, 1807." Pritzel 4549. PH.

171. Kaempfer, Engelbert. *Icones selectae plantarum quas in Japonia collegit.* London, 1791. "From Sir Joseph Banks." Pritzel 4565; TL–1 593. PH.

172. Kalm, Peter. *Travels into North America.* J. R. Forster transl. 3 vols. Warrington and London, 1770. Sabin 36989; TL–2 3493. APS.

173. Keith, William. *History of the British Plantations in America. Part I. Virginia.* London, 1738. Sabin 37240. APS.

174. Labatt, Samuel Bell. *An Address to the medical Practitioners of Ireland, on the subject of the Cow Pock.* Dublin, 1805. cf. Countway 6: 2. APS.

175. Lacépède, Bernhard Germain Etienne, comte de. *Histoire naturelle des Quadrupedes ovipares et des serpens.* 2 vols. Paris, 1788–1789. Dryander 2: 14; cf. Blacker-Wood 4: 560. PH.

176. Lacépède, Bernhard Germain Étienne, comte de. *Discours . . . du Cours de Zoologie donné dans le Museum National d'Histoire naturelle.* Paris, an IX [1800]. APS.

177. Lacépède, Bernard Germain Étienne, comte de. *Sur l'Histoire des Races ou principales Variétés de l'Espèce humaine.* [Paris, 1800]. APS.

178. Laet, Johannes de. *Novus orbis.* Antwerp, 1633. Sabin 38557. APS.

179. Lamarck, Jean B. de, and A. P. de Candolle. *Flore française.* Paris, 1805. Pritzel 1468; TL–1 641; TL–2 987. PH.

180. Lambe, William. *Additional Reports on the Effects of a peculiar Regimen in Cases of Cancer, Scrofula, Consumption, Asthma, and other chronic Diseases.* London, 1815. "London, July 29, 1815 BSB." Countway 6: 27. APS.

181. Landivar, Raphaele. *Rusticatio Mexicana.* 2d ed. Bologna, 1782. Sabin 38839. APS.

182. La Rochefoucauld-Liancourt, F. A. F. *Travels through the United States of North America.* London, 1799. Sabin 39056. APS.

183. Latreille, Pierre André. *Histoire naturelle des Salamandres de France.* Paris, 1800. To BSB from F. A. Michaux. APS.

Latrobe, Christian Ignatius. *See* G. H. Loskiel.

184. Laws, John. *On the Rationale of the Operation of Opium on the Animal Economy*. Wilmington, Delaware, 1797. Evans 32364. APS.

185. Lawson, John. *History of Carolina*. London, 1714. Sabin 39452. APS.

186. Leon Pinelo, Antonio Rodriguez de. *Epitome bibliotheca oriental y occidental, nautica y geografica*. 3 vols. Madrid, 1737–1738. Sabin 40053. APS.

187. L'Escluse, Charles de. *Curae posteriores . . . Accessit seorsim Everardi Vorstii de Clusii*. [Antwerp], 1611. Pritzel 1761; TL–2 1150. PH.

188. LeLarge de Lignac, Joseph Adrien. *Lettres à un Américain sur l'histoire naturelle . . . de monsieur de Buffon*. Hamburg, 1751. Dryander 5: 14 has Joseph Albert LaLande de Lignac; Sabin 9066 and 41054. PH.

189. Leon, Solomon de. *Dissertatio medica inauguralis de inflammatione*. Leiden, 1790. PH.

190. Le Page du Pratz, A. S. *History of Louisiana . . . transl. from the French . . . a new edition*. London, 1774. BSB loaned to M. Lewis, June 1803, "conveyed by me to the Pacific Ocean," returned 9 May 1807. Sabin 40124. LCo.

191. Lery, Jean de. *Historia navigationis in Brasiliam quae et America dicitur*. [Geneva], 1586. Sabin 40153. LCo.

192. *Lettsom, John Coakley. *Natural History of the Tea-Tree*. London, 1772. Samuel Paterson, Edinburgh bookseller, sent copy 21 Nov. 1808. Pritzel 5252.

193. Lettsom, John Coakley. *Memoirs of John Fothergill*. 4th ed. London, 1786. Cf. DNB 11: 1015 on 4th ed. BSB purchased from E. Earle & Co., 21 Feb. 1811, for $3.25. Ewan Lib. (Emily Cheston copy).

194. *Lettsom, John Coakley. *Naturalist's Companion*. 3d. ed. London, 1799. Author presentation to BSB (undated incompl. letter at APS).

195. l'Heritier, Charles Louis. *Stirpes novae aut minus cognitae*. Paris, 1784. 2 fascicles only. Fasc. 2 gift of "Mr. Andrew Michaux," 8 Feb. 1793. Pritzel 5268; TL–2 4484. PH.

196. Lightfoot, John. *Flora scotica*. 2 vols. London, 1777. Pritzel 5308; TL–2 4522. PH.

197. Ligon, Richard. *True and Exact History of the Island of Barbadoes.* London, 1673. "Extremely rare in the US," *Phila. Med. Phys. Jour.* 1 (1805) Pt. 2: 134. Wolf 140; Sabin 41058. APS.

198. Linnaeus, Carolus. *Fauna svecica.* Stockholm, 1746. Gift from J. Acrelius to Wm. Bartram; to BSB 18 Feb. 1794. Soulsby 1151. PH.

199. Linnaeus, Carolus. *Flora zeylanica.* Stockholm, 1747. Pritzel 5422; Soulsby 420; TL–2 4740. PH.

200. Linnaeus, Carolus. *Flora svecica.* 2d ed. Stockholm, 1755. "Received Nov. 13, 1812 BSB." Pritzel 5414; Soulsby 409. TL–2 4731. PH.

201. Linnaeus, Carolus. *Systema naturae.* 9th ed. Leiden, 1756. Pritzel 5404; Soulsby 57; TL–2 4742. PH.

202. Linnaeus, Carolus. *Naturlyke historie of intvoerize beschrying der dieren plasten en mineralslen.* 37 vols.! Amsterdam, 1761–1785. Soulsby 73; TL–2 3080. PH.

203. Linnaeus, Carolus. *Systema vegetabilium.* 14th ed. Göttingen, J. A. Murray, 1784. BSB loaned to C. F. H. Denke 26 April 1811. Pritzel 5404; Soulsby 583. PH.

204. Linnaeus, Carolus. *Termini botanici.* Hamburg, 1787. Pritzel 5433; cf. Soulsby 2194. PH.

205. Linnaeus, Carolus. *Philosophia botanica.* 3d ed. Berolini, 1790. Pritzel 5426; Soulsby 449; TL–2 4760. PH.

206. Linnaeus, Carolus. *Praelectiones in ordines naturales plantarum.* Hamburg, 1792. Soulsby 620. PH.

207. Linnaeus, Carolus. *Flora lapponica exhibens plantes per Lapponiam.* London, 1792. Borrowed by the US Exploring Expedition from Pennsylvania Hospital. Soulsby 281. PH.

208. Linnaeus, Carolus. *Species plantarum.* 4th ed.: 4 vols. in 9; Berlin, 1797–1805. Lib. lacks vols. 5 and 6. Soulsby 512. PH.

209. Linnaeus, Carolus [by Stewart, Charles]. *Elements of Natural History.* London, 1801–2. Vol. 1 only. Soulsby 153. PH.

210. L'Obel, Matthias. *Stirpium adversaria nova.* London, 1570 [1571]. TL–2 4906. PH.

211. Lockette, Henry Wilson. *On the warm Bath.* Philadelphia, 1801. Presented by author. APS.

212. Logan, James. *Experimenta et meletemata de plantarum generatione nec non canonum.* Leiden. 1739. Bookseller's ticket "Libreria de Malleu, Valencia." Pritzel 5582; Sabin 41796. PH.

213. *Loskiel, George Henry. *History of the Mission of the United Brethren among the Indians in North America.* London, 1794. BSB had purchased, see letter to Heckewelder, 11 February 1798. Sabin 42110. APS.

214. Loureiro, Joao de. *Flora cochinchinensis.* 2 vols. Lisbon, 1790. Pritzel 5637; TL-2 5038. PH.

215. Ludwig, Christian Gottlieb. *Institutiones historio physicae regni vegetabilis.* 2d ed. Leipzig, 1757. Pritzel 5664; TL-2 5069. PH.

216. *Mackenzie, Alexander. *Voyages from Montreal.* 2 vols. [Probably Philadelphia, 1802.] BSB loaned to Heckewelder in 1804.

217. Mackenzie, Colin. *On the Dysentery.* Philadelphia [1797]. Presented by the author to BSB. Countway 6: 457. APS.

218. M'Mahon, Bernard. *American Gardener's Calendar.* Philadelphia, 1806. Sabin 43560. PH.

219. Malpighi, Marcello. *Opera omnia.* London, [1686]–1687. Bound with *Opera posthuma . . . ejusdem vita.* London, 1697. "London, August 1815." Pritzel 5764. PH.

220. Marschall von Bieberstein, Friedrich August. *Flora taurico-caucasica.* 2 vols. Kharkov, 1808. "From Dr. Fisher at Gorenki near Mores." Pritzel 5831; TL-2 5453. PH.

221. Marshall, Humphry. *Arbustum americanum.* Philadelphia, 1785. Interleaved. Pritzel 5834; Sabin 44776. PH.

222. Martyn, Thomas. *Thirty-eight Plates with Explanations.* London, 1788. 38 pls. incl. frontis. Pritzel 5928; TL-2 5569. PH.

223. Martyn, Thomas. *The Language of Botany: being a Dictionary of Terms.* 2d ed. London, 1796. Pritzel 5930. PH.

224. May, Frederick. *On the animating principle, or anima mundi . . . and tetanus or lock-jaw.* Boston, 1795. For degree B. M., Harvard, 3 July, 1795. APS.

225. Melsheimer, Frederick Valentine. *Catalogue of Insects of Pennsylvania. Part I.* Philadelphia, 1806. Presentation copy to BSB. Sabin 47470. LCo.

226. Michaux, André. *Histoire des chênes de l'Amérique.* Paris, an IX [1801]. Pritzel 6194. PH.

227. Michaux, André. *Flora boreali-americana.* 2 vols. Paris, 1803. Pritzel 7611; Sabin 48691; TL–2 5958. PH.

228. Michaux, André. *Quercus or Oaks, from the French of Michaux . . . by Walter Wade.* Dublin, 1809. Sabin 48701; TL–2 5957. PH.

229. *Minot, George Richards. *History of the Insurrections in Massachusetts.* Worcester, 1788. Sabin 49324.

230. Mirbel, C. F. D. de. *Exposition de la théorie de l'organization végétale.* 2d ed. Paris, 1809. Author presentation copy. Pritzel 6290. PH.

231. Molina, Giovanni Ignatio. *Compendio de la historia geographica, natural y civil del Rayno de Chile.* Madrid, 1788–1795. See letter Wm. Hamilton to BSB 21 Dec. 1802. Sabin 49889. LCo.

232. Monardes, Nicolas. *Joyfull News out of the New found World.* 1597. Sabin 49944. BPL.

233. Moore, Samuel W. *On the medical Virtues of the white Oxide of Bismuth.* New York, 1810. Presentation to BSB. Degree M.D. from Columbia College. APS.

234. Morison, Robert. *Hortus regius Blesensis.* London, 1669. "Ex Libris Jacobi Bobart," and "Ex Bibl. J. Banister, ex dono [1679] in Virginia." Wm Byrd bookplate. Pritzel 6462; Wolf 155. PH.

235. Mott, Valentine. *Experimental Inquiry into the chemical and medical Properties of the Statice limonium of Linnaeus.* New York, 1806. APS.

236. Müller, Otho Frederik. *Zoologica Danica.* 4 vols. Copenhagen, 1788. (3 vols. in 1, lacks vol. 4.) Blacker-Wood 5: 414; Wood 475. PH.

237. Muñoz, Juan Bautista. *Historia del Nuevo-mundo.* Tom. I. Madrid, 1793. Sabin 51343. APS.

238. *Munro, Alexander. *Structure and Physiology of Fishes explained and compared with those of Man and other Animals.* Edinburgh, 1785. Account. PH purchase 3 May 1817.

239. Muratori, Ludovico Antonio. *Relation of the Missions of Paraguay.* London, 1759. Sabin 51422. APS.

240. Newton, James. *A compleat herbal . . . of [plants] not to be found in the herbals.* Amsterdam, 1764. Cf. Pritzel 6689 and TL–2 6779. Amsterdam ed. Unknown to B. Henrey. *Hort. Lit.*, London, 1975.

Norberto, J. I. *See* Aucourt e Padilha

241. Oeder, Georg Christian. [*Flora Danicae*] *Icones plantarum sponte nascentium in regnis Daniae et Norvegiae.* 9 vols. Copenhagen, 1770–

1816. See BSB letter 12 June 1803 to C. L. Willdenow. Pritzel 6799; TL–2 8001: 17 vols. Vol. 9 (1813) was the last received. PH.

242. Pallas, Peter Simon. *Novae species quadrupedum*. Erlangen, 1778. Blacker-Wood 7: 122. APS.

243. Palm, Johan Jakob. *Dissertatio gradualis sistens nonullas de lapide obsidiano*. London [1799]. APS.

244. Paris. *Histoire de la Société Royale de Médecine*. Années 1777 & 1778. Paris, 1780. PH.

245. Paris. *Actes de la Société d'Histoire Naturelle de Paris*. Vol. 1, pt. 1. Paris, 1792. Gift of L. Bosc. PH.

246. Peña, Pierre, and Mathias de Lobel. *Stirpium adversaria nova*. London, 1570 [i.e. 1571]. Plates missing. Pritzel 7029. PH.

247. Pendleton, James J. *Materials for an Alphabet to the Science of Medicine*. Philadelphia, 1804. "For Dr. Barton with the highest respects of the author." PH.

248. Pennant, Thomas. *British Zoology*. Vols. 1–4. Warrington, 1776–1777. Daines Barrington's copy with his notes. Blacker-Wood 7: 210; Dryander 2: 25. LCo.

249. Pennant, Thomas. *Indian Zoology*. 2d ed. London, 1790. Gift of Pennant; BSB loaned to Wm. Bartram 30 December 1792 (Barton Papers, 1: 8, Historical Society of Pennsylvania). Dryander 2:30. PH.

250. Pennant, Thomas. *Arctic Zoology*. 2d ed. London, 1792. Vol. 1 only. from the author. Dryander 2: 34. PH.

251. Pennant, Thomas. *Natural History of Quadrupeds, Birds, Fishes and Insects*. 2 vols., 3d ed. London, 1793. 1 plate. Presentation copy. Blacker-Wood 7: 210; Dryander 2: 47. PH.

252. Pennant, Thomas. *Literary Life of the late Thomas Pennant*. London, 1793. Presentation copy, "recd May 14, 1794." "For my Son, Thomas Pennant Barton BSB 1804." Dryander 1: 175; Blacker-Wood 7: 210. BPL.

253. Perrault, Claude. *Suite des Mémoires pour servir a l'Histoire Naturelle des Animaux*. 3 vols. Amsterdam & Leipzig, 1780. Cf. Dryander 2: 276. PH.

254. Petty, William. *Several Essays in Political Arithmetick*. London, 1699. Wm. Byrd bookplate. Wolf 167. APS.

255. *Philibert, J. C. *Dictionnaire abrégé de botanique*. Paris, 1803. F. A. Michaux purchased for BSB 11 January 1811 (misc. papers, receipts). Pritzel 7120.

256. Piso, Guilielm, and Georg Marcgrave. *Historiae rerum naturalium Brasile. Medicina Brasiliensi . . . cum appendice*. Antwerp, 1648. Pritzel 7157; Sabin 63028. PH.

257. Plaz, Anton Wilhelm. *Organicarum in plantis partium historia physiologica*. Leipzig, 1751. Pritzel 7188. PH.

258. Plenk, Joseph Jacob, ritter von. *Physiologia et pathologia plantarum*. Vienna, 1794. Pritzel 7202. PH.

259. Plescheef, Sergey. *Survey of the Russian Empire . . . from the Russian with considerable additions by James Smirnove*. London, 1792. LCo.

260. Plukenet, Leonard. *Phytographia*. [3 vols.] 2d ed. London, 1769. 4 pts. with separate t.p.'s. Z. Collins listed 3 vols. $40.00. Pritzel 7212; TL–2 8064. PH.

261. Plumier, Charles. *Traité des fougeres de l'Amérique*. Paris, 1705. "BSB Aug. 1807." Pritzel 7216; Sabin 63458. PH.

262. [Pownell, Thomas.] *Administration of the Colonies*. 4th ed. London, 1768. Sabin 64817. APS.

263. Priestley, Joseph. *Considerations on the Doctrine of Phlogiston and Decomposition of Water*. Philadelphia, 1796. Presentation to the hospital by BSB. Evans 31049. PH.

264. Priestley, Joseph. *Rudiments of English Grammar*. New ed. London, 1798. APS.

265. Pulteney, Richard. *Account of the Writings and Life of Linnaeus*. London, 1781. Soulsby 3628. APS.

266. Pulteney, Richard. *Revue général des écrits Linné*. 2 vols. London & Paris, 1789. "Number 118 BSB." Soulsby 3629. PH.

267. Pulteney, Richard. *Historical and Biographical Sketches of the Progress of Botany*. 2 vols. London, 1790. "BSB 1790." Pritzel 7265. PH.

268. Ramsay, David. *An Oration on the Cession of Louisiana to the United States, . . . 12th of May, 1804, in St. Michael's Church, Charleston, South Carolina*. Charleston, 1804. APS.

269. Ray, John. *Synopsis methodica stirpium britannicarum*. 3d ed. London, 1724. Pritzel 7438. PH.

270. Ray, John. *L'Histoire naturelle, éclaircie dans une de ses parties principales. l'Ornithologie.* Paris, 1767. Blacker-Wood 7: 591. PH.

Ray, John. *Ornithology. See* Willughby, F.

271. [Rede, Leman Thomas.] *Bibliotheca americana; or a chronological Catalogue . . . upon the Subject of North and South America.* London, 1789. Sabin 5198. APS.

272. *Reeve, Henry. *On the Torpidity of Animals.* London, 1809. "I lately purchased . . . ," *Philosophical Magazine* (London) 35 (1810): 241.

273. Reimarus, Johann Albrecht Heinrich. *De animalium inter naturae regna statione et gradibus. Oratio pro suscipiendo manere professoris physices et historiae naturalis.* Hamburg, 1796. APS.

274. [Reisch, Gregor.] *Margarita philosophica nona cui insunt sequentia.* Strassburg, 1512. Sabin 69127. PH.

275. Richard, Louis Claude Marie. *Demonstrations botaniques.* Paris, 1808. Pritzel 7606; TL–2 9154. PH.

276. Rittenhouse, David. *Oration delivered . . . 1775.* Philadelphia, 1775. APS.

277. Robson, Joseph. *Account of six years Residence in Hudson's Bay.* London, 1752. Sabin 77259. APS.

278. Rochefort, Charles de. *Histoire naturelle et morale des Iles Antilles de l'Amérique.* Lyon, 1667. Sabin 72317. PH.

279. Rohde, Michael. *Monographiae cinchonae generis tentamen, fragmentum ex materia medica.* Göttingen, 1804. Cf. Pritzel 7731. APS.

280. Romans, Bernard. *Concise Natural History of East and West Florida.* New York, 1776. Sabin 72993. APS.

281. Rumpfius, Georg Eberhard. *Herbarium amboinense.* [Reissue of 1741–1750.] 6 vols. Amsterdam, 1750–1755. Pritzel 7908. PH.

282. Saint-Simon, Maximilien Henri, Marquis de. *Dos Jacinthes, de leur anatomie, reproduction et culture.* Amsterdam, 1768. Pritzel 7996; TL–2 10.092. PH.

283. Salazar, Thomas de. *Tratado del uso de la Quina.* Madrid, 1791. APS.

284. Salerne, [François]. *L'Histoire naturelle . . . l'Ornithologie . . . des Oiseaux.* Paris, 1767. Blacker-Wood 8: 59. PH.

285. Salisbury, Richard Anthony. *Paradisus londinensis.* 2 vols. London, 1805–1808. Composite work with artist Wm. Hooker who contributed 117 pls. in color. Copy examined in 1952, but missing since. Pritzel 8003. PH.

286. Sargent, Winthrop. *Papers in Relation to the Conduct of Governor Sargent.* Boston, 1801. APS.

287. Saussure, Horace Benedict de. *Voyages dans les Alpes.* 4 vols. Neuchâtel, 1779–1796. 2 of 4 vols. Dryander 1: 242. APS.

288. Sayre, Francis Bowes. *On the Causes which produce a Predisposition to phthisis pulmonalis, and the method of obviating them.* Trenton, 1790. APS.

289. Schaeffer, Jakob Christian. *Isagoge in botanicum expeditionem.* Regensburg, 1759. Pritzel 8110; TL–1 753. PH.

290. Schaeffer, Jakob Christian. *Elementa ornithologica iconibus.* Regensburg, 1774. Blacker-Wood 8: 104. PH.

291. Schaeffer, Jakob Christian. *Museum ornithologicum.* Regensburg, 1789. Bound with *Elementa ornithology.* PH.

292. Schaeffer, Jakob Christian. *Icones insectorum.* 3 vols. in 1. Regensburg, 1791. "Dec. 18, 1807." PH.

293. Schneider, Johann Christian. *Historia amphibiorum naturalis.* 2 vols. Jena, 1799–1801. PH.

294. *Schoepf, Johann David. *Materia medica americana.* Erlangen, 1787. PH missing, Z. Collins paid $1.00. Pritzel 8316.

295. Schoepf, Johann David. *Reise durch einige der mittlern und südlichen Vereinigten Nordamerikanischen Staaten.* 2 vols. Erlangen, 1788. Sabin 77757. APS.

296. Schreber, Johann Christian Daniel. *Beschreibung der Gräser.* 2 vols. in 1. Leipsig, 1766–[1779]. "BSB August 1807." Pritzel 8395; TL–1 1195. PH.

297. Schulze, Johann Heinrich. *See* Blanckaert, Steven.

298. Scopoli, Joannes Antonius. *Deliciae florae et faunae insubricae . . . Austriaca.* 3 vols. in 1. Pavia, 1786–1788. See letter BSB to Targioni-Tozzetti 25 June 1809 (APS). Pritzel 8558; TL–1 1217. PH.

299. Senebier, Jean. *Physiologie végétale.* 5 vols. Geneva, 1800. Pritzel 8609. PH.

300. Sharaf Al-Din, Yazdi. *Histoire de Lemur . . . Empereur des Moguls et Tartaras.* 4 vols. Delft, 1723. LCo.

301. Shaw, George, and J. F. Stephens. *General Zoology or Systematic Natural History.* 6 vols. in 12. London, 1800–1804. Wood 566; Blacker-Wood 8: 262. PH.

302. Shaw, George. *Zoological Lectures delivered at the Royal Institution.* 2 vols. London, 1809. Blacker-Wood 8: 263. PH.

Sibthorp, John. *See* J. E. Smith.

303. Sloane, Hans. *Catalogus plantarum quae in insula Jamaica.* London, 1696. On half-title: "Ex dono Domini Gilbert Ransom"; on t.p. "John Furner" and "Benjamin Smith Barton." Sabin 82166; Pritzel 8722. PH.

304. Smart, Christopher Goldsmith. *The World Displayed, or a Curious Collection of Voyages and Travels . . . from the Writers of all Nations.* Vol. 1 of 8 vols. 1st Amer. ed. Philadelphia, 1795. John Bartram, Jr.'s copy. Sabin 105486. LCo.

305. Smeathman, Henry. *Some account of the termites, which are found in Africa and other hot climates.* London, 1781. "BSB April 4, 1808." Dryander 2: 273. PH.

306. Smith, James Edward. *Plantarum icones hactens ineditae.* 3 fasc. in 1 vol. London, 1789–1791. From Pennant 27 May 1797. Pritzel 8734. PH.

307. Smith, James Edward. *Flora lapponica.* Editio altera. London, 1792. Pritzel 5410; TL–1 1239. PH.

308. Smith, James Edward. *Tracts relating to Natural History.* London, 1798. Inscribed from the author to BSB. Pritzel 8740. PH.

309. Smith, James Edward. *Flora brittanica.* 3 vols. London, 1800–1804. Pritzel 8742; TL–1 1241. PH.

310. Smith, James Edward, and John Abbot. *Natural history of the rarer lepidopterous insects of Georgia.* 2 vols. London, 1797. Dryander 5: 36; Sabin 25. PH.

311. Smith, James Edward, and John Sibthorp. *Florae graecae prodromus.* London, 1806. Vol. 1 only. "November 1809." Pritzel 8744. PH.

312. Smith, Samuel. *History of the colony of Nova-Caesaria or New Jersey.* Burlington, 1765. Evans 10166; Sabin 83980. APS.

313. Smith, Thomas. *Essay on Wounds of the Intestines.* Philadelphia, 1805. Cf. Countway 9: 296. APS.

314. Smith, William. *History of Virginia.* Williamsburg, 1747. APS.

315. Smith, William. *History of the Province of New York.* 2d ed. Philadelphia, 1792. Sabin 84568. APS.

316. Solis y Ribadeneyra, Antonio de. *History of the Conquest of Mexico by the Spaniards.* London, 1724. Sabin 86487. LCo.

317. Soulavie, Jean-Louis Giraud. *Histoire naturelle de la France Méridionale.* 7 vols. Paris, 1780–1784. Dryander 1: 237. PH.

318. Spalding, Lyman. . . . *On the Production of Animal Heat.* Walpole, N.H., 1797, Austin, 1804; Countway 9: 371. APS.

319. *Spallanzani, Lazzaro. *Expériences sur la digestion de l'homme et de différentes espéces d'animaux.* 1st ed. Geneva, 1783. transl. by Jean Senebier. Cf. Countway 9: 372. Argosy Catalogue 706 (1982): 492.

320. Spallanzani, Lazzaro. *Tracts on the Natural History of Animals and Vegetables.* 2d ed. 2 vols. Edinburgh, 1803. cf. Pritzel 8811. PH.

321. Sparrman, Andreas. *Museum Carlsonianum.* 4 parts in. 1 vol. Stockholm, 1786–1789. No BSB sig. Account, PH purchase 3 May 1817. Dryander 2: 118; Blacker-Wood 8: 446. PH.

322. Sprengel, Kurt. . . . *Historia rei herbariae.* 2 vols. Paris & Amsterdam, 1807–1808. Pritzel 8866. PH.

323. Staehlin Storcksburg, Jacob von. *Account of the new northern Archipelago lately discovered by the Russians in the Sea of Kamtschatka.* London, 1774. APS.

324. Stahl, Georg Ernst, and Joannes Juncker. *Conspectus medici medicinae theoretico-practicae . . . primarios morbos.* Halle, 1718. Cf. Countway 9: 413. PH.

Stewart, Charles. *See* Linnaeus, Carolus.

325. Stiles, Ezra. *United States elevated to Glory and Honour.* 2d ed. Worcester, 1785. Sabin 91750. APS.

326. Stuart, James. *On the salutary Effects of Mercury, in malignant Fevers.* Presented by the author. Philadelphia, 1798. APS.

327. Sullivan, James. *History of the District of Maine.* Boston, 1795. Sabin 93499. APS.

328. Svenka, K. *Analecta transalpina: Epitome commentariorum Regiae Scientarum Academiae Suecicae, 1739–1746.* 2 vols. Venice, 1762. PH.

329. Swartz, Olaf Peter. Dianome Epidendri generis Linn. *Nov. Act. Reg. Soc. Sci. Upsala* 6 (1800). Cf. Pritzel 9067. PH.

330. Tachenius, Otto. *Hippocrates chimicus, qui novissimi viperini.* Venice, 1666. Cf. Countway 9: 617. PH.

331. *Thornton, Robert John. *New Illustration of the Sexual System of Carolus von Linnaeus.* London [1799]–1807. Vol. 1. BSB wrote J. C. Lettsom asking he subscribe for him, 15 July 1803. Soulsby 772; Pritzel 9235.

332. Thunberg, Carol Peter. *Dissertatio . . . de Protea.* Upsala, 1781. Pritzel 9264. PH.

333. Thunberg, Carol Peter. *Dissertatio . . . de Erica.* Upsala, 1785. Pritzel 9272. PH.

334. Thunberg, Carol Peter. *Dissertatio . . . de Ficus.* Upsala, 1786. Pritzel 9273. PH.

335. Thunberg, Carol Peter. *Dissertatio . . . de Moraea.* Upsala, 1787. Pritzel 9275. PH.

336. Thunberg, Carol Peter. *Dissertatio botanica de Acere.* Upsala, 1793. BSB acknowledged receipt 5 Sept. 1796. Pritzel 9283. PH.

337. Thunberg, Carol Peter. *Dissertatio . . . utili atque Jucunda.* Upsala, 1793. Pritzel 9285. PH.

338. Thunberg, Carol Peter. *Dissertatio . . . de Hermannia.* Upsala, 1794. Pritzel 9286; TL–1 1299. PH.

339. *Thunberg, Carol Peter. *Icones plantarum japonicarum.* Upsala, 1794. Sent to BSB 8 Aug. 1797. Dryander 3: 184.

340. Thunberg, Carol Peter. *Prodromus plantarum capensium.* Upsala, 1794. Received pt. 1, 5 Sept. 1796. On verso of t.p. "From the author to Benjamin Smith Barton." Pritzel 7261. PH.

341. Thunberg, Carol Peter. *Dissertatio . . . de Diosma.* Upsala, 1797. Pritzel 9288. PH.

342. *Thunberg, Carol Peter. *Flora capensis.* 3 vols. Upsala, 1807–1813. 8 Aug. 1797. Pritzel 9262.

343. Tineo, Joseph. *Synopsis plantarum horti botanici Academiae Regiae Panormitanae.* Palerma, 1799. Pritzel 9365. PH.

344. Tournefort, Joseph Pitton de. *Corollarium institutionum rei herbariae.* 3 vols. Paris, 1719. Mss. notes by Wm. Byrd II. Wolf 215; Pritzel 9428. PH.

345. Tournefort, Joseph Pitton. *Voyage into the Levant.* London, 1741. 3 vols. Dryander 1: 123; Pritzel 9426. APS.

346. Trew, Christoph Jacob, and G. D. Ehret. *Plantae selectae.* Nürnberg, 1750. "Paid Mr. Dupef [?] 26 dollars." Pritzel 9499; TL–1 1328. PH.

347. Vallancy, Charles. *Grammar of the Irish Language.* Dublin, 1781. APS.

348. Venegas, Manuel. *Noticia de la California.* 3 vols. Madrid, 1757. Sabin 98848. APS.

349. Ventenant, Etienne Pierre. *Tableau du règne végétal selon la méthode de Jussieu.* 4 vols. Paris, 1794. Pritzel 9729. PH.

350. Vesalius, Andreas. *Humani corporis fabrica.* Basel, 1555. Signed, but not in account of purchases. PH.

351. *Vespucius, Americus. [Suggested:] *Quatuor Americi Vesputii navigationes.* St. Die, 1507. Letter BSB to John Eliot, 4 July 1801: "I have recd . . . very curious and interesting old book" which he plans to translate and annotate. Sabin 99354. Perhaps no. 274.

352. *Vieillot, Louis Jean Pierre. *Histoire naturelle des oiseaux de l'Amérique Septentrionale.* 2 vols. Paris, 1807. Blacker-Wood 9: 118.

353. Villette, Charles, marquis de. *Oeuvres* [with pref. by A. P. Leonier Delisle]. Printed on paper made from the bark of the lime [linden] tree, followed by twenty specimens of paper made from various grasses, leaves, and bark. Gift of Michel G. St. Jean de Crèvecoeur, May 1789. Presented by Mrs. Barton 6 Dec. 1816. APS.

354. Volney, Constantin François de Chasseboeuf, comte de. *Tableau du climat et du sol des Etats-Unis d'Amérique.* Paris XII [1803]. Jefferson, Washington 20 June 1804, letter to BSB, is sending copy (Barton corres. f. 72. HSP). BSB loaned to Heckewelder 4 March 1805. Sabin 100692.

355. *Volney, Constantin, F de C. *Rapport . . . sur l'ouvrage russe de . . . Prof. Pallas, intitulé Vocabulaires comparés des langues de toute la terre.* Paris, [1806]. Presented by author. APS.

Vorstius, A. E. *See* L'Escluse, Charles de.

356. Wade, Walter. *Plantae rariores in Hibernia inventae.* Dublin, 1804. Presented by the author. Pritzel 9893. PH.

357. Wade, Walter. *Sketch of Lectures on Meadow and Pasture Grasses, delivered in the Botanical Gardens, Glasnevin,* Dublin, 1808. Presented by the author. APS.

Wade, Walter. *See* André Michaux, *Oaks,* no. 228.

358. Walter, Thomas. *Flora Caroliniana.* London, 1788. Pritzel 9978; Sabin 101198; TL–1 1388. PH.

359. *Weaver, Thomas. *Treatise on the external Characters of Fossils.* Dublin, 1805. Transl. of Abraham Gottlob Werner's . . . *Fossilien.* Leipzig, 1774. Account of PH purchase 3 May 1817.

360. Weber, Friedrich. *Nomenclator entomologicus secundum . . . Fabricii.* Chillon & Hamburg, 1795. Presented by author. Dryander 5: 32. PH.

361. White, Gilbert. *A Naturalist's Calendar.* London, 1795. PH.

362. Willard, Moses. *Observations on the Remitting Fever which prevailed in the City of Albany in the Summer & Autumn of 1809.* Albany, 1810. APS.

363. Williams, Samuel. *Natural and Civil History of Vermont.* 2d ed. 2 vols. Burlington, 1809. Sabin 104350. APS.

364. Willughby, Francis. *Ornithology of Francis Willughby by John Ray.* London, 1678. Dryander 2: 111; Blacker-Wood 9: 258; Wolf 232. PH.

365. Willughby, Francis. *Historia piscium.* London, 1686. "For his honourd friend Mr Banister, M[artin] L[ister] Westminster, in the old Palace Yard Sept. 7, [16]87." Wm. Byrd bookplate; gift of Sarah Zane Wolf. PH.

366. Wilson, Alexander, *American Ornithology.* 9 vols. Philadelphia, 1808–1814. Blacker-Wood 9: 259. PH.

367. Wilson, Joseph Nicholas. *Dissertatio medica inauguralis, de tetano.* Edinburgh, 1788. Presented by author. APS.

368. Withering, William. *Systematic Arrangement of British Plants.* 4th ed. 4 vols. London, 1801. Pritzel 10360; TL–1 1440. Hist. Sci. Library, Cornell Univ.

369. Woodward, John. *Essay toward a Natural History of the Earth.* London, 1695. Inscribed on flyleaf: "bought at Philad'a at the auction of books of that most valuable library of the late Col. Byrd [III] of

Westover in Virginia November 1781. Du Simitiere." Wm. Byrd [II] bookplate. Wolf 234, another William Byrd copy at LCo. PH.

370. Yates, Christopher C. *Essay on the bilious Epidemic Fever, prevailing in the state of New-York . . . a letter from Dr. James Mann . . . a dissertation from Dr. John Stearns.* 2d ed. Albany, 1813. Presentation by the authors to BSB. APS.

371. Yeates, Thomas Pattinson. *Institutions of Entomology.* London, 1773. APS.

372. Zeisberger, David. *Essay of a Delaware-Indian and English Spelling Book for the use of the Schools of the Christian Indians on Muskingum River.* Philadelphia, 1776. Evans 15228; Sabin 109229 or 106300? LCo.

Foreign Membership of Biological Scientists in the American Philosophical Society During the Eighteenth and Nineteenth Centuries

BENTLEY GLASS

INTRODUCTION

FOR persons who may be interested in the history of science during the eighteenth and nineteenth centuries the question of contemporary fame and international recognition has considerable interest, but appears to have been essentially unexamined. Whitfield J. Bell, Jr., who is honored for his long service to the American Philosophical Society as its Librarian and Executive Officer, has during his life been particularly interested in the history of science in America during the period covered by the existence of the American Philosophical Society. From its beginning, the American Philosophical Society was devoted to the natural sciences and their applications to the improvement of human society through "the promotion of useful knowledge." Benjamin Franklin's "Proposal for Promoting Useful Knowledge among the *British Plantations* in *America*," issued on 14 May 1743, suggested that the American Philosophical Society always contain as a nucleus of members "a Physician, a Botanist, a Mathematician, a Chemist, a Mechanician, a Geographer, and a general Natural Philosopher," in addition to its necessary officers.[1] Some nine foreign members were elected during this first period of the Society's life (the only biologist being Buffon), but the "Brief History" records that "interest soon lagged . . . and by 1746 the Society was moribund, if not dead."[2]

Twenty years later a similar Society was organized, the American Society for Promoting Useful Knowledge, and in a spirit of rivalry the American Philosophical Society was revived. On 2 January 1769 the two societies were merged, taking the name of the "American Philosophical Society, Held at Philadelphia for Promoting Useful Knowledge." Franklin was chosen its president. Its prototype was the British Royal Society, conjoined with the Society of Arts of London as its model for promoting practical concerns. Although the American Revolution interrupted the Society's activities for a few years, in 1779 a new period of activity commenced. It occurred to me that it might be of value to see what recognition was given in America, and specifically by the American Philosophical Society, to foreign biological scientists who are now recognized as having been the greatest contributors to scientific discovery during the latter half of the eighteenth century and the entire nineteenth century. Which ones, indeed, were recognized by election to foreign membership in the Society, and what others of their contemporaries were passed over? Was the American Philosophical Society broadly cognizant of scientific advances in Europe and other countries of the world during that century and a half? Can plausible reasons be given for cases of neglect?[3]

1. Benjamin Franklin, "A Proposal for Promoting Useful Knowledge among the *British Plantations* in *America*," 14 May 1743, William Smith Mason Collection, Yale University Library.

2. "A Brief History of the American Philosophical Society," *Year Book 1980* of the American Philosophical Society (Philadelphia: American Philosophical Society, 1981), 35.

3. For a fuller account of the APS see Brooke Hindle's *The Pursuit of Science in Revolutionary America 1735–1789* (Chapel Hill: Institute for Early American History

From the 1980 *Year Book* of the American Philosophical Society the listing of former foreign members of the Society was obtained, and those who might be described as biological scientists, or important in the development of biology, were identified. The list was subdivided chronologically into three parts, comprised of biologists whose primary significance fell in the eighteenth century, in the first half of the nineteenth century, or in the second half. From a former publication of my own, "Milestones and Rates of Growth in the Development of Biology," an extensive table of the great discoveries in biology during these same periods served as a basis for a list of scientists who might reasonably have been considered for election to foreign membership in the American Philosophical Society.[4] To this listing I added certain very notable biologists who were not included in the basic listing because they contributed no single major "milestone." They were, nevertheless, greatly reputed as leaders of science in their own times, not only by contemporaries in their own countries but internationally. Any such list, of course, is inescapably subjective, yet I trust it represents reasonably well the greatest leaders and discoverers in the biological sciences.[5] This list was then compared with the list of foreign biologists actually elected to the A.P.S.

---

and Culture, 1956) and "The Rise of the American Philosophical Society, 1766–1787" (unpublished doctoral dissertation, University of Pennsylvania, 1949), and Whitfield J. Bell, Jr.'s "As Others Saw Us: Notes on the Reputation of the American Philosophical Society," *Proceedings of the American Philosophical Society* 116 (1972): 269–278, and his introduction to "Account of the American Philosophical Society by John Vaughan (1841)," (Philadelphia: Friends of the Library, 1972).

4. Bentley Glass, "Milestones and rates of growth in the development of biology," *Quarterly Review of Biology* 54 (1979): 31–53.

5. In making up this list I consulted especially the *Dictionary of Scientific Biography*, C. C. Gillispie, ed. (16 vols. to date, New York: Charles Scribner's Sons, 1970–); Erik Nordenskiöld, *The History of Biology* (New York: Tudor Publishing Company, 1942); Charles Singer, *A History of Biology* (revised edition, New York: Henry Schuman, 1950); William Coleman, *Biology in the Nineteenth Century* (New York: John Wiley and Sons, 1975); and Joseph S. Fruton, *Molecules and Life* (New York: Wiley-Interscience, 1972).

One wonders what persons, if any, were nominated but failed to achieve election. The records of the Society, however, do not preserve that information; by specific direction of the membership those records were considered confidential and were destroyed. I found only one anomalous record. Francis Galton of England was actually elected to membership, but for unspecified reasons declined. His name is included in the appropriate table of foreign scientists elected to membership.

Three biological foreign members (Linnaeus, Daubenton, and Lavoisier) had been elected by 1779, but most of the foreign members representing eighteenth-century science were elected after that year. The resident members at this period who were perhaps most cognizant of the biological sciences were physicians or naturalists (such as John and William Bartram, J. J. Audubon, and Caspar Wistar), together with a few generalists such as Benjamin Franklin who knew something about everything. How the foreign members were actually selected the records do not show, as the Society's policy of secrecy in such matters extended to the destruction of all records dealing with elections. Only in the second half of the nineteenth century does one run across any names of well-known biologists among the resident members. These included Louis Agassiz, who was elected as a foreign member but came to the United States as a permanent resident in 1846, Asa Gray, William K. Brooks, Alpheus S. Packard, Samuel H. Scudder, Charles S. Minot, E. D. Cope, Othniel C. Marsh, E. G. Conklin, Edmund B. Wilson, and Jacques Loeb. Some of them may have nominated foreign biologists for membership, but there is a record only of E. G. Conklin's activity in this respect. The roster of distinguished European scientists actually elected to membership in the Society is consequently rather surprising, even though, as will be shown hereafter, many equally worthy or even superior possibilities were overlooked.

A comment also seems needed in regard to the identification of certain foreign biologists as having been elected as foreign associates of the National Academy of Sciences of the United States. The National Academy was founded in 1863, so foreign elections to it pertain only to the last third of the nineteenth century. In making a comparison with similar elections to the American Philosophical Society, it is significant that during that period the National Academy of Sciences remained a small body, and its election of foreign associates corresponded in number rather closely to the number of scientists elected to foreign membership in the American Philosophical Society. The N.A.S. had only 22 living foreign associates in 1900. Here, then, is an approximate standard for the comparison of foreign elections by the two organizations. It is especially important to keep this equivalence in mind in these present times, when the National Academy of Sciences has so much larger a membership, and one that is restricted to a much narrower range of disciplines than that of the American Philosophical Society.

Tables 1 to 3 are arranged in the following manner. In the column of persons elected to the American Philosophical Society, the individuals are arranged in order of date of election (date in parentheses). Other dates are those of the discoveries and contributions of outstanding significance to science. In the column of persons not elected, the dates are all simply of the latter kind, and the order is that of the date of the person's most significant contributions.

Following each table are thumbnail sketches of all persons listed; they provide some idea of the nature and relative significance of each man's contribution to biology. Here the arrangement, for the sake of convenient reference, is alphabetical. In order to provide a second, and perhaps less subjective, evaluation of each scientist's standing, the number of columns devoted to the biography and bibliography of each person included in the *Dictionary of Scientific Biography* appears just after the birth and death dates. For example, Claude Bernard (elected) received 19¼ columns; Louis Pasteur (elected), 131 columns; Robert Koch (not elected), 30³/₄ columns. The sketches also indicate conferral of Nobel Prizes; although these prizes began only in the twentieth century, the early awards honored scientists for work done mainly in the nineteenth century. Election as a foreign associate of the National Academy of Sciences (U.S.A.) is also noted for persons in Tables 2 and 3, beginning with 1865. Even in the year 1900, there were only 22 living Foreign Associates of the National Academy of Sciences.

There is unavoidably some overlapping between biologists of the late nineteenth and early twentieth centuries. I included for consideration here all those biologists whose most significant work, or at least much of their very significant work, occurred prior to 1900, even though they might well have been considered for election as a foreign member of the Society only in the twentieth century.

DISCUSSION

For the eighteenth-century elections and non-elections, several points need to be kept in mind. The only foreign biologist elected in the first period of activity of the American Philosophical Society (1744–46) was Buffon, who was indeed a worthy choice. Upon the resumption of the Society's activity in 1769, the election of such great European scientists as Stephen Hales, Réaumur, and Maupertuis was precluded by their deaths. The same might be said of Haller, who died in 1777. Malthus might be excluded from the present consideration because he was primarily a contributor to social thought rather than a biologist, although his influence on Darwin and Wallace accounts for his inclusion in the general list.

As for the remainder of the non-elected great, or near great, it seems rather surprising that Michel Adanson was overlooked when Daubenton was elected, for France produced no greater botanist than Adanson in the eighteenth century; his experimental refutation of Linnaeus's late idea of the origin of new species, by means of hybrids between existing species, was brilliant. True, the latter work was virtually forgotten by all biologists of the nineteenth and twentieth centuries until it was rediscovered in 1959. Nevertheless, Adanson ranked with Daubenton in his fundamental contributions to Buffon's great *Natural History*, Daubenton providing the basic studies of animals and Adanson those of plants.

German scientists, in comparison with those of France, Britain, or even Sweden, received little regard in the eighteenth-century elections. Wolff, Koelreuter, and Scheele clearly rank as the peers of any of the persons actually elected. Even more amazing is the total neglect of biological scientists from Switzerland and Italy. Trembley's discovery and studies of the behavior and reproduction of the so-called "plant-animal" Hydra and Bonnet's discovery of parthenogenesis in plant lice were the talk of the European intelligentsia. And what are we to think of the failure to recognize the greatness of Lazzaro Spallanzani, by far the greatest experimental biologist of his time? Or of Galvani, whose very name has become a common English term, "galvanic," because of his extraordinary discoveries of the effects of electrical shocks on the contraction of muscles? It will be well, therefore, to keep in mind this blindness toward the greatness of the science in countries other than England, France, and Northern Europe (excluding Germany) as we proceed to examine the record of elections in the nineteenth century.

In the first half of the nineteenth century some very odd elections, and especially non-elections, transpired. Great Britain was represented by only a single election, that of the very distinguished zoologist and anatomist Richard Owen. Ignored was the equally distinguished botanist and plant morphologist Robert Brown, who presided for many years as chief curator over Sir Joseph Banks's famous natural history collections, and who was justly famous for pointing out that every living cell possesses a nucleus as an indispensable organelle. The elections of French foreign members included the leader of French chemistry, J. B. Dumas, who also, with Prévost, in a youthful foray into biology, observed that frog's eggs are fertilized by spermatozoa, a step in the understanding of animal reproduction that even Spallanzani, in his experiments of the second half of the eighteenth century, had failed to achieve. Yet Dumas was accompanied in the honor extended to him by the American Philosophical Society by three quite ordinary scientists, whereas scientists like Lamarck, Cuvier, Magendie, and other notables were passed over. The omission of Cuvier is particularly hard to envisage, for his bust was enshrined in the hall of the Philosophical Society, and many of his associates, students, and colleagues were elected, for example, Louis Agassiz, Ducrotay de Blainville, and Flourens. Can it be, as Whitfield Bell has suggested to me on several occasions, that everyone simply supposed that Cuvier *must* have been elected at some earlier date, and that no one checked to see? By 1819, when Ducrotay de Blainville was elected, it was indeed possibly taken for granted that Cuvier had been honored earlier by the Society. (We may note that Cuvier died prior to the election of Louis Agassiz as a member.) A further conjecture is that Cuvier's outright hostility to Lamarck and to Geoffroy Saint-Hilaire prevented the election of either of those great early evolutionists.

The six German biologists elected to foreign membership in the Society in this half-century were certainly all of the highest caliber. Even so, there were about three times as many others of equal distinction who were not elected. Among these, the illustrious zoologists von Baer, Leuckart, Kölliker, and the physiologist DuBois-Reymond were all elected to the National Academy of Sciences of the United States, like Wöhler, von Liebig, and Helmholtz. (Johannes Müller and Humboldt died too soon to achieve that election, but so, for that matter, did Gall, Meckel, Pander, Rudolphi, and Rathke.) Especially surprising, in view of the election of Theodor Schwann to membership, seems the failure to elect M. G. Schleiden, his botanical collaborator in enunciating the Cell

Theory. Each of them harbored erroneous ideas of how new cells are formed, so in that respect they were equally meritorious. Perhaps it was the later work of Schwann in developing methods of sterile culture for bacteria that kept up his fame. Because of the lack of any documentation of nominations for the election of foreign, or indeed of any, members in those years, one can do no more than guess at the reasons for what appears to be clear discrimination.

The eighteenth-century neglect of Swiss scientists of distinction in these elections was apparently remedied in the early nineteenth century; but the blindness with respect to Italy persisted. Two elections of Swedish biologists of moderate qualifications seem best explained as a residue of respect for compatriots of the great Linnaeus. Austria, in this period, joined the ranks of the disregarded, inasmuch as a first-rate microscopic anatomist and physiologist, the Czech (Bohemian) Purkinje and the Viennese physician Semmelweiss joined the ranks of the overlooked.

The second half of the nineteenth century might be supposed to exhibit a more balanced and judicious election of foreign members to the Society, for by that time the Society was larger, embraced a greater variety of members in different disciplines who might be expected to make nominations of persons for foreign membership, and in general scientific communication had advanced greatly and assumed a worldwide embrace. In fact, when one considers the large number of young scientists who travelled eagerly to Europe to engage in graduate and postdoctoral studies under the acknowledged masters of the biological disciplines, one would hardly expect to find the previous degree of oversight and national or disciplinary blindness. That, however, is not what the record of actual elections reveals.

The elections of foreign members heavily favored England and France. In spite of the phenomenal rise of German biological science to superiority in many fields, the number of Germans elected to foreign membership in the Society scarcely exceeded the number of French, although, in spite of the overwhelming distinction of Claude Bernard and Louis Pasteur, French science was definitely losing its leadership to the Germans.

First, the British. Surely no one will cavil at the elections of Darwin, Wallace, Huxley, or Hooker. Sclater, Lankester, and Foster, all three elected, fall into a second category, certainly matched or exceeded in distinction by Graham and Overton, who were not. Of the two representatives of medicine elected, Osler was perhaps the most distinguished physician of his time, but he made no particular scientific discoveries. Lister, too, the father of anti-sepsis in surgery, found his methods quickly superseded by superior aseptic methods. On the other hand, Bruce and Ross made notable discoveries of the way in which insect vectors, flies or mosquitoes, transmit by their bites the respective agents of sleeping sickness and of malaria. Neither Bruce nor Ross seems to have been proposed for election. Even more disconcerting is the discovery that Charles S. Sherrington, one of the greatest neurobiologists of all time, was not elected to foreign membership in our Society. Sherrington's work on reflexes and nervous coordination was done in the 1890s, somewhat preceding that of Ivan Pavlov, who was elected a foreign member in 1932. And of course he, too, like Pavlov, was a Nobel laureate and a foreign associate of the National Academy of Sciences. In addition, although those writings came later, in the twentieth century, he was a distinguished philosopher of science. Why, then, was he not elected along with Pavlov, in the early decades of this century? The record provides no answer.

The French scientists elected to foreign membership in the American Philosophical Society stand out as a generally better group of choices than the British. True, Brown-Séquard and Broca have declined in esteem because of their espousal of erroneous opinions, respectively, about rejuvenation by means of glandular extracts and about craniology as a guide to intelligence. Yet they also made solid contributions to science. Nevertheless, Boussingault's insight into nitrogen fixation, Laveran's independent discovery of the plasmodium that causes malaria, Bordet's work on complement in immunity, and Bertrand's discovery of the role of trace metals in enzyme function rank far more significantly today. Boussingault, Bordet, and Bertrand were each elected to the National Academy of Sciences, and Laveran was a Nobel laureate. Boucher de Perthes, discoverer of some of the caves decorated by Stone Age Man in France, appears a very weak choice by comparison.

German biological science, in view of its overwhelming dominance in the latter half of the nineteenth century, seems gravely slighted in the elections. Although some excellent choices were made, the elections of Anton Dohrn and of Adolf Engler suggest personal friendships more than scientific eminence. There were, moreover, quite astonishing omissions. How could it be that Louis Pasteur was elected, whereas Robert Koch, his peer in bacteriology and in establishing the Germ Theory of Disease, was not? There was, furthermore, a remarkable galaxy of cytologists who worked out, by common effort, the intricate details of mitotic cell division, fertilization of the egg cell by a sperm cell, the fusion of the male and female nuclei in the egg, and the behavior of the chromosomes during mitosis and fertilization. These giants included Walter Flemming and Eduard Strasburger, Oskar Hertwig, Otto Bütschli, N. Pringsheim; in Belgium, Edouard van Beneden; and in Switzerland, Hermann Fol. Wilhelm Roux, who together with Oskar Hertwig was first to see the implications that the chromosomes are the carriers of the material basis of heredity, and who founded experimental embryology, was also passed over. Theodor Boveri, who established firmly the concepts of the persistence of the chromosomes between cell divisions and their individuality in the control of hereditary traits, was neglected.

For the failure to recognize these great pioneers in the study of cells in heredity, the blame, I fear, must fall squarely on the head of our own member Edmund B. Wilson. As the foremost American working in that field, and one whose notable book *The Cell in Development and Inheritance* reveals a solid familiarity with the pioneers in cytology, Wilson, who was elected to the American Philosophical Society in 1888, had ample opportunity to nominate any, or all, of these men. Instead, he appears to have nominated exactly no one —a negligence hard to condone. Whatever nominations did occur in these areas were made, according to the Society's records, by E. G. Conklin, the Society's President in later years. Conklin, though himself not a chromosomal cytologist, was familiar with the work done in that field and nominated August Weismann, who was in fact more of a theorist in the fields of heredity and evolution than a meticulous observer, since his eyesight became very poor early in his scientific career.

Notable plant scientists who were ignored include Sachs (a member of the N.A.S.), DeBary, Schwendener, and Engelmann. Noted zoologists overlooked were Leuckart and Gegenbaur, both members of the National Academy of Sciences. The great embryologists of the period included not only Wilhelm Roux, already mentioned, but also Richard Hertwig (N.A.S.), who with his brother Oskar was an early proponent of the germ layer theory of animal development. Physiology and biochemistry were illuminated by the work of Kronecker, Hoppe-Seyler, Pflüger, and Buchner. The last-named isolated and purified the first enzyme. Mysterious is the strange failure to elect Emil von Behring, who won a Nobel Prize for the work he did with S. Kitasato (elected) in the development of antitoxins to diphtheria and tetanus. Perhaps there was simple justice was at work here, for Kitasato, though elected a foreign member of the Society, failed to share in the Nobel Prize. Even more astonishing was the failure to elect Paul Ehrlich, the illustrious colleague of Robert Koch, who pioneered in the discovery of stains for refractory bacteria, in the development of antitoxins, and in the theoretical analysis of immune reactions. Ehrlich, who won a

Nobel Prize and was also elected to the National Academy of Sciences, did much of his famous work in the twentieth century, but already by 1900 he was a world leader in bacteriology and immunology. Loeffler, who discovered the first animal virus, also went unnoticed. It is of course quite understandable that Friedrich Miescher, who was first to isolate and characterize nucleic acids as phosphorus-containing organic molecules, was not elected, for the great significance of his work, like that of Gregor Mendel in heredity, was not recognized until the twentieth century, long after these pioneers had died.

Sweden continued to win attention in the elections of the Society's foreign members, although its representation could scarcely be regarded as illustrious in comparison with the Germans who were passed over. Italy was once again sadly neglected. Both Golgi, a Nobel laureate and one of the greatest of all students of the microscopic anatomy of the brain, and Grassi, one of the pioneers in the elucidation of the transfer of malaria by mosquitoes and observer of the action of quinine on the growth of the plasmodia causing malaria, were never—it seems—considered for election. It is interesting that a new country, Spain, previously unrepresented in the Society's elections, was honored by the election of Ramón y Cajal, whose work so brilliantly followed up that of Golgi.

The most serious lapse of all, in the late nineteenth-century elections of the American Philosophical Society, was a total blindness toward the rising eminence of biology in Russia. No Russian was ever honored by the Society with foreign membership until 1932, when Ivan Pavlov was elected. But Pavlov's work of distinction fell almost entirely within the twentieth century and so is excluded from the present consideration. In the nineteenth century, however, there were a number of very distinguished Russian life scientists, many of whom were trained or did their most notable work while in laboratories in France or Germany. Among the earliest of these was Sechenoff, who by his work on reflexes laid the foundation for Pavlov's later work. Sechenoff was in the laboratory of Claude Bernard when he first analyzed behavior into conscious and unconscious reflexes. Later he went back to Russia and founded there the study of neurophysiology and experimental psychology.

Two Russian zoologists made contributions of great importance. Both bore the name of Kovalevsky. A. O. Kovalevsky was a notable embryologist who studied invertebrate animal development, especially that of the tunicates and Amphioxus, invertebrates that he was the first to discern have embryonic relationships to the vertebrate animals. V. O. Kovalevsky was the famous paleontologist who worked out the evolutionary reduction of the toes in fossil horses. It is indeed odd that the former was not nominated for election by such a member as E. G. Conklin, or the latter by O. C. Marsh, E. D. Cope, or later by Henry Fairfield Osborn. The Society's botanists ignored Timiryazeff, and in bacteriology and immunity no one put forward the name of Vinogradsky, who worked with Pasteur and discovered active nitrogen fixation by the soil bacterium Azotobacter, and went on to discover and study the iron and sulfur bacteria of the soil. Similarly, Metschnikoff, discoverer of phagocytosis by white blood cells, a Nobel laureate and a pioneer in the study of immunity, and Ivanovski, discoverer of the very first virus, that of tobacco mosaic disease, failed to win regard from the Society.

All in all, the record of the Society's elections of foreign members among eighteenth- and nineteenth-century biologists presents many enigmas. Some of them seem to derive from an overstrong tendency to look only toward Western Europe when considering nominations. Yet even within that area, astonishing neglect, or possibly bias, shows up. Even the election of British members is spotty and sometimes ill-considered. In the case of German science, entire burgeoning fields were neglected, apparently because of the negligence of one, or a very few, competent members of the

Society to make nominations of persons whose work was well-known to them. Sometimes, as in the cases of Cuvier and Sherrington, the oversight seems inexcusable. Or again, as in the case of Robert Koch and Paul Ehrlich, perhaps nationalistic feelings and wartime bias played a part.

In the Society today, we have evolved a new machinery to take better care of nominations. Each of the five "Classes" of members has its own Class Membership Committee, which is expected to prevent the kind of oversight and negligence evident in the period before 1900. The personnel of the Class II Membership Committee, that for the Biological Sciences, changes with every year through overlapping terms of service, and appointments represent different fields of biological activity. It will be interesting to see whether, by the end of the present century, the Society's membership has done conspicuously better in selecting and honoring foreign members. The task is almost infinitely more difficult now than in the earlier periods, because there are hundreds of competent scientists today, in all countries, where there were but scattered individuals before. It will indeed take systematic, careful effort to avoid the oversights made in earlier times.

TABLE I

Biologists of the Eighteenth Century Elected,
or not Elected, to Foreign Membership
in the American Philosophical Society

A date in parentheses indicates year of election to the APS.
Other dates are of notable contributions to science.

Elected			Not Elected	
Great Britain				
Joseph Priestley	1771	(1785)	Stephen Hales*	1727/53
Joseph Banks		(1787)	James Lind	1752
John Hunter		(1787)	Thomas Robert Malthus	1798
Erasmus Darwin		(1792)		
Edward Jenner	1798	(1804)		
France				
Georges Louis LeClerc de Buffon		(1744–46)	René F. de Réaumur*	1734–42/52
Louis J. M. Daubenton		(1775)	Pierre Louis Moreau de Maupertuis*	1745/51/52

* The asterisks designate those scientists who died before 1769, the year of the union of the American Philosophical Society and the American Society for Promoting Useful Knowledge. Until that year the American Philosophical Society was dormant except in its earliest years (1744–46), and the only foreign biologist elected in the early period was Buffon.

Elected		*Not Elected*	
Antoine Lavoisier	1780 (1775)	Michel Adanson	1763/72
		Felix Vicq d'Azyr	1786
Germany			
Johann F. Blumenbach	(1798)	Christian Friedrich Wolff	1759
		Johann Gottlieb Koelreuter	1761
		Carl Wilhelm Scheele	1776/80/85
		C. K. Sprengel	1793
Netherlands			
Jan IngenHousz	1779 (1786)		
Pieter Camper	(1789)		
Sweden			
Carolus Linnaeus	1735 (1769)		
Switzerland			
		Abraham Trembley	1744
		Charles Bonnet	1745
		Albrecht von Haller	1753
		Jean Senebier	1788
Italy			
		Lazzaro Spallanzani	1780/83
		Luigi Galvani	1791

Thumbnail Sketches of Each Biological Scientist's Notable Contributions
The number with a "c" appended which follows the birth and death dates indicates the number of columns in the Dictionary of Scientific Biography *devoted to the subject.*

Adanson (1727–1806) 2¹/2 c. Elected to Académie des Sciences, 1759; to Royal Society, 1761. Pioneer study of African mollusks. Great *Famille des plantes*, 1763–64. Studied regeneration in animals. Contributed the section on plants to Buffon's *Histoire naturelle*. By means of carefully designed experimental studies, refuted Linnaeus's claim that new species may arise by hybridization.

Banks (1743–1820) 8¹/2 c. Naturalist on the cruise of the *Endeavour* with Captain Cook; described many new species, especially of plants. President of the Royal Society 1778–1820.

Blumenbach (1752–1840) 3¹/2 c. Founder of anthropology, based on comparative anatomy; student of human races (1778).

Bonnet (1720–1793) 3¹/2 c. In youth, studied regeneration in worms; discovered parthenogenesis in plant lice. Wrote *Traité d'insectologie* at age 25. Studied the functions of the leaves of plants. With increasing blindness, became an experimental theorist and champion of preformation in the origin of new individuals. Advanced a theory of "palingenesis": there must be some sort of structural basis for the inheritance of the characteristics of a species within the egg or bud from which a new individual arises.

Buffon (1707–1788) 11¹/2 c. Known especially for the great *Histoire naturelle*, 1747–79, containing among other sections the "Epoques de la Nature." Embraced evolution of the natural world by means of an inheritance of acquired characteristics, migration, and adaptation of forms to new environments. Rejected Linnaeus's view that families are a natural category; really, he insisted, only *species* exist in nature. Rejected preformation in favor of epigenesis in the formation of new individuals.

Camper (1722–1789) 2 c. During his lifetime acclaimed one of the most famous scientists of Western Europe. Made numerous discoveries in anatomy, such as the air spaces in birds' bones that lighten their structure.

Darwin, Erasmus (1731–1802) 7 c. Physician and writer of scientific poetry, especially the *Zoonomia* (1794) with its evolutionary theme. Also *Phytologia*, 1800, a horticultural and agricultural poem.

Daubenton (1716–1800) 7 c. Elected to Académie des Sciences, 1744. Wrote careful descriptions of animal morphology and anatomy for the *Encyclopédie* and for Buffon's *Histoire naturelle*. Extended the comparative method to the study of plants.

Galvani (1737–1798) 4¼ c. Originated the study of animal electricity. Used frog nerve-muscle preparations to study irritability and response by contractions of muscles.

Hales (1677–1761) 25 c. The founder of plant physiology. *Vegetable Staticks*, 1727. Also investigated animal blood pressure; *Haemostaticks*, 1733.

Haller (1708–1777) 13 c. The leading comparative anatomist and physiologist of his time. Invented the injection of blood vessels with colored substances to make them readily visible. Hemodynamics; lymph vessels; structure and function of the human diaphragm; the nervous system, especially the function of the cerebellum; sensibility and irritability. A strong preformationist. Developed botanical nomenclature.

Hunter (1728–1793) 5 c. Surgeon and anatomist. Wrote *The Natural History of Human Teeth*, 1771, 78. Founder of a famous anatomical museum in London.

IngenHousz (1730–1799) 11 c. Photosynthesis; proof that light is required for the renewal of air by plants, and that in the dark they respire like animals.

Jenner (1749–1823) 5 c. Immunization against smallpox by vaccination with pus from cowpox, 1798.

Koelreuter (1733–1806) 3½ c. Studied the function of pollen grains in plant fertilization, by insects or by wind. Made many plant hybrids, including the first hybrid between different species, 1761. Distinguished between fertile and infertile hybrids. Held that hybrids are not new species, because in backcrosses to the parent species the hybrids tended to revert to the parental forms.

Lavoisier (1743–1794) 50½ c. Proved animal respiration to be a form of combustion, that is, an oxidation, 1780. Determined products of respiration to be animal heat, carbon dioxide, and water.

Lind (1716–1794) 4 c. Discovered that citrus fruits (oranges, lemons) would prevent scurvy, 1753–57. The beginning of vitaminology.

Linnaeus (1707–1778) 14½ c. Wrote the *Systema Naturae*, 1735, foundation of scientific binomial nomenclature of living organisms and of modern classification. Wrote on higher categories, 1751. At first convinced that species never change, later was convinced that new species can originate by hybridization of existing species: evolution within the limits of the "order."

Malthus (1766–1834) 8¼ c. A political economist, Malthus is in this list because of his *Essay on Population*, 1798, which formed the stimulus to Darwin's theory of natural selection.

Maupertuis (1698–1759) 7 c. Best known as a mathematician and physicist, Maupertuis also interested himself in biology. Confuted preformation by emphasizing biparental heredity. Made first important study of a human inherited trait, a pedigree of polydactyly. Advocate of evolution ('transformism') by mutation, migration and isolation, and a sort of natural selection.

Priestley (1733–1804) 17 c. Demonstrated the "revivification" of breathed air (i.e., will again support animal life) by means of the action of green plants when in the light, 1771–2.

Réaumur (1683–1757) 16½ c. *Histoire des insectes*, 1734–42, emphasizing their habits and behavior. Studied bees extensively; also regeneration in animals. Developed the technology of artificially incubating hen's eggs. Made an experimental study of the gastric digestion of meat by a pet hawk by enclosing the meat in a perforated metal capsule that could be swallowed and regurgitated at intervals, 1752.

Scheele (1742–1786) 13½ c. Better known as a chemist, Scheele was a founder of biochemistry, discovering and isolating, from 1776 on, uric acid, lactic acid, citric acid, tartaric acid, malic acid, and glycerol among other important substances.

Senebier (1742–1809) 1½ c. Proved the necessity of chlorophyll for the assimilation of carbon by plants in photosynthesis.

Spallanzani (1729–1799) 28½ c. Disproof of spontaneous generation of microorganisms, 1765–80. Artificial fertilization of frog's eggs with semen from a male frog, 1780.

Sprengel (1750–1816) 7½ c. In *Das entdeckte Geheimnis* . . . Sprengel treated the nature of fertilization in plants, distinguishing flowers pollinated by wind from those pollinated by insects, and the role and adaptations of nectaries in relation to insects. He was much admired and cited by Charles Darwin, for emphasizing the importance of cross-pollination.

Trembley (1710–1784) 4 c. Discovered *Hydra*, a "plant-animal," and studied its reproduction and behavior, especially its ability to regenerate lost parts, 1744. Also discovered, in the protozoan *Stentor*, reproduction by fission, 1744.

Vicq d'Azyr (1748–1794) 7¼ c. Studied the comparative anatomy of vertebrates, especially that of the brain. Made first good classification of animal functions, 1786.

TABLE 2

Biologists of the First Half of the Nineteenth Century
Elected, or not Elected, to Foreign Membership
in the American Philosophical Society

A date in parentheses indicates year of election to the APS.
Other dates are of notable contributions to science.

	Elected		Not Elected	
Great Britain				
Richard Owen	1840	(1845)	T. A. Knight	1806
			Charles Bell	1811
			William Buckland	1824
			Robert Brown	1833; 1864–9
France				
H. M. Ducrotay de Blainville		(1819)	M. F. Xavier Bichat	1798–1802
M. J. P. Flourens	1842	(1825)	J. B. Lamarck	1802/09
Felix A. Pouchet		(1848)	J. Le Gallois	1811
Jean Baptiste Dumas	1824	(1860)	F. Magendie	1811
			Georges Cuvier	1812–17/38
			P. Pelletier & J. B. Caventou	1817
			J. B. F. Turpin	1826
			A. T. Brongniart	1828
			E. Geoffroy Saint-Hilaire	1830
			H. J. Dutrochet	1832
			Felix Dujardin	1835

	Elected		Not Elected
Germany			
Alexander von Humboldt	(1804)	F. J. Gall	1811
Johannes P. Müller	1837–40	J. F. Meckel	1811
Friedrich Wöhler	1828	H. C. Pander	1817
Justus von Liebig	1842	Karl Ernst von Baer	1817/27
Theodor Schwann	1839	Jakob Henle	1830/31
Hermann L. F. von Helmholtz	1847/50	C. G. Ehrenberg	1838
		M. G. Schleiden	1839
		Robert Remak	1841
		A. von Kölliker	1841
		Emil du Bois-Reymond	1848
		Wilhelm Hofmeister	1849
		C. F. Gaertner	1824
		Julius Robert Mayer	1842/45
		Hugo von Mohl	1846
		Carl A. Rudolphi	1819
		M. H. Rathke	1832/41/42
Switzerland			
Augustin P. de Candolle	(1841)	J. L. Prévost	1824
Louis Agassiz	(1843)		

	Elected	Not Elected
Sweden		
A. A. Retzius	(1813)	
Carl A. Agardh	(1835)	
Austria (including Czechoslovakia)		
J. E. Purkinje		1830–41
I. P. Semmelweiss		1850
Italy		
G. B. Amici		1824
Agostino Bassi		1835

Thumbnail Sketches of Each Biological Scientist's Notable Contributions

The number with a "c" appended which follows the birth and death dates indicates the number of columns in the Dictionary *of Scientific Biography devoted to the subject.*

Agardh (1785–1859) 2½ c. Contributed to the taxonomy of algae, 1817–24, and developed a natural system of plant classification, 1817–26. A Naturphilosophe.

Agassiz (1807–1873) 5 c. Under Cuvier, became an authority on fishes, fossil animals, and geology. His *Poissons fossiles* especially famous, 1833. Also his theory of glaciation. To America in 1846. Strongly opposed Darwinism. NAS.

Amici (1786–1868) 4½ c. Inventive microscopist who increased good optical resolution to 1000X. Discovered the palisade parenchyma in leaves, and growth of pollen tubes through the flower's pistil, 1824.

von Baer (1792–1876) 7 c. In 1827, discovered the true mammalian egg. Also discovered the embryonic notochord, and how the neural groove of the embryo folds over and unites to make the neural tube, whence come the five primary brain vesicles and spinal cord. With Pander, in 1817, identified the three embryonic germ layers. Held development to be epigenetic. Noted the great similarity of embryos of vertebrates very different as adults. Opposed Darwinism. NAS.

Bassi (1773–1856) 2½ c. Discovered the cause of silkworm disease to be a fungus, of importance in the understanding of contagious disease. Used experimental inoculations to prove causation of the disease by the fungus. Advanced the Germ Theory of Disease; recommended disinfection.

Bell (1774–1842) 2 c. Demonstrated surgically that each division of a peripheral nerve derives its functional specificity from the part of the brain to which it is connected, 1811; also by cutting anterior and posterior roots of the spinal nerves, showed former are motor in function, latter are not.

Bichat (1771–1802) 3 c. Anatomist who founded the science of histology by distinguishing 21 different types of tissues in the animal body and noting their properties, 1801.

Blainville (1777–1850) 2½ c. Cuvier's deputy at the Musée naturelle. Based classification of animals on principles exemplifying the Creator's design, and into series corresponding to the Great Chain of Being, distinguishing levels of sensibility.

Brongniart, A. T. (1801–1876) 4¼ c. Pioneer paleobotanist; classification and distribution of fossil plants 1822, 28, 37. Used a modern classification and accepted geological succession, but believed in fixity of species. Described the formation of tetrads in male sporogenesis, pollen tubes, embryo sac in the seed, and distinguished between the fertilized egg cell and the seed. Cuvierian.

Brown (1773–1858) 10½ c. Considered the "first" among the botanists of his time. Discovered Brownian movement and the streaming of protoplasm in plant cells. Affirmed that every cell has a nucleus and that it is an essential part of the cell. Found that gymnosperms lack an ovary, and bear naked seeds in their cones. From his early voyages on the *Investigator* to Australia, described about 2000 new species. In charge of Banks's collections from 1831. Described pollination in orchids and growth of pollen tubes to convey nuclei to the ovule. Royal Society member.

Buckland (1784–1856) 12 c. English follower of Cuvier, identifying fossil hyenas, bears, and various marine shelled invertebrates. Discovered the first dinosaur, 1824; coprolites of ichthyosaurs; and Megatherium, a giant primitive mammal. Adherent of the Great Chain of Being, contributor to the Bridgewater Treatises, and opponent of Lyell's uniformitarianism in geology.

Candolle (1778–1841) 5 c. Pioneer work on freshwater algae and succulent plants. Monographs on eight major families of flowering plants; *Prodromus*, 1824–39. Contributed to knowledge of lichens, medicinal botany, pharmacology, phytogeography, history of botany.

Caventou (1795–1877) 3½ c. Jointly with Pelletier, discovered chlorophyll (1817). Extracted many alkaloids from plants, including strychnine, brucine, veratrine, quinine, caffeine. Chemist and toxicologist.

Cuvier (1769–1832) 12½ c. Acknowledged "first" in zoology in his time. *Le règne animal*, 1817; *Histoire des poissons*, 1828–32, 9 vols. Authority on classification. Developed principle of the correlation of parts whereby from available skeletal parts the nature of missing parts could be predicted. Formulated the "balance of nature" concept. Strongly opposed the transformism (evolutionary theory) of Lamarck and later Geoffroy St. Hilaire.

du Bois-Reymond (1818–1896) 10½ c. Electrophysiologist; studied animal electricity, 1842–3; injury currents; muscle contraction; polarization; electrotonus, shock; electrical character of the nerve impulse, 1848. Also contributed to history of science. NAS.

Dujardin (1801–1860) 7 c. Protozoologist; studied rhizopods such as amoeba, and the shelled foraminifera. Developed concept of "sarcode," synonym for protoplasm, and studied its vacuoles and behavior in white blood cells and worms as well as protozoans, in coelenterates, intestinal worms, insects, echinoderms. A pioneer in the colloidal chemistry of protoplasm.

Dumas (1800–1884) 11½ c. In youth, with Prévost, found proof that frog's egg is fertilized by spermatozoa, 1824. Primarily a chemist, studied animal fat. NAS.

Dutrochet (1776–1847) 4 c. Physiologist and embryologist; studied development of eggs within the ovary, and fetal envelopes. Gas exchange, proving that respiration in plants and in animals is the same. Function of stomata, 1832. Observed that only *green* plants absorb CO_2 and transform light into chemical energy. Studied excitability and motility; osmosis and diffusion. Observed that mushrooms are the fruiting bodies of mycelium. Anticipated the Cell Theory. Rejected the concept of "vital force."

Ehrenberg (1795–1876) 9 c. First to observe the fission of protozoans, 1838; regarded protozoans as complete organisms, with organs comparable to those in higher animals.

Flourens (1794–1867) 3¼ c. Physiologist and deputy of Cuvier. Founder of *Comptes rendus*. Studied physiology of brain function by ablation of parts and opposed the phrenology of Gall. Work on nervous system: perception and transmission; excitation of muscle contraction; coordination. Effects of lesions of the semicircular canals leading to disturbed equilibrium. Located the respiratory center in the medulla oblongata. Also studied the reunion of nerves, role of the periosteum, and anesthetic properties of chloroform.

Gaertner (1772–1850) 2³/₄ c. Made many plant hybrids. Discovered dominance of some parental traits and noted other "regularities."

Gall (1758–1828) 12 c. Neuroanatomist and psychologist. Conceived the nervous system to be a hierarchical series of separate ganglia designed on a unified plan. Distinguished white from gray matter, and postulated the plurality and independence of cerebral organs. General doctrines poor; neuroanatomy good. From recognition of the cerebral localization of functions he went into phrenology. Helped to establish psychology as a biological science.

Geoffroy Saint-Hilaire (1772–1844) 7 c. Work on marsupials, 1796. Studied Egyptian mammals, recent and mummified. A stout supporter of transformism = evolution, contra Cuvier. As paleontologist, established the fossil reptile Teleosaurus. Proposed two fundamental principles of evolutionary biology: the principle of homology; the principle of balance (when one organ hypertrophies, neighboring organ is reduced).

Helmholtz (1821–1894) 24¹/₂ c. Physiologist: energetics, physiological acoustics and optics; hydrodynamics; electrodynamics. Established the Conservation of Energy Law, 1847. Determined the velocity of the nerve impulse. Formation of acid in contracting muscle. Invented the ophthalmoscope. NAS.

Henle (1809–1885) 5¹/₄ c. First study of histology based on extensive microscopic examination; relation of tissues to cells; especially work on epithelia of the intestinal villi, where he distinguished pavement epithelium from "cylindrical" (= columnar) epithelium. Great textbook, 1841. Also theory of "miasma" as basis of contagion.

Hofmeister (1824–1877) 8¹/₂ c. Botanist. Described how plant embryo comes from a pre-existing cell in embryo sac of ovule. Also described pollen formation. Discovered and emphasized the alternation of generations in seed plants.

Humboldt, A. (1769–1859) 13 c. Travels in South America, 1799–1804; 60,000 specimens, 6300 new species, maps, measurements, magnetism, meteorology, climatology, mineralogy, geology, oceanography, zoology, ethnography, history, linguistics, plant geography and physiognomy of regions visited. Detected connection between the Orinoco and Amazon Rivers. Ascended the volcano

Chimborazo. Travel journal of 34 vols. Elected to APS in 1804, at conclusion of S. and Central American travels and visit to USA. Author of *Kosmos*, 1845–47.

Knight (1759–1838) 3½ c. Discovered geotropisms in plants (responses by differential growth to gravity). Observed the decline in vigor of horticultural stocks propagated by grafts repeatedly, and emphasized need for sexual reproduction to maintain vigor. Genetic studies of crosses between varieties of peas, noting dominance of certain parent traits and reappearance of the recessive trait in second generation. Forerunner of Mendel.

Kölliker, A. von (1817–1905) 6½ c. Observed origin of spermatozoa in special sperm-producing cells of the testes. Studied development in cephalopods. Observed the division of red blood cells and the formation of blood in embryo's liver. Distinguished the smooth muscle fibers. Worked on the formation of the skull; distinguished 5 kinds of tissues; noted that the egg is a single cell; attributed a hereditary function to the cell's nucleus. NAS.

Lamarck (1744–1829) 20 c. *Flore françoise* (1779) introduced dichotomous keys for identification. Major work on invertebrate animals, 1801–22. In *Philosophie zoologique*, 1809, developed his theory of evolution, based on natural tendency to increase in complexity, indirect influence of the environment on an animal, effects of use and disuse and changes of habits, and inheritance of acquired characteristics.

Le Gallois (1770–1814) 4¼ c. Physiologist; recognized respiratory center in the medulla oblongata; anticipated the concept of internal secretions and hormones. Conducted experiments on animals; recognized the metameric organization of the spinal cord; distinguished gray matter from white matter.

Liebig (1803–1873) 37½ c. Notable early biochemist. Noted that the gas exchange of hemoglobin is related to the oxidation of iron. Great textbook *Organische Chemie* (in applications to Agriculture and Physiology), 1840. Studied use of mineral fertilizers in agriculture. Conceived of respiration as combustion, all animal heat coming from breakdown of fats, carbohydrates, and proteins. Identified the nitrogen cycle in nature. NAS.

Magendie (1783–1855) 10 c. In 1811 distinguished the functions of the dorsal and ventral roots of the spinal nerves. The Bell-Magendie Law, 1822. 1809–13, studied the experimental action of drugs on animals, the beginning of modern pharmacology.

Mayer, J. R. (1814–1878) 10½ c. Studied oxidation of foods as source of animal heat; established Conservation of Energy Law as applied to animals. A critic of vitalism and believer in evolution, he noted the embryonic resemblances of animals and proposed recapitulation. Textbooks of pathological anatomy, 3 vols., and comparative anatomy, 6 vols., 1821–31.

Meckel (1781–1833) 3¼ c. Comparative anatomist, embryologist. An early student of birth defects. First to note the striking resemblances of animal embryos that develop into quite different sorts of adults. Formulated a prototype of the "Law of Recapitulation."

von Mohl (1805–1872) 2¼ c. Botanist who helped to establish the concepts of protoplasm and the cell. Plant cells comprised of membrane, utricle, protoplasm, nucleus, and cellular fluid; formed by growth and production of a partition wall.

Müller, Johannes P. (1801–1858) 14½ c. Discovered that different stimuli to the same sense organ produce the same sensation; the dorsal and ventral spinal nerve roots in the frog; and that there is no direct communication of afferent and efferent nerve fibers. Studied cyclostomes; the placenta in sharks; the lancelet; and pluteus larvae of sea urchins, noting their change from bilateral to radial symmetry in metamorphosis. A vitalist and initially a "Naturphilosophe." Many important students, e.g., Schleiden, Schwann, Henle, Virchow, Haeckel.

Owen, R. (1804–1892) 7 c. Studied pearly nautilus and other cephalopods. Identified monotremes as mammals although egg-laying. Primate anatomy, from aye-aye to gorilla. Worked up Darwin's South American fossils. Classic studies of invertebrates, fishes, higher vertebrates; fossil reptiles, and Archaeopteryx. Concepts of special and serial homology, contrasted with analogy. Attracted by Naturphilosophie; attacked Darwin's *Origin of Species*, especially the theory of natural selection. NAS.

Pander (1794–1865) 2¼ c. Notable embryologist, with von Baer identifying the three germ layers of the embryo. Believer in animal evolution, Lamarckian.

Payen (1795–1871) 1½ c. Discovery that starch is converted to sugar by diastase, a substance extracted from barley seeds (actually diastase is a mixture of enzymes).

Pelletier (1788–1842) 3¼ c. With Caventou, in 1817 isolated chlorophyll. With Magendie, isolated emetine from ipecac. Also isolated ambrein, strychnine, brucine, veratrine, cinchonine, quinine, and caffeine. Later collaborated with Dumas on alkaloids.

Persoz (1805–1868) 1¼ c. With Payen, isolated diastase; with Biot, obtained dextrin by hydrolytic action of mineral acids on starch or cane sugar.

Pouchet (1800–1872) 2 c. Known chiefly for prolonged contest with Pasteur over spontaneous generation of bacteria, which Pouchet supported.

Prévost (1790–1850) 2 c. With Dumas, discovered that frog's egg is fertilized by spermatozoa. Described the segmentation of the developing frog's egg. Studied the biochemistry of digestion and the formation of the circulatory organs and blood in amphibian embryos, and of the heart in the chick embryo.

Purkinje (1787–1869) 8¾ c. Many discoveries in microscopic anatomy, including the shift from perception of light by retinal cones to rods, in dim light; physical properties of sensory organs; formation of reflex images; vertigo from drugs; early development of the bird's egg within ovary, including discovery of the germinal vesicle of the ovum; identified the Purkinje cells of the cerebellum and the Purkinje fibers of the heart. Relation of tissues to cells. Also studied plant structures.

Rathke (1793–1860) 2¼ c. A notable embryologist, discoverer of the gill slits and gill arches in embryo birds and mammals; Rathke's pouch of the oral cavity that becomes the pituitary gland; and the pronephros, or embryonic kidney. Studied the regression of the gills and tails in tadpoles metamorphosing into frogs. Made first description of the lancelet.

Remak (1815–1865) 4³/₄ c. Observed the amitotic division of blood cells. Observed nuclear division in the cleavage of the frog's egg. Studied the fine structure of nervous tissue, and identified the ganglion cells. Found that the medullary canal (neural tube) is formed from the ectoderm. Identified the respective derivatives of the ectoderm, mesoderm, and endoderm.

Retzius, A. A. (1796–1860) 4¹/₂ c. Studied the development of elasmobranchs, lamprey, and Amphioxus. Collaborated with Müller and Purkinje.

Rudolphi (1771–1832) 2³/₄ c. Noted for study of the parasitology of intestinal worms. Discovered the intestinal villi.

Schleiden (1804–1881) 5¹/₂ c. With Schwann, proposed the original Cell Theory, although he held wrong idea of how new cells are formed. Notable textbook of botany in 1842–3, but held wrong idea of pollen tube. Scientific popularizations in 1848.

Schwann (1810–1882) 11¹/₂ c. With Schleiden, enunciated the original Cell Theory: the cell a mass of protoplasm surrounding a nucleus; but held an erroneous idea of how new cells are formed. Measured the contractility of muscle following same stimulus but under different loads. Discovered pepsin. Held fermentation to be a product of living organisms. Observed that yeast cells originate by budding. Became a mystic.

Semmelweis (1818–1865) 7 c. Proved beyond doubt that puerperal fever is transmitted to women in labor by the hands of physicians or midwives.

Turpin (1775–1840) 2¹/₂ c. One of the first botanists to study cell division (1826) and to identify the plant cell as a "utricle" containing vesicles and granules. Produced a Flora parisienne. Adopted Goethe's idea of the leaf as the archeypal plant organ.

Wöhler (1800–1882) 10¹/₂ c. Synthesis of urea from ammonium cyanate (1828), first production in laboratory of an organic compound. Worked with Liebig. Held that eventually all organic substances will be produced in the laboratory. NAS.

TABLE 3

Biologists of the Second Half of the Nineteenth Century
Elected, or not Elected, to Foreign Membership
in the American Philosophical Society

*A date in parentheses indicates year of election to the APS.
Other dates are of notable contributions to science.*

Elected		Not Elected	
Great Britain			
Charles Darwin	1858–9/68/71 (1869)	George Newport	1851
Joseph D. Hooker	1855 (1869)	Thomas Graham	1861
Thomas Henry Huxley	1856–63/66 (1869)	George J. Romanes	1882/83
Alfred Russel Wallace	1858–9/69/76 (1873)	J. W. Dawson	1888
Philip L. Sclater	1858 (1873)	C. S. Sherrington	1892/98
William Osler	(1885)	C. E. Overton	1893
Joseph Lister	1867 (1897)	David Bruce	1893
Michael Foster	1859/69 (1897)	Ronald Ross	1898
Edwin Ray Lankester	1859/69/76 (1902)		
Francis Galton	1869/83/93 (1903)		
France			
Charles E. Brown-Séquard	1852/56 (1854)	J. B. Boussingault	1851–5/64–9
Claude Bernard	1849/52/55/56/59 (1860)	G. Thuret	1854
J. Boucher de Perthes	(1863)	Alphonse Laveran	1880
		Jules Bordet	1895

	Elected		Not Elected	
Wilhelm P. Schimper	1836/48	(1866)	Gabriel Bertrand	1897
Paul Broca	1861/66–9	(1872)		
Louis Pasteur	1857/60–79/79–85	(1885)		
Marcellin Berthelot	1885/95	(1895)		

Belgium

Henri Milne-Edwards	1860/68–74	(1860)	Edouard van Beneden	1883–87

Netherlands

Hugo de Vries	1889/96	(1903)	M. H. Beijerinck	1888

Switzerland

			Hermann Fol	1877–79

Germany

Heinrich G. Bronn		(1860)	K. F. W. Ludwig	1852
Rudolf Virchow	1855/58/63	(1862)	N. Pringsheim	1854/55/58
K. T. E. von Siebold	1854/63	(1869)	R. Leuckart	1854/63
Ernst Haeckel	1866	(1885)	Carl Gegenbaur	1861
Anton Dohrn		(1903)	Max Schultze	1861
Wilhelm Waldeyer	1891	(1904)	Julius von Sachs	1860s
Adolf Engler		(1906)	H. A. DeBary	1868
August Weismann	1875/1883–92	(1906)	Simon Schwendener	1868
Wilhelm Pfeffer	1873–77	(1909)	T. W. Engelmann	1868–70
			Friedrich Miescher	1869

Elected	Not Elected	
	G. Fritsch & E. Hitzig	1870
	Walther Flemming	1870s
	Eduard Strasburger	1870s/83–5/88
	Wilhelm His	1874
	Anton Schneider	1873/4
	Robert Koch	1876/84
	Oskar Hertwig	1877–9/83–5
	Wilhelm Roux	1883–5/88
	Hugo Kronecker	1871
	Otto Bütschli	1876/89
	Felix Hoppe-Seyler	1866/76–78
	Edward Pflüger	1872/75/93
	Moritz Traube	1858/74–78
	Paul Ehrlich	1885
	Hermann Hellriegel	1888
	Theodor Boveri	1888–95
	Emil von Behring	1890
	Richard Hertwig	1884/90
	H. Henking	1891–2
	Eduard Buchner	1896–7
	Friedrich Loeffler	1897

Sweden

M. G. Retzius   1875–6   (1912)

	Elected		Not Elected	
Austria			Johann Gregor Mendel	1865
Italy			Carlo Golgi	1871/3
			G. B. Grassi	1898
Spain (1932)	S. Ramón y Cajal	1890		
Russia			I. M. Sechenoff	1863
			A. O. Kovalevsky	1866–77
			Constantin Timiryazeff	1868–70
			V. O. Kovalevsky	1870–73
			Sergius Vinogradsky	1876–85/87/94
			Elie Metschnikoff	1884
			Dmitri Ivanovski	1892
Japan (1914)	S. Kitasato			

Thumbnail Sketches of Each Biological Scientist's Notable Contributions
The number with a "c" appended which follows the birth and death dates indicates the number of columns in the Dictionary of Scientific Biography *devoted to the subject.*

Altmann (1852–1901) Not in *D.S.B.* Discoverer of mitochondria, 1894, structures now known to be the "powerhouses of the cell" and essential cellular organelles. Held a theory of the granular structure of protoplasm.

Behring, von (1854–1917) 7 c. With Kitasato, developed antibodies to tetanus and diphtheria; antitoxins, 1890; serum therapy, 1893. Nobel prize.

Beijerinck (1851–1913) 4 c. Discovered Rhizobium, the root nodule bacterium of legumes, 1888, isolated and cultured it. Also Azotobacter, a nitrogen-fixing bacterium. Developed auxanography, 1889, and enrichment culture. Concept of role of microorganisms in cycles of matter in nature.

Beneden, van (1846–1910) 3 c. 1883–7, described union of two "half-nuclei" after fertilization in Ascaris, each gamete bearing complementary sets of chromosomes. Discovered reduction to haploidy (= single set of chromosomes) in the maturation of the egg and polar body. Described role of centrosome in cell division.

Bernard (1813–1878) 19¼ c. Major physiologist of period. Discovered formation of glycogen from sugar in liver, 1849; vasoconstrictor nerves, 1852; role of pancreas in digestion, 1849; vasomotor regulation of blood supply, 1855, and concept of the constancy of the "internal milieu." Glands of internal secretion, 1859; pancreatic juice and digestive regulation, 1856.

Berthelot (1827–1907) 17 c. Fixation of nitrogen by soil bacteria, 1885. Isolation of nitrogen-fixing bacteria, 1893. NAS.

Bertrand (1867–1962) 2½ c. From isolation of an oxidase, 1896, found importance of trace metals, e.g., manganese, as necessary functioning part of the oxidative enzyme. Metal is "cofactor" of enzyme. NAS.

Bordet (1870–1961) 2 c. Discovered the role of complement in immune reactions, 1895. Often necessary to ensure reaction of antibody with antigen. NAS.

Boucher de Perthes (1788–1868) ¼ c. Archeologist; discoverer of Stone Age cultures in France.

Boussingault (1802–1887) 1½ c. Discovered that plants absorb nitrogen from soil, carbon dioxide from the air, 1852–5; that nitrogen is fixed in salt form by legumes. Demonstrated that only cells with chloroplasts assimilate CO_2, 1864–9. NAS.

Boveri (1862–1915) 9 c. Showed that an enucleated egg can be fertilized by a spermatozoon, and will then yield a haploid type with purely paternal inheritance, 1889–95. Established the persistent identity and individuality of the chromosomes. Discovered meiosis in animals (Ascaris), 1887–90. Proposed the Chromosome Theory of Heredity, 1902. NAS.

Broca (1824–1880) 2¼ c. Discovered the cortical localization of motor functions, especially the speech center, in the brain, 1861. His interest in craniology led him into ethnology and anthropology.

Bronn (1800–1862) 2 c. Paleontologist, known primarily for his great *Tierreich*, an attempt to provide a complete coverage of the Animal Kingdom. An opponent of Lamarckian evolution, Bronn accepted the gradual transition of forms.

Brown-Séquard (1817–1894) 3 c. Discovered crossing of the nerve pathways in the spinal cord, 1849. While in U.S., studied the vasoconstrictor nerves and demonstrated the artificial production of epileptic seizures through lesions in the nervous system, 1856. Opposed the idea of cerebral localization of functions. Showed that removal of the adrenal glands is fatal, 1856. Late in life, advocated rejuvenation by means of extracts from the testes.

Bruce (1855–1931) 6 c. Proved that the tsetse fly's bite transfers the agents of trypanosomiasis (nagana), 1895–7.

Buchner (1860–1917) 6 c. Extracted "zymase" from yeast and purified first enzyme, 1897. Demonstrated fermentation can occur in vitro.

Bütschli (1848–1920) 4¾ c. Studied cell division, conjugation, and fertilization in ciliates. Determined the order of the successive stages in mitosis; described the fertilization cone arising where a spermatozoon enters the egg, and concluded that only one sperm normally enters the egg. Observed the fusion of male and female pronuclei in snail eggs after fertilization; described the presence of "rodlets" on the nuclear plate of the dividing cell. Put forward an alveolar theory of the structure of protoplasm, 1889–92.

Darwin, C. (1809–1882) 24¼ c. Voyage of the *Beagle* around the world, and observations as naturalist; collections of fossils and many animal specimens. *Origin of Species by means of Natural Selection,* 1858–9, forced scientific world to accept reality of organic evolution. *Variation in Animals and Plants under Domestication,* 1868. *Descent of Man,* 1871, and sexual selection. Authority on the classification and morphology of barnacles. By common consent, the greatest of all naturalists.

Dawson (1820–1899) 4½ c. Discovered, 1859, the earliest fossil land plant, a psilophyte, of the Devonian Period. An anti-evolutionist.

DeBary (1831–1888) 6½ c. Studied lichens, and found them to be composites of an alga and a fungus; developed the concept of symbiosis, 1869–79. Sexual reproducion of fungi; alternation of generations in rust fungi, alternate host the barberry.

Dohrn (1840–1909) Not in *D.S.B.* Marine zoology. Founder and director of the Marine Zoological Station in Naples.

Dubois (1858–1940) Not in *D.S.B.* Discovery of Pithecanthropus, now known as *Homo erectus,* a fossil type of early man, found in Java, 1892. First Homo not recognized as *H. sapiens.*

Ehrlich (1854–1915) 19¾ c. With Koch, 1882, developed staining of tubercle bacilli; vital staining of bacteria, 1885–91; fluorescein staining; diazo reaction. Used ricin and abrin as antigens, Prepared bacterial toxins and antitoxins. With von Behring, improved diphtheria antitoxin. Developed side-chain theory of immune reactions, 1889–1906. Nobel Prize. NAS.

Engelmann (1843–1909) 3½ c. The action spectrum of chlorophyll: red most efficient, violet less, rest of spectrum negligible, 1882. Action of flagella in protozoa. Myogenic induction of stimuli in heart, and velocity of conduction in heart muscle. Chemotaxis in bacteria. Sensitivity of protozoa to light and color.

Engler (1844–1930) Not in *D.S.B.* Geography of recent and fossil plants. Joint author of widely used Engler-Prantl work on plant taxonomy. NAS.

Flemming (1843–1905) 4½ c. Described mitotic cell division in animal cells; distinguished the doubled threads of chromatin and stated that chromosomes divide longitudinally, 1882. Specified the constancy of chromosome number in each species.

Fol (1845–1902) 4 c. Described the fusion of male and female pronuclei in fertilization of the animal egg, 1877–9. Also adduced proof that a single sperm is the agent of fertilization.

Foster (1836–1907) 10¹/₄ c. Physiologist. Noted that heartbeat could be either myogenic or neurogenic; demonstrated that a snail heart lacking ganglia could generate a heartbeat myogenically, 1859. Same for vertebrate (frog) heart in lower two-thirds of the ventricle, 1869. Biographer of Claude Bernard. History of physiology in 16th to 18th centuries.

Fritsch (1838–1927) 3 c. Functional localization in the cerebral cortex (dog), 1870.

Galton (1822–1911) 2¹/₂ c. Pioneer in biometrical statistics, anthropometrics, experimental psychology, and heredity in humans. Developed measures of regression and correlation. Studied the inheritance of talent. Father of eugenics. Introduced use of fingerprinting as unique test of individuality, 1893.

Gegenbaur (1826–1903) 12 c. Identified ovum as a modified cell, 1861. Described life cycles of various marine invertebrates, 1855. Defined homologies of vertebrate limbs and cranial bones, and used comparative anatomy to indicate phylogeny. A Darwinist. NAS.

Golgi (1843–1926) 4 c. Notable studies of the microscopic structure of the nervous sysem; discovered the glial cells, 1871; developed the superb Golgi stain for nerve cells, 1873. Determined the functional differences of axons and dendrites. Found the Golgi bodies in nerve cells. 1885–93, worked on the tertian and quartan types of malarial parasites. Nobel Prize.

Graham (1805–1869) 5¹/₂ c. Defined osmosis and osmotic force, 1854. Father of colloid chemistry, 1861. Studied dialysis and sol-gel reversible changes.

Grassi (1854–1925) 4 c. With Bignami, determined the transfer of malaria by mosquitoes, 1898; identified Anopheles as the mosquito carrier. Studied termites, worms, flies; and identified *Plasmodium vivax*, parasite of malaria. Found that quinine suppresses the growth of the parasites in the blood.

Haeckel (1834–1919) 9¹/₂ c. Monographs on radiolaria, medusae, siphonophores, sponges, echonoderms, 1862–89. Wrote *Generelle Morphologie der Organismen*, 1866. Leading German Darwinist, arguing for a monophyletic descent of all plants and animals. Proposed the Recapitulation Law, 1872; accepted inheritance of acquired characteristics; unreliable interpreter of history of biology.

Hellriegel (1831–1895) 1³/4 c. With Wilfarth, identified the symbiotic root bacteria on legumes that fix nitrogen as nitrate, 1888.

Henking (1858–1942) 2 c. Discovered the sex chromosomes, although he failed to discern their relation to sex determination, 1891. Also observed the conjugation of chromosomes, two by two, during meiosis.

Hertwig, O. (1849–1922) 6 c. Discovered the fusion of the male and female pronuclei in the egg, the central event in fertilization, 1877–9. Proposed that heredity resides in the chromosomes, 1883–5. Discovered meiosis in animal germ cells, 1890. Noted that only one sperm is required for fertilization in the sea urchin egg, others being prevented from entering by formation of a vitelline membrane, 1875. Identified the parallelism of spermatogenesis and oogenesis, 1890. Proposed the Germ Layer Theory of development.

Hertwig, R. (1850–1937) 2 c. In embryology, contrasted the development of the two-layered coelenterates with that of the three-layered higher animals. Proposed theory of the coelom. Found that fragmented eggs, both nucleate and non-nucleate portions, can be fertilized. Described artificial parthenogenesis. Studied the life cycles of protozoans. NAS.

His (1831–1904) 5 c. Embryologist, studying products of the germ layers and mesodermal cavities; distinguished epithelia and endothelia. Identified the origin of nerve fibers from nerve cells in the central nervous system. Determined the existence in the animal embryo of predetermined organ-forming regions, 1874. Rejected Haeckel's Law of Recapitulation. Constructed the first microtome.

Hitzig (1838–1907) 2¹/2 c. With Fritsch, determined that specific muscles move when particular parts of the cerebral cortex are stimulated electrically, 1870. Contralateral responses. Removal of certain areas produced (dog) muscle weakness. Studied the visual cortex, 1874.

Hooker (1817–1911) 8 c. Botanist on the voyage of the *Erebus*, plant geographer (New Zealand, Tasmania, Antarctica). Flora of Himalayan India, especially rhododendrons, 1849–55. Paleobotany. Close friend of Darwin. NAS.

Hoppe-Seyler (1825–1895) 3 c. Physiologist. Chemical studies of blood and urine; absorption spectrum of hemoglobin, 1862. Discovered oxyhemoglobin, and that carbon monoxide displaces oxygen from oxyhemoglobin. Also studied porphyrins, lecithin, cholesterol, and noted that chlorophyll resembles porphyrin in chemical structure.

Huxley (1825–1895) 17 c. Naturalist on the voyage of the *Rattlesnake*. Described the invertebrates, 1850–54; vertebrates, 1854–58. Subdivided Pisces (fishes) into Agnatha, Chondrichthyes, and Osteoichthyes, 1866. Darwin's "Bulldog." Identified the first Neanderthal skull as a new human type or species; wrote "on fossil remains of man"; *Man's Place in Nature*, 1856–63. Leader in science education in Great Britain. NAS.

Ivanovski (1864–1920) 3¼ c. Discovered the first filtrable virus, tobacco mosaic virus, although thought it a bacterium, 1892. Obtained virus in form of crystalline particles.

Kitasato (1852–1931) 3¼ c. With von Behring, discovered the first known antibodies, the diphtheria and tetanus antitoxins, 1890. Earlier work on the cholera bacillus, the anaerobe Clostridium (for which he developed a pure culture technique), and tetanus bacillus. In Robert Koch's laboratory 1886–91. Developed tuberculin, 1890. Studied the bubonic plague organism (*Pasteurella pestis*), 1894.

Koch (1843–1910) 30¾ c. In study of anthrax, first to show that a specific bacterium causes a specific disease, 1876. Discovered bacterial spores. Studied etiology of wound infections, 1877–80. Stated the famous Koch Postulates for positive proof of the cause of a disease by a bacterium: the Germ Theory of Disease, 1884. Introduced sterile techniques and use of nutrient gelatin for cultures. Studied the tuberculosis bacillus, 1881–4. Relation of cholera, in India, to water supply and sewage disposal. Also studied agents of leprosy and plague; but Koch's tuberculin preparation held suspect. Trained Ehrlich, Behring, Kitasato, Kossel, Wassermann, and others. NAS.

Kovalevsky, A. O. (1840–1901) 5 c. Embryologist of invertebrates, studying development of Amphioxus and tunicates from the gastrula stage, 1866–77, as well as coelenteraes, echinoderms, and worms. Found evidence of evolutionary descent in embryonic stages, Amphioxus and tunicates related to vertebrates.

Kovalevsky, V. O. (1842–1883) 1¼ c. Paleontologist, especially the evolution of horses, 1872. Noted the reduction of the digits in horses and ungulates during the course of evolution. Related changes in the teeth and skull to changes in the particular food plants consumed.

Kronecker (1830–1914) 3 c. Identified the acid formed in muscles during contraction as lactic acid, 1871. Studied the all-or-none nature of contraction in heart muscle, in the isolated heart, 1873–4; and the refractory period of heart muscle following contraction. Studied mountain sickness. NAS.

Lankester (1847–1929) 1¾ c. Studied homologies in spiders, scorpions, and limulids. Systematized the study of embryology. A strong Darwinian and anti-Lamarckian. NAS.

Laveran (1845–1922) 2 c. One of the discoverers of the agent of malaria, the plasmodium, 1880. Nobel Prize, 1907.

Leuckart (1822–1898) 3¾ c. Distinguished coelenterates from echinoderms, 1848. Worked out the life cycles of intestinal parasites (Pentastomum, Taenia, liver fluke, trichina), 1854–68. Studied the Sporozoa and Coccidia. NAS.

Lister (1827–1912) 27½ c. Notable surgeon, who introduced antisepsis with carbolic acid, 1867. Introduced catgut for ligatures. Studied fermentation by *B. lactis*, 1877. NAS.

Loeffler (1852–1915) 6 c. Discovered the first animal virus, that of foot-and-mouth disease, 1897. With Koch, 1879–83; discovered infectious agent of glanders. Developed pure culture of diphtheria bacillus, 1884; also agents of swine erysipelas and of plague. Worked on *Salmonella typhimurium*, 1888–92.

Ludwig (1816–1895) 4¼ c. Physiologist; studied the dependence of glandular secretion on the nerve supply. Wrote *Lehrbuch der Physiologie*, 1852. Invented the mercurial blood pump, 1859, and the kymograph, 1846. NAS.

Mendel (1822–1884) 14 c. From experimental crossing of varieties of peas, determined the laws of heredity: segregation; dominance and recessivity; independent assortment of different pairs of contrasting traits, 1865.

Metschnikoff (1845–1916) 8½ c. Discovered phagocytosis in starfishes, 1882, and developed theory of protection against disease by phagocytosis on the part of white blood cells, 1883. Theory of immunity, 1901. Nobel Prize, 1908.

Miescher (1844–1895) 2³/4 c. From pus cells, obtained a new kind of organic compound containing phosphorus, "nucleins," 1869. From salmon sperm, obtained nuclein combined with protamine, 1874.

Milne-Edwards (1800–1885) 2³/4 c. Studied physiology of crustaceans, 1860. *Histoire naturelle des crustacées*, 3 vols., 1834–40. Corals, 1858–60. Mammals, 1868–74. Comparative physiology and anatomy of animals. Law of the division of labor between organs (within organisms). NAS.

Newport (1803–1854) 3 c. Identified fertilization as the entry of a spermatozoon into the egg (frog), 1851; its point of entry determining the location of the first cleavage plane and thus the median plane of the embryo.

Osler (1849–1919) Not in *DSB*. Famed physician; angina pectoris. History of medicine. *Principles and Practice of Medicine*, 1892.

Overton (1865–1933) 2½ c. Identified the alternation of diploid and haploid generations in plants, 1890–93. Studied cell permeability and fat solubility and proposed theory of the lipid structure of plasma membranes. Studied cellular transport against the concentration gradient of the solutes; narcosis; loss of muscle irritability upon diffusion outward of sodium ions, reversible.

Pasteur (1822–1895) 131 c. Fermentation (lactic; alcoholic) identified as a microbial process, 1857–60. Disproof of spontaneous generation of bacteria, 1861–4. Studies on the cause of silkworm disease; anthrax. Germ Theory of Disease, 1878/80. Study of immunity and production of vaccines: anthrax, 1881; rabies, 1881–6. NAS.

Pavlov (1849–1936) 8½ c. Studied circulation of the blood, physiology of digestion. First development of conditioned reflexes in digestion (dog), 1897–1903. Nobel Prize, 1904. NAS.

Pfeffer (1845–1920) 6¹/4 c. Plant physiologist; studied osmosis, 1873–77; periodic movements of plant organs. Wrote *Pflanzenphysiologie*, 1881. Many students from USA. NAS.

Pflüger (1829–1910) 5¹/4 c. Physiologist. Studied gas exchange in animals; defined the respiratory quotient. Studied electrotonus, 1858; metabolism of nutrients, divided into proteins, fats, carbohydrates.

Pringsheim (1823–1894) 8 c. Botanist and plant physiologist. Discovered fertilization in lower plants by spermatozoids. Contrary to Schleiden, held that cell division is the basic method of cell multiplication. Confirmed Thuret on fertilization by spermatozoids and found them in Vaucheria, Oedogonium (actual observation of penetration of spermatozoid into the egg), 1856–8. Noted the alternation of generations.

Ramón y Cajal (1852–1934) 5 c. Histological study of the central nervous system. Determined that there are separate neurons with connections (synapses) over which transmission of nerve impulses moves from one neuron to another. Also that the nerve impulse always moves from the terminal arborizations of sensory neurons to the dendrites of a c.n.s. neuron, then to cell body, and thence to axon. Nobel Prize, 1906. NAS.

Retzius, M. G. (1842–1919) 4¹/4 c. Studied anatomy of nervous system and connective tissues and the comparative anatomy of human and ape brains. NAS.

Romanes (1848–1894) 7¹/2 c. Studied the neurocontractility of medusae; mental evolution. Wrote *Animal Intelligence*, 1882; *Mental Evolution in Animals*, 1883; *Mental Evolution in Man*, 1888. As evolutionist, accepted inheritance of acquired characteristics; emphasized physiological selection; the reproductive isolation of varieties; and the role of isolation in evolution.

Ross (1857–1932) 4 c. Determined, parallel with Grassi, that mosquitoes transmit malaria, that Anopheles is the specific vector, Plasmodium the parasite. Work done on bird malaria. Nobel Prize, 1902.

Roux (1850–1924) 11¹/4 c. Proponent of experimental embryology. Heredity seen to reside in the chromosomes, 1883. Killing of one blastomere in embryo led to idea of the natural selection of parts within an organism; adaptation of skeletal parts to stresses; injury experiments, 1881–3; formation of half-embryos. Concept of mosaic development. Associated with "Entwicklungsmechanik." NAS.

Sachs (1832–1897) 5½ c. Observed that chlorophyll is formed only when plant is in light, and that starch is first product of photosynthesis, 1860s. Studied growth in roots and shoots, geotropism, phototropism, hydrotropism. Attacked Darwin's Theory of Natural Selection. *History of Botany*, 1875. NAS.

Schimper (1808–1880) 2 c. Studied mosses. Wrote *Bryologia Europaea*.

Schneider (1831–1890) 4 c. Observed that in the flatworm Mesostomum the chromosomes split longitudinally during mitosis, 1873. Introduced carmine as stain for chromosomes.

Schultze (1825–1874) 4½ c. Defined the cell as consisting of protoplasm plus nucleus, 1861. Found chlorophyll occurring in an animal (Turbellaria). Studied cell division.

Schwendener (1829–1919) 2 c. Like DeBary, identified lichens as composites of a fungus and an alga, 1869. Studied phyllotaxy, 1878.

Sclater (1829–1913) 3 c. Ornithologist. Developed foundations of terrestrial zoogeography: six regions, 1858.

Sechenoff (1829–1905) 3 c. Studied theory of cerebral behavior mechanisms in laboratory of Claude Bernard. Analyzed all conscious and unconscious behavior into reflexes, such as sensation resulting in movement, 1863. Founded in Russia the study of neurophysiology and psychology, foundation of work of Pavlov. Discovered periodic and spontaneous fluctuations of electrical potentials in the brain, "brain waves," 1881.

Sherrington (1857–1952) 15¼ c. Developed the concepts of nervous integration and reflex coordination, 1898. Studied knee jerk reflex, 1891–2. Mapped the motor pathways in the lumbrosacral plexus, 1892. Studied sensory nerves in muscles, 1894; cutaneous distribution of the posterior spinal roots, motor functions of the spinal cord, compounding of reflexes, inhibitory and excitatory actions at synapses, and the reciprocal innervation of antagonistic muscles, 1892–8. Especially important was the development of the concept of the synapse, 1897. Life work and ideas summarized in *The Integrative Action of the Nervous System*, 1906. Nobel Prize, 1932. NAS.

Siebold (1804–1885) 4³/4 c. Like Leuckart, studied the life cycles of intestinal parasites, especially cestodes, 1854/63. Discovered parthenogenesis in honeybees, 1850–56. Studied freshwater fishes of Europe.

Strasburger (1844–1912) 6³/4 c. Notable work on the mitotic division of plant cells in the 1870s. Recognized role of nucleus and chromosomes in heredity, 1884. Observed the fusion of the gametic nuclei in an alga and in flowering plants, 1877/84. Observations of mitoses in embryo sacs; discovered reduction of chromosome number to haploid state, 1888, and meiosis. NAS.

Thuret (1817–1875) 2³/4 c. Discovered the spermatozoids of Fucus, 1844, and their role in fertilization, 1854.

Timiryazeff (1843–1920) 3 c. Identified the action spectrum of chlorophyll, 1871. Assimilation of light by plants; highest activity in red light, 1877. Demonstrated that First Law of Photochemistry and Conservation of Energy hold true in photosynthesis. Found second absorption band, 1890. Observed relation of amount of photosynthesis to intensity of light.

Traube (1826–1894) 3³/4 c. Known especially for his work on oxygen-carrying ferments, 1858; semipermeable membranes, 1864/67; and artificial models of the cell, 1875.

Tschistiakoff [Chistiakov] (1843–1877) Not in *D.S.B.* Simultaneously with Schneider, observed that chromosomes replicate during mitosis by longitudinal splitting, 1873/4.

Virchow (1821–1902) 9 c. Developed a sociological theory of disease, 1848– , holding that bacterial agents are only one factor in disease. Founder of cellular pathology, 1858; famous aphorism, "Omnia cellula e cellula." Study of causation of tumors, 1863. After 1870, worked mainly in anthropology; held that there is "no pure German race." NAS.

Vinogradsky (1856–1953) 3³/4 c. With Pasteur until 1905. Discovered Azotobacter and its active nitrogen fixation, 1885. Worked out the morphology of iron and sulfur bacteria and their life cycles, 1887. Microbial chemosynthesis and auxanography, 1894. Also developed pure culture technique for the anaerobe *Clostridium pastorianum*. After 1900 developed concept of chemoautotrophism in bacteria, the oxidation of hydrogen sulfide being analogous to respiration in respect to supply of energy for the organism.

de Vries (1848–1935) 18²/3 c. *Intracellular Pangenesis*, 1889, as theory of heredity and development. Discovered genetic segregation, 1896, and rediscovered and confirmed the work of Mendel, 1900. Studied osmosis and plasmolysis in plant cells, 1890, curling of tendrils, turgor and isotonicity, plant protoplasm. Began work on mutations, 1895. NAS.

Waldeyer (1836–1921) 3¹/4 c. Developed doctrine of the neuron, 1891. Coined the term "chromosome." Studied lymphoid tissues, especially the tonsils. NAS.

Wallace (1823–1913) 15 c. Simultaneously with Darwin, proposed Theory of Natural Selection, 1858. Traveled as naturalist in the Amazon with Bates, 1848–53. Great works: *The Malay Archipelago*, 1869; *Geographic Distribution of Animals*, 1876; *Island Life*, 1880. Noted the sharp boundary of animal zones in the Malayan archipelago, Wallace's Line; explained the evolution of mimicry.

Weismann (1834–1914) 13¹/4 c. Experiments to disprove inheritance of acquired characteristics. Inferred the necessity of a chromosomal reduction division in the maturation of germ cells. Established the doctrines of the isolation and the continuity of the germ plasm, 1883–5; 1892. A stout neo-Darwinist, explaining evolution by means of natural selection alone. NAS.

Louis Agassiz as an Early Embryologist in America

Jane M. Oppenheimer

A T the time the American colonies established themselves as a
Republic, embryology did not yet exist as a coordinated discipline. Men of medicine had been recording observations on embryos or fetuses at least since the time of Hippocrates, and they
have continued to do so ever since. Presumably colonial physicians
and midwives, naturalists and farmers, looked at fetuses of man
and of his domestic animals and thought about them. Few if any
records reveal what they thought they saw, or what they thought
about what they thought they saw. Those that have been preserved
are often teratological in interest. Thus I am unable, for the purposes of this volume, to concentrate on the colonial world that has
so interested Whitfield Bell. It is a privilege, however, to pay
tribute to him for all he has done for us as a scholar and as a
functionary of this Society, even though, as an embryologist, I
must do so by dwelling on the wrong century.

Our protagonist here—not always a hero—is Louis Agassiz. He
was born in 1807, during Jefferson's presidency, although not as
one of his constituents, since he was originally of Swiss nationality.
Jefferson lived until 1826, thus his life and Agassiz's overlapped in
time for most of two decades. But not in space: Agassiz did not
come to America until 1846. He remained here, except when traveling, until he died in 1873; he became a naturalized American
citizen in 1861.

Whitfield Jenks Bell, Jr., has at least one personal reason to be interested in some of Louis Agassiz's work. Among Agassiz's contributions to embryology was his *Embryology of the Turtle*, included along with a work on the classification and zoology of American Testitudinata (turtles and tortoises) in the first monograph forming part of Volumes I and II of his *Contributions to the Natural History of the United States of America*.[1] In the preface to Volume I, Agassiz gratefully acknowledged that he had received specimens from, among others, J. W. P. Jenks of Middleboro, Massachusetts;[2] Jenks received a bachelor of arts degree from Brown University in 1838, and in the early 1870s became a professor of zoology there. He is also listed among the subscribers to the published work. John Whipple Potter Jenks (1819–1894) and Whitfield Jenks Bell, Jr., had a common ancestor in Joseph Jenks (Jencks) who was in Massachusetts in 1643, but their lines diverged in the third generation, around 1700.[3] What purports to be J. W. P. Jenks's own account of the challenges, vicissitudes, and success of getting turtle eggs to Agassiz within three hours of their being laid was presented in an entrancing story first published in 1910 in the *Atlantic Monthly*;[4] it was republished in an Anniversary Number of the *Atlantic* in 1932 (the year that I completed my metamorphosis into an embryologist), and I remember well the literary sensation that it then caused.

Whit Bell, as a historian of the American Philosophical Society, may find greater interest in Louis Agassiz's connections with this Society. Agassiz was elected a Foreign Member in 1843. This was only the beginning of what was to become a truly extended family

1. L. Agassiz, *Contributions to the Natural History of the United States of America*, vols. 1–4 (Boston: Little, Brown and Co., 1857–1862). Ten volumes were planned, but only four appeared. *North American Testitudinata* is in volume 1: 233–452. *Embryology of the Turtle* constitutes the whole of volume 2.

2. Ibid., 1: xii.

3. Memorandum Whitfield J. Bell, Jr., to JMO, 21 June 1984.

4. D. L. Sharp, "Turtle Eggs for Agassiz," *Atlantic Monthly* 105 (1910): 156–164; 150 (1932): 537–545.

membership in the Society. His second wife, Elizabeth Cabot Cary Agassiz, was elected in 1869. She was the second woman in the history of the Society chosen for membership, the first as a Resident Member. His first child, Alexander, was elected as a Resident Member in 1875. Alexander was the son of Louis Agassiz's first wife, née Cécile Braun, who died in 1848 of tuberculosis. In 1850 Louis married Elizabeth Cary. Alexander had come to America in 1849; his two sisters emigrated shortly after the second marriage, and the second Mrs. Louis Agassiz was a warm and devoted mother to all three children. To expand further the unique multiple record of this family's membership in the American Philosophical Society, it may be added that Alexander Braun, the brother of Agassiz's first wife, was admitted to Foreign Membership in 1862.

Historians of biology and medicine agree that embryology had not become an organized science—or even a science at all—until early in the nineteenth century, when the first two volumes of Karl Ernst von Baer's *Entwickelungsgeschichte der Thiere*[5] were published. Agassiz became interested in the subject a-borning even before 1828 when von Baer's first volume appeared. From 1827 to 1831 he was instructed in it, formally and informally, by Ignaz Döllinger. During those years, while a student at the University of Munich, Agassiz lived, along with other students, in Döllinger's house. It was Döllinger who had the first clear insight as to how the morphogenesis of complex organisms might be investigated scientifically, and he also developed the technical skills to make such study possible; it was he who had previously been the preceptor of von Baer. Agassiz places Döllinger accurately in the history of embryology in the last of the *Lowell Lectures* that he delivered in Boston in 1848–49: "Döllinger, the Physiologist, traced the growth of the young Chicken within the egg, and for the first time showed how im-

---

5. K. E. von Baer, *Entwickelungsgeschichte der Thiere*, 2 vols. (Königsberg: Borntraeger, 1828, 1837); the second volume was incomplete.

portant for physiological investigations it would be to understand the manner in which organs were formed, in order to arrive at more precise conclusions upon their functions." Although Agassiz claimed that Döllinger had published nothing on this subject, he asserted that "those who are conversant with the history of Embryology, acknowledge him as the first and most eminent among those who have devoted themselves to these investigations. Indeed, his pupils have, under his directions, carried out his views, and developed the new science up to the point at which it has arrived at the present day."[6]

Agassiz could not have had better beginnings in embryology. He always remembered with deep gratitude his experience with Döllinger and "the influence that learned and benevolent man had upon my studies and early scientific application." Agassiz appreciated the introduction to "what was then known of the development of animals, prior to the publication of the great work of von Baer; and from his lectures I first learned to appreciate the importance of Embryology to Physiology and Zoology."[7]

We shall return shortly to Agassiz's views on embryology as a part of physiology, but meantime let us remain briefly with another aspect of Döllinger's influence on Agassiz: he not only introduced him to microscopy, he also taught him the value of using good instruments. Specifically, Döllinger enlightened him as to the virtues of Fraunhofer's microscope, an instrument with improved objectives, made of flint glass, that greatly decreased both chromatic and spherical aberration. Agassiz bought one in 1828, and

6. L. Agassiz, *Lowell Lectures on Comparative Embryology* . . . (Boston: Henry Flanders & Co., 1849). Quotation from p. 96. Döllinger has long been spoken of reverently by historians of embryology but little studied, a balance that has recently been at least partially redressed by T. Lenoir in *The Strategy of Life* . . . (Dordrecht, Holland and Boston: D. Reidel Publishing Co., [1983]), 65–71 et passim.

7. L. Agassiz, *Essay on Classification*, in *Contr. Nat. Hist. U.S.A.* 1: 1–232. Quotation from p. 220.

ultimately became an accomplished microscopist, at least for a while.

To return to Agassiz's categorization of embryology as a branch of physiology: this coincided with the views of his contemporaries. When he considered embryology as physiological, he was thinking about it as the study of particular mechanisms at play during the development of specific organs. He was concerned with those "gradual changes, although complicated, and at first sight so mysterious, [that] follow laws which are uniformly the same in each department of the Animal Kingdom,"[8] as he put it in a textbook of zoology published jointly in 1848 with Augustus A. Gould. But this was not Agassiz's own primary concern. He did not attempt to investigate or explain such specific manifestations of change. His own fundamental interest, almost from the beginning, was in what the comparison of observed changes in various organisms might reveal of the mutual relationships of these animals as expressible in schemes of so-called natural classification. Such arrangements had previously been based upon adult anatomical rather than developmental characters. As Agassiz put it in 1848, "[e]mbryological investigations have been particularly made with reference to Physiology—that is, with reference to the mode of formation of the various organs which exist in animals, and not with reference to ascertaining their natural relations among themselves."[9] It was the ascertainment of such postulated natural relations that obsessed Agassiz. This was the contribution, in his view, that embryology could make to zoology as contrasted with physiology.

His theoretical interest ultimately became focused upon parallels between developmental changes in embryos and historical changes in animal succession as surmised from the fossil record,

8. L. Agassiz and A. A. Gould, *Principles of Zoology: . . . Part I. Comparative Physiology* (Boston: Gould, Kendall and Lincoln, 1848). [Two volumes were planned, but only the first appeared.] Quotation from p. 110.
9. *Lowell Lectures*, 7.

and their combined bearing on the formulation of schemata of natural classification.[10] In Agassiz's mind, by the way, such parallels were totally unrelated to what was to become evolution theory. When the *Origin of Species* appeared, he was totally opposed to Darwin's concepts, and he remained so throughout his life. He was a firm believer in special creation, indeed, in a continuing series of special creations. To him the significance of schemes of natural classification lay in their exemplification of the plan of the Creator. He never abandoned his conviction that they reflected a plan, or plans, devised by the superior Intelligence of a Divine Being. Agassiz remarked that "All organized beings exhibit in themselves all those categories of structure and of existence upon which a natural system may be founded, in such a manner that, in tracing it, the human mind is only translating into human language the Divine thoughts expressed in nature in living realities." Furthermore, he explained, "The combination in time and space of all these thoughtful conceptions exhibits not only thought, it shows also premeditation, power, wisdom, greatness, prescience, omniscience, providence."[11]

Agassiz's ideas have been frequent subjects for discussion. Because of his own particular emphases, however, most biographers and historians concerned with his embryological ideas have discussed them mainly in connection with their relationships to classification. Questions with respect to his actual knowledge of what passed for embryological fact in his time have arisen less frequently. How well-versed was he in the embryological literature? How accurate were his own observations? How original were his interpretations of what he saw? Did he express any new ideas influential

---

10. For the best recent discussion, especially with reference to Agassiz's studies on invertebrate development, see M. P. Winsor, *Starfish, Jellyfish, and the Order of Life* . . . (New Haven and London: Yale University Press, 1976).

11. *Essay on Classification*, 135.

within embryology itself? These topics encompass our principal interest here.

Agassiz's reputation as a research scientist rests primarily on his early technical studies and publications in ichthyology, geology, and paleontology. Before he came to America he had published a distinguished monograph on Brazilian fishes (1829), a set of volumes on fossil fishes (1833–1844), a four-volume monograph on living and fossil echinoderms (1838–1842), and an important study on glaciers and glaciation (1840).

In addition, in 1845, a publication appeared in Neuchâtel under his authorship entitled *Histoire Naturelle des Poissons d'Eau Douce de l'Europe Centrale*.[12] Below Agassiz's name and the main title on the title page appears a single subtitle which reads: "*Embryologie des Salmones* par C. Vogt." Accompanying a volume of text is an atlas containing magnificent colored illustrations of adult fishes, and engravings in black and white of eggs, embryos, hatchlings, and fry of a European Coregonid, *Coregonus palaea*. Coregonids are white-fishes, small-mouthed fresh-water fishes related to primitive sal-monid types; our American whitefish of the Great Lakes is *Coregonus clupeiformis*.

Agassiz later objected to the appearance of Vogt's name as au-thor of the *Embryologie des Salmones*; this was not the only dissension in which he sparred with former direct associates and colleagues over possible plagiary. When Agassiz cited the *Embryologie des Sal-mones* in a long bibliography of works on embryology published in 1857 (see p. 410 below), he himself attributed its authorship to Vogt, but in so doing he specified that "these investigations were made under my direction and supervision."[13] He thought that he,

---

12. L. Agassiz, *Histoire Naturelle des Poissons d'Eau Douce de l'Europe Centrale* (Neuchâtel: Petitpierre, 1845). Two volumes were planned, but only the first appeared. In later notes, it will be referred to as Agassiz-Vogt.

13. *Essay on Classification*, 81.

Agassiz, should have been named as principal author.[14] Agassiz had suggested to Vogt that he undertake the study, and Agassiz supervised the work during its progress. In his preface, Vogt wrote that when the work began in 1839, Agassiz participated in it fully. But as it proceeded, Agassiz became increasingly involved in other activities. As a result Vogt carried the heavy labor of the long and arduous continuing observations that are required for the sort of complete study of teleost development that the *Embryologie des Salmones* became. No matter who deserves the credit, the publication of the work, together with the dispute over its authorship, attests to the fact that Agassiz still in 1857 took himself seriously as an embryologist, whether or not others did.

Of the monographs published while Agassiz was living in Europe, the volume on the Coregonids was the only one to enter into detail on embryology. Agassiz's choice of teleost (bony fish) eggs for the first detailed investigation of morphogenesis in which he was involved merits attention. The eggs of most teleost species develop externally, and are produced in large numbers. If not overcrowded in their containers, or otherwise subjected to inappropriate environmental conditions, they develop normally in the laboratory—and could have done so in whatever quarters substituted for laboratories in the period of which we are speaking. The eggs and embryos and their coverings are extraordinarily transparent. Teleost eggs, once their coverings are removed, are very fragile, and are therefore used in few laboratories today. In the

---

14. The controversy between Agassiz and Vogt is discussed more fully by E. Lurie, *Louis Agassiz: A Life in Science* (University of Chicago Press, 1960). This is the fullest biography of Agassiz available, but equally interesting is E. C. Agassiz, ed., *Louis Agassiz. His Life and Correspondence*, 2 vols. (Boston: Houghton Mifflin, 1885). Neither Lurie nor Mrs. Agassiz discuss fully Agassiz's work in vertebrate embryology. M. E. Bettes, *Embryology and the Natural Classification of Animals: Louis Agassiz (1807–1873)* (Master's thesis, Bryn Mawr College, 1977), in addition to discussing Agassiz's ideas on classification, presents more details of his strictly embryological ideas than do most other authors.

nineteenth century when embryos were more often used for observation than for experiment, they were more popular for study. Agassiz's interest in teleost development went as far back as his time with Döllinger; Vogt reported that he had seen sketches of young developing perch eggs that Agassiz had made in 1831.[15] Among Agassiz's immediate contemporaries, von Baer had described the development of a carp (1836)[16] and Rusconi (1836)[17] that of the tench. Thus these works appeared before the Agassiz-Vogt description of the embryology of the whitefish. They were referred to in the *Embryologie des Salmones*, as were several studies of later stages of development in viviparous blennies.

Pictorial illustrations are extremely important for the exposition of embryological findings. The engravings in *Embryologie des Salmones* portray accurately many aspects of the development of the eggs and embryos at all periods from before fertilization through hatching and beyond. One bizarre illustration misrepresents the position of four of the segmentation products at what we know as the eight-cell stage, but this is an exception (fig. 105, pl. 5 in the original book). Figure 1 on the plate on page 402 of this chapter, photocopied from Agassiz and Gould's 1848 textbook on zoology, reproduces a reasonably faithful copy of an excellent figure of a whitefish hatchling published originally in the *Embryologie des Salmones* (fig. 86, pl. 30 in the original book). Furthermore, the Agassiz-Vogt figures for the whitefish illustrate far more clearly than had the earlier figures of von Baer for the carp and of Rusconi for the tench a structure produced by an accumulation of what we know as cells that migrate and gather together, some time after

15. Vogt in Agassiz-Vogt, 31.
16. K. E. von Baer, *Untersuchungen über die Entwickelungsgeschichte der Fische, nebst einem Anhange über die Schwimmblase* (Leipzig: Friedrich Christian Wilhelm Vogel, 1835).
17. M. Rusconi, "Ueber die Metamorphosen der Eier der Fische," *Müllers Archiv für Anatomie, Physiologie und wissenschaftliche Medizin,* Jahrg. 1836: 278–288.

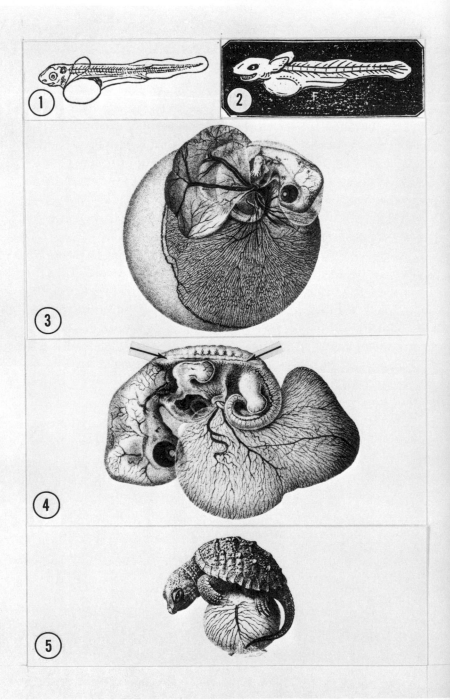

segmentation, to give rise to the embryo proper (as contrasted with the epithelia covering its yolk). This group of cells, now known as the embryonic shield in teleosts is, to be sure, easier to discern in the living eggs of some teleost species than in those of others.

The principal new contribution to knowledge made by the *Embryologie des Salmones* was its minutely detailed description of a long, closely-staged series of developmental phenomena in a particular teleost whose formative phases had not been described before. No new interpretations were introduced here; no new principles of development were derived. The accuracy of the observations and the excellence of the workmanship in the preparation of the plates, however, far exceeded that in earlier publications on teleost development. In this respect the volume was significant in establishing high standards for the new science of embryology. Zoological and anatomical illustrations were typically magnificent in the mid-nineteenth century, as they had been for several centuries before. Presenting splendid pictures was not an innovation by Agassiz and Vogt, but they were first to accomplish it in an original study of teleost development.

Figure 1. Whitefish hatchling; photocopy of an illustration from *Principles of Zoology*, 118. The figure in the *Principles* is a reasonably accurate copy of Agassiz-Vogt's fig. 86, pl. 30.

Figure 2. Whitefish hatchling; photocopy of an illustration from *Lowell Lectures*, 7. This is an altered adaptation of the same figure from Agassiz-Vogt reproduced in *Principles of Zoology* as shown here in fig. 1. Among other changes, the shape of the mouth has been altered and grinding teeth (non-existent in whitefish of all ages) have been added. The shape of the eye is different, and a crescentic highlight replaces the spherical lens. The shape of the eye and lens are correct in the Agassiz-Vogt figure and in fig. 1 here.

Figures 3, 4, and 5. Copies of figures from plate XV in *Embryology of the Turtle*, which illustrates vitelline and allantoidean circulation in developing *Chelydra serpentina*. Fig. 3 (fig. 12 in the original) represents an embryo in an egg laid 10 June, opened 1 August 1855; the allantois has been reflected to show the underlying vitelline vessels. Fig. 4 (fig. 13 in the original) is an embryo removed from its shell; the egg was laid 18 June, opened 31 July 1855. The straight rod, indicated here by arrows added to my copy of the figure, is not a structure normally seen in this position in vertebrate embryos at a comparable stage of development. Fig. 5 (fig. 3 in the original) is an unhatched specimen shown after removal of the amnion and the allantois, taken from an egg that was laid 1 June and opened 1 September 1855.

When Agassiz came to America in 1846, he probably still thought of himself as a research investigator in embryology, rather than as entrepreneur with respect to this new subject of investigation. In 1847 and 1848, he made some observations on the development of an American starfish. The study of echinoderm development was exciting great interest at this time. The metamorphosis of larva to adult began to come under active study from 1835 on; the crowning discovery, by Johannes Müller, that pluteus larvae metamorphosed to become adult brittle stars or sea urchins (observed 1845, announced at a meeting 1846),[18] had greatly excited Agassiz. He was distressed that his own observations on the development of echinoderms were published not under his own name but under that of his former colleague Edouard Desor.[19] Desor also added his name, unjustifiably according to Agassiz, as co-author of an important catalog of families, genera, and species of echinoderms published in Paris (1846–1847).[20] Agassiz later wrote that this was "one of the most extraordinary pieces of plagiarism I know of."[21]

Soon after Agassiz's arrival in America, his emphasis on embry-

18. J. Müller, "Ueber die Larven und die Metamorphose der Ophiuren und Seeigel," *Abhandlungen der königliche Akademie der Wissenschaften*, Berlin 1846 [1849]: 273–312. Also published separately in 1848, according to Winsor, *Starfish, Jellyfish, and the Order of Life*, 106 n 20.

19. E. Desor [no title], *Proceedings of the Boston Society of Natural History* 3 (1851): 13–14, 17–18.

20. L. Agassiz and E. Desor, "Catalogue raisonné des familles, des genres et des espèces de la classe des echinodermes," *Annales des Sciences Naturelles. Zoologie.* 3d ser. 6 (1846): 305–374; 7 (1847): 129–168; 8 (1847): 5–35, 355–380.

21. *Essay on Classification*, 97 fn. Part I of the *Catalogue* (1846) is the only work by Agassiz listed in the bibliography for the *Origin of Species* drawn up by R. C. Stauffer, *Charles Darwin's Natural Selection* . . . (Cambridge, England: Cambridge University Press, 1975). Darwin, however, did misattribute to Agassiz, in the first two editions of the *Origin*, a remark on embryology made by von Baer, and thus must somehow have thought of Agassiz in connection with embryos; the error was corrected in the third edition. For details, see J. M. Oppenheimer, "An Embryological Enigma in the Origin of Species," in B. Glass, O. Temkin, and W. L. Straus, Jr., eds., *Forerunners of Darwin: 1745–1859* (Baltimore: Johns Hopkins University Press, 1959), Chapt. IX, 292–322.

ology as phenomenological research began to diminish and he became an ardent popularizer of the subject. By invitation, he delivered twelve popular lectures on Comparative Embryology at the Lowell Institute in Boston in 1848–49. The anonymous author of the preface strongly emphasized that embryology was a new science whose findings Agassiz wished to introduce to Americans, and Agassiz, near the conclusion, stated that his "object has been to bring the present knowledge which is possessed upon Embryology, into one point of view."[22] The point of view: "I think I have particularly been able to show that classification in its details may be improved by Embryological evidence."[22] The evidence that he adduced was derived from studies on a great variety of forms; he devoted separate chapters to Radiates, Articulates, Mollusca, and Vertebrates; this was Cuvier's classification. Among the Radiates he included echinoderms, jellyfishes, and medusae. The Articulates included crustaceans, insects, and worms.

As previously mentioned, pictorial illustrations are especially important when presenting embryological research data. They are also indispensable in teaching, a fact which Agassiz knew well and fully exploited. This is evident from portraits depicting him against the background of a blackboard containing chalk drawings.[23] The diagrams of teleost development in the published version of the *Lowell Lectures* are extremely crude. This is evident, for instance, when the drawings of teleost embryos accompanying the *Lowell Lectures* are compared with those in the *Embryologie des Salmones* from which they were adapted. Figure 2 in the plate accompanying this chapter, photocopied from the *Lowell Lectures*, reproduces an adaptation of the same outline drawing of the whitefish hatchling

22. *Lowell Lectures*, 103.
23. Two portraits of Agassiz at the blackboard, one a photograph, the other an engraving made from a sketch, are reproduced in Lurie, *Louis Agassiz*. The reproductions are found near the end of a group of illustrations without page numbers located between pp. 84 and 85 of the text.

shown in figure 1. In the illustration to the *Lowell Lectures* a leering eye and a toothy grin have been gratuitously added as pure invention. Was the new portrayal a joke on Agassiz's part introduced to amuse the audience? An appropriate portion of the transcript of the lectures reads: "Let me, with a reference to a few diagrams, show what I mean. Here are the various stages of the growth of a fish. See here . . . the egg in the earliest condition. . . . Next we see it still further advanced. There are afterwards successive changes taking place. . . . The transverse divisions are introduced until we see a little fish is coming. [Laughter]."[24] "[Laughter]" is nowhere else indicated in the published lectures. Was it a reaction to Agassiz's ploy of giving a little fish an impish expression by adding the leer and the grin? The *Principles of Zoology* specifies that teeth are not yet formed in hatchlings of the age shown. The shape of a hatchling's teeth, once they do form, does not remotely resemble that of the grinding teeth shown in the caricature; they are sharp pointed recurved trapping teeth. Was the device of giving the hatchling an unfishy expression evidence of the beginning of the transformation of Agassiz from a research scientist to a different *persona*? An engraving that portrays him teaching in his short-lived school on Penikese Island[23] shows on the blackboard beside him a chalk drawing of an adult fish that is as highly overstyled and inaccurate, indeed as false a representation of a real fish, as is the ludicrous whitefish hatchling of the *Lowell Lectures*. Agassiz could draw well if he chose to do so. Lurie has reproduced some of his pencil drawings of hatching turtles;[25] these are deliberately sketchy, but very graceful and in no way exaggerated or distorted.

Not all of his lectures indulged in frivolity and inaccuracy. When in 1873 he presented a very dogmatic lecture at much the same popular level as the Lowell Lectures to the State Board of Agriculture in Barre, Massachusetts, on "The Structure and Growth of

24. *Lowell Lectures*, 7.
25. Reproduced among the same group of illustrations described in n 23 above.

Domesticated Animals,"[26] accurate reproductions of Bischoff's illustrations of rabbit development and of Rathke's viviparous blenny embryos were included in the publication, but there were no jokes. By now Agassiz was again taking himself seriously.

The *Principles of Zoology* appeared shortly before the delivery of the Lowell Lectures. It attempted to do for all of what was then considered physiology what the Lowell Lectures attempted for embryology. Its subtitle was "Touching the structure, development, distribution, and natural arrangement of the races of animals, living and extinct. With numerous illustrations. For the use of schools and colleges." The "numerous illustrations" to which the subtitle refers are greatly superior to those of the *Lowell Lectures* (compare again my figure 1 with my figure 2). It is curious, by the way, that the *Principles of Zoology* omit credit to Vogt for the illustrations of teleost development, but their original provenance is acknowledged in the *Lowell Lectures* (p. 96), and they are there attributed to Vogt without invidious comment. The *Principles of Zoology* attempted to cover the development of a number of vertebrates, not only that of teleost fishes. The *Lowell Lectures*, as we have seen, and later the *Essay on Classification*, which expounded at their fullest Agassiz's ideas on the potential contribution of embryology to classification, attempted to cover the development of all animals insofar as it was then known, not only that of the animal groups on which Agassiz himself had worked.

The breadth of his knowledge, and particularly its sources, are important considerations. Before coming to this, however, it remains to describe the *Embryology of the Turtle*, an extensive, if not exactly impeccable, contribution to embryological research. *Embryology of the Turtle*[27] is Part III, the final Part, of the First Monograph of the *Natural History of the United States of America*; it consti-

26. L. Agassiz, "The Structure and Growth of Domesticated Animals," *American Naturalist* 7 (1873): 641–657.
27. See n 1 above.

tutes the whole of Volume II. Its text covers pages 451 through 643; 33 plates follow.[28] Two of the plates contain illustrations of adult turtles acknowledged as copies of illustrations originally made by a different naturalist; the remaining illustrations were made "from nature," as is stated on the plates and in the explanation of the plates. In the explanation of the plates the illustrations are attributed to Henry James Clark, to Auguste Sorel, or to both. None are credited to Agassiz himself. Clark was responsible for the greater number of the strictly embryological figures, and in the preface to the *Natural History* Agassiz expressed gratitude to him for the preparation of microscopic illustrations. However, following a familiar pattern, in 1860, well after the volume containing *Embryology of the Turtle* had been published, there was great controversy between Agassiz and Clark as to which of them had been primarily responsible for much of their joint enterprise on the embryology of the turtle.

The illustrations to *Embryology of the Turtle* can only be described as sumptuous. For examples of the skill with which they were executed, several are reproduced here (figs. 3, 4, and 5). These were drawn by Sorel, but are representative of all. Their delicacy of detail is astonishing. Are they accurate? Not always, in the judgment of embryologists. For instance, no such structure exists in any vertebrate embryo as the long straight rod indicated by arrows in the figure reproduced in my figure 4. Sorel or Clark or Agassiz may well have seen what seemed to be a bar in that position. If they did, they were no doubt looking at an artefact. The representations made by Sorel and Clark, on plate X, of segmentation stages are magnificent, with outstanding tactile value, as historians of art used to call it. Some of the illustrations show geometric patterns that we recognize as predictable; others depict patterns that are highly irregular and unusual for eggs of this type. We need not

28. Lurie gives the number as 27. The final plate bears the number 33 but plate 7a is interposed after plate 7, and plates 9a–e after plate 9.

question whether the artists drew what they saw; clearly they did. As embryologists we may doubt whether the irregularly segmenting eggs would have developed into normal turtles; perhaps they might have, but perhaps not. It did not occur to Agassiz to raise the question. Then again, the turtle embryo has what is called an embryonic shield (different from the structure in teleost embryos that carries the same name). It is well portrayed on plate XI. But over fifty years ago it was pointed out that its anterior end is interpreted as posterior, and vice versa.[29] *Embryology of the Turtle* contributes much to natural history concerning the reproductive habits of turtles, and presents useful data concerning differences among eggs and among hatched specimens of different taxonomic groups, but it sometimes falters on details of embryonic morphogenesis.

It is inappropriate to categorize early work as right or wrong, as good or bad. But it is evident that the accuracy of Agassiz's study on the development of the turtle does not equal that on the development of the whitefish. Many possible extenuating reasons may explain the differences between the studies. The eggs of turtles are fertilized internally, unlike those of the Coregonids, and begin their development before they are laid. The shells of turtle eggs are not transparent, in contrast to those of Coregonids; this is true also of the embryos themselves. Under the conditions that prevailed when these studies were made, it would have been possible to watch successive changes as a single particular fish embryo developed from one stage to the next. This would hardly have been possible at the time for the turtle. Furthermore, Clark was not Vogt, and the Agassiz of the 1850s was no longer the Agassiz of the 1830s and the early 1840s. He was too involved in many other activities over and

---

29. G. C. Davenport, "Agassiz's Work on the Embryology of the Turtle," *American Naturalist* 32 (1898): 187–188. Gertrude C. Davenport was Mrs. Charles B. Davenport, B.S. Kansas, 1889, who worked both on turtle embryology and on heredity in man.

above research. The Agassiz who had begun his American life as an embryologist was by now only in some respects the equal of the Agassiz who had started to be an embryologist in the European home of Ignaz Döllinger.

Still, Agassiz made one major contribution to American embryology. This was his publication, in a series of long detailed footnotes to the *Essay on Classification*, of a very full list of references[30] to previous writings on embryology. He justified his publication of the extensive list by saying: "The limited attention, thus far paid in this country to the study of Embryology, has induced me to enumerate more fully the works relating to this branch of science, than any others in the hope of stimulating investigations in that direction."[31] The total number of works referred to is over three hundred and eighty, for the nineteenth century alone.

The figure is approximate, because it is not always clear, when two references to different periodicals are listed under a single title, whether or not the identical article was published twice. The count excludes translations of listed articles already referred to in their original languages; it excludes the items described in an additional short list of textbooks, handbooks, and monographs. It omits publications that appeared before the nineteenth century. It includes some references to articles on anatomical features and to possible taxonomic features that in Agassiz's mind related to embryology. The articles listed appeared originally in France, Belgium, Germany, Austria, Switzerland, Italy, the British Isles (England, Scotland, Ireland), Scandinavia (Norway, Denmark, Sweden), Poland, Russia (Moscow, St. Petersburg), also in Schleswig and in Estonia, separate places if not separate countries in the nineteenth century. Twenty American publications were listed, including three short notes by Agassiz himself and seven articles by Joseph Leidy.

---

30. *Essay on Classification*, fns from p. 68 through p. 85. On twelve of these pages the footnotes occupy more space on the pages than does the text.
31. Ibid., 68 n 2.

The American articles were, with three exceptions, published in the 1850s. The three exceptions were the two publications already mentioned, that appeared under the authorship of Desor and Agassiz in 1848 and 1849 respectively, and an article by Charles Meigs, M.D., a Philadelphia obstetrician, that appeared in 1849 describing the reproductive organs and fetus of a dolphin.

The articles listed were grouped according to the organisms to which they referred. A comparison of the numbers for the groups of organisms in which Agassiz was interested because of his own work suggests that the list was mainly compiled before Agassiz came to America: there were 41 items listed for fishes (a few of them regarding non-teleost fishes); 19 for echinoderms; and 6 for reptiles (3 snakes, 3 turtles). There was only a single reference to a work on turtles in the monograph on the embryology of turtles, published the same year as the *Essay* in which the long general bibliography was published.

Citing references to possible sources of knowledge is of course not equivalent to the acquisition of wisdom. It is hard for us now to admire the acumen of a would-be biological generalist who was convinced that Vorticella is a larval Bryozoan, Paramecium the larva of a flatworm, even though others at the time held similar opinions. Agassiz included in the long bibliography just described an 1849 article by Siebold on one-celled plants and animals (so called in the title) and even an 1855 article by his brother-in-law Alexander Braun on the unicellularity (designated by that very word in Latin) of some algae. What did cells mean to Agassiz?

In other words, how did he relate to embryology the discoveries made during his time in areas of biology related to embryology? The realization that what we know as cells were the units which constitute all organisms developed gradually. Cell theory, however, was enunciated in its most influential general terms by Theodor Schwann in 1838, just four years before the publication of the *Embryologie des Salmones*, ten before the publication of the *Principles of*

Zoology. The *Principles* is unequivocal with respect to the cellular constitution of embryos: "In order to understand the successive steps of embryonic development, we must bear in mind that the whole animal body is composed of tissues, whose elements are cells. . . . These cells are much diversified in the full grown animal; but, at the commencement of embryonic life, the whole embryo is composed of minute cells of nearly the same form and consistence."[32] So much is current, but the continuation of the passage seems less meaningful today. As an excellent microscopist using a fine instrument, Agassiz's personal observations were accurate indeed. An illustration in the *Principles* provides an accurate pictorial representation of an ovarian egg as a cell: "Primitive [ovarian] eggs . . . are nearly the same in all animals, and are in fact merely little cells containing yolk substance, . . . including other similar cells, namely, the germinative vesicle . . . and the germinative dot."[33] The accompanying illustration presents a spherical egg with small yolk granules or globules evenly distributed within it. A nucleus, Agassiz's germinative vesicle, eccentric in position, is included within it; the nucleus of an egg in this stage is still called a germinal vesicle today. Within it, Agassiz's "germinal dot" is our nucleolus. Agassiz's interpretation that germinative vesicle and germinative dot are cells within cells proved incorrect, but he formulated these ideas when the manner of origin of new cells by what we know as cell-division had not yet been described, and the constitution and function of the nucleus were still total mysteries. The progressive breaking up by successive divisions of the non-yolky portions of teleost eggs that we call segmentation or cleavage of blastomeres is beautifully illustrated in the *Embryologie des Salmones.* It appears a bit more crude in the *Principles of Zoology,* and very rough and sketchy in the *Lowell Lectures.* The quality of comparable figures in

32. *Principles of Zoology,* 110.
33. Ibid., 104.

the *Embryology of the Turtle* has been already discussed. Agassiz specifically stated in the *Lowell Lectures* that "the egg [shows] . . . a successive division into cells."[34] But the statement refers to a figure that does not show division of what we know as the cytoplasmic portion of the egg. Instead this illustration represents a pre-segmentation stage in which what Agassiz interprets as cells are merely lipo-protein droplets that constitute the yolk in teleost eggs of the stage in question. It was commonly accepted at the time that what were then called granules in the yolk were separate cells. Nonetheless, in 1844, when Agassiz was still in Europe, Albert von Kölliker[35] recognized that the segmentation products were cells that gave rise to the other cells of the embryo. Agassiz referred to the appropriate book of von Kölliker more than once in his footnotes and in the long bibliography, but never in connection with the direct derivation of all embryonic cells from the segmentation products of the young egg as cells themselves. As an embryologist, Agassiz made no pioneering forays into the new areas of embryology that were awaiting exploration.

In sum, how do we judge Agassiz as an early embryologist in America? It is evident that when he came here he brought embryology with him from Europe; from this point of view he was not only *an* early embryologist in America, but the first one of all. The extensive and important bibliography that he published here, however, seems to have been based mainly on his European experience. The sometimes flawed *Embryology of the Turtle* was his most extensive original American production in embryology. At least it did open up a field which others could enter.

Agassiz was, however, to make a strong mark on American zoology, if not so vivid a one in American embryology. As his strictly research interests in embryology diminished, his role as educator

---

34. *Lowell Lectures*, 98.
35. A. von Kölliker, *Entwickelungsgeschichte der Cephalopoden* (Zürich: Meyer und Zeller, 1844).

grew. He became an enormously popular lecturer, on embryology as well as on other zoological subjects. He drew wide attention, of audiences of varied intellectual level, to both new and old aspects of the biological scene. Perhaps it was in part because as a youth he enjoyed a taste, in embryology, of what science can be at its best, that he was later able to develop into the scientific statesman (or politician?) who greatly assisted zoologists by participating in the foundation of the Harvard Museum of Comparative Zoology, and benefited all by his involvement in the establishment of the National Academy of Sciences.

Index

415

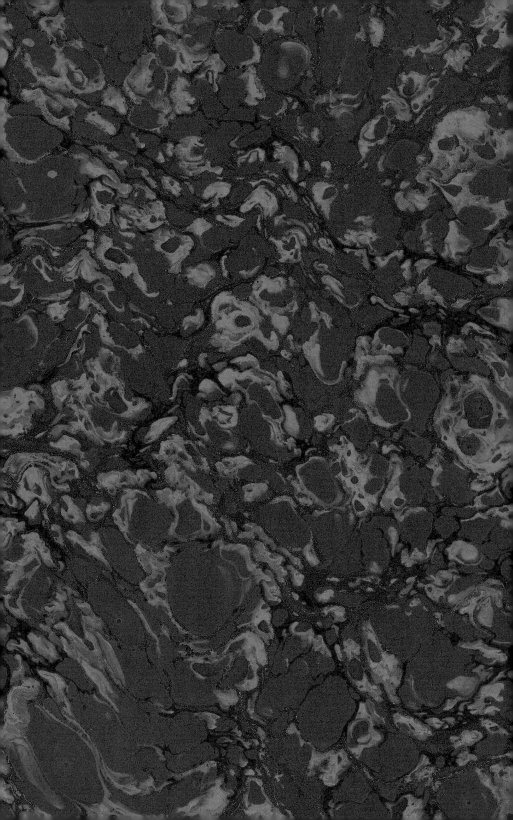